T0220784

Medicine and Biomedical Sciences in Modern History

Series Editors
Carsten Timmermann
University of Manchester
Manchester, UK

Michael Worboys
University of Manchester
Manchester, UK

The aim of this series is to illuminate the development and impact of medicine and the biomedical sciences in the modern era. The series was founded by the late Professor John Pickstone, and its ambitions reflect his commitment to the integrated study of medicine, science and technology in their contexts. He repeatedly commented that it was a pity that the foundation discipline of the field, for which he popularized the acronym 'HSTM' (History of Science, Technology and Medicine) had been the history of science rather than the history of medicine. His point was that historians of science had too often focused just on scientific ideas and institutions, while historians of medicine always had to consider the understanding, management and meanings of diseases in their socio-economic, cultural, technological and political contexts. In the event, most of the books in the series dealt with medicine and the biomedical sciences, and the changed series title reflects this. However, as the new editors we share Professor Pickstone's enthusiasm for the integrated study of medicine, science and technology, encouraging studies on biomedical science, translational medicine, clinical practice, disease histories, medical technologies, medical specialisms and health policies.

The books in this series will present medicine and biomedical science as crucial features of modern culture, analysing their economic, social and political aspects, while not neglecting their expert content and context. Our authors investigate the uses and consequences of technical knowledge, and how it shaped, and was shaped by, particular economic, social and political structures. In re-launching the Series, we hope to build on its strengths but extend its geographical range beyond Western Europe and North America.

Medicine and Biomedical Sciences in Modern History is intended to supply analysis and stimulate debate. All books are based on searching historical study of topics which are important, not least because they cut across conventional academic boundaries. They should appeal not just to historians, nor just to medical practitioners, scientists and engineers, but to all who are interested in the place of medicine and biomedical sciences in modern history.

This series continues the Science, Technology and Medicine in Modern History series.

Miguel García-Sancho • James Lowe

A History of Genomics across Species, Communities and Projects

palgrave
macmillan

Miguel García-Sancho
Science, Technology and
Innovation Studies
University of Edinburgh
Edinburgh, UK

James Lowe
Science, Technology and
Innovation Studies
University of Edinburgh
Edinburgh, UK

ISSN 2947-9142   ISSN 2947-9150   (electronic)
Medicine and Biomedical Sciences in Modern History
ISBN 978-3-031-06132-5   ISBN 978-3-031-06130-1   (eBook)
https://doi.org/10.1007/978-3-031-06130-1

This Palgrave Macmillan imprint is published by the registered company Springer Nature Switzerland AG.
The registered company address is: Gewerbestrasse 11, 6330 Cham, Switzerland

Acknowledgements

This book is the culmination of over five years' work as part of the multi-disciplinary 'TRANSGENE: Medical translation in the history of modern genomics' project, which was funded by the European Research Council (ERC) under the European Union's Horizon 2020 research and innovation programme, grant agreement No. 678757. Without this support it would not have been possible to produce this book. There have been several full-time members of staff and multiple collaborators associated with the project in addition to ourselves: Giuditta Parolini, Erika Szymanski, Mark Wong, Rhodri Leng, Jarmo de Vries and Rodrigo Liscovsky Barrera all worked on the project, while Ann Bruce, Niki Vermeulen and Gil Viry were invaluable collaborators. Jarmo de Vries also undertook a careful copy-editing of the manuscript, and provided insightful comments on the contents and writing style.

More broadly, we would like to thank colleagues within and outside the University of Edinburgh for their invaluable feedback, some of it at conferences and other events at which we presented our work. The members of the Advisory Board of our project, along with Robert Cook-Deegan, devoted considerable time to providing us with guidance and commenting on our ongoing work.

We are grateful to all those who have provided us with access to archives or agreed to be interviewed, including all those who have informed our analyses but have not been cited in the book. Furthermore, we appreciate the additional clarifications and answers to our queries provided by many of our interlocutors, including Alan Archibald, Jon Beever, David Bentley, Martin Bobrow, Horst Domdey, Horst Feldmann, Mark Guyer, Barbara

Harlizius, Kerstin Howe, Zhi-Liang Hu, Mark Johnston, Karl Kleine, Peter Li, Jane Loveland, Brigitte Obermaier, Jane Peterson, Peter Philippsen, Christine Renard, Claire Rogel-Gaillard, Lawrence Schook, Ross Sibson, Rolf Stucka and Robert Waterston.

We thank the series editors, Michael Worboys and Carsten Timmermann, as well as Molly Beck, Lucy Kidwell, Ruby Panigrahi, Eliana Rangel and Petra Trieber at Palgrave Macmillan and Springer Nature for guiding and informing us through the publication process. Their comments and those of the anonymous reviewer of an earlier version of the manuscript were highly valuable in helping us to refine the text and argument of the book. Naturally, any errors that remain belong only to ourselves.

James Lowe would like to thank Miguel García-Sancho for giving him the opportunity to participate in such a fascinating project, and for his stimulation and support in what has been an enjoyable and productive collaboration. He would also like to express his gratitude for all the school, college and university teachers who have helped him get to the position today where he is co-authoring this book, and to Sabina Leonelli and Staffan Müller-Wille for their exemplary role as his PhD supervisors from 2010 to 2015, as well as their ongoing support. They and many others made Egenis at the University of Exeter a nurturing academic environment for him and others at such a crucial stage. Finally, he would like to thank his family and friends, in particular his parents, Doug and Jackie. One could not wish for a better start in life than the one they gave him, and he treasures their love, encouragement and company. On his part, this book is dedicated to them.

Miguel García-Sancho would like to thank James Lowe for his loyalty, enthusiasm, insight and exemplary dedication during the life of the project. He is also indebted to all those who helped him throughout his career, especially at Imperial College, the University of Manchester, the Spanish National Research Council and the University of Edinburgh. Most importantly, he would like to dedicate this book to Cate, Clarissa, Emma and Winnie for creating a home filled with love, support, patience and inspiration.

Praise for *A History of Genomics across Species, Communities and Projects*

"A comparative historical analysis of the breadth of genome projects set up and carried out around the Human Genome Project is so far missing. *A History of Genomics across Species, Communities and Projects* fills this gap. This comparative study of yeast, pig, and human genome projects is a well-researched and impeccable piece of socio-historical scholarship that gives a balanced picture of the coming into being of genomics, a new field of research with a huge future impact on the biological and biomedical sciences, and on society as a whole."

—Hans-Jörg Rheinberger, *Max Planck Institute for the History of Science, Germany*

"The history of genomics, García-Sancho and Lowe argue, is more than just the history of the Human Genome Project. Diving deeply into the history of the yeast and pig genomic project next to those of the human, the authors show how multifaceted and varied the field of genomics is. What is regarded as a reference sequence, how it is turned into a useful resource and who participates in the effort changes from species to species. These insights also change our understanding of the Human Genome Project. The book is an important addition to the historiography of genomics."

—Soraya de Chadarevian, *University of California, USA*

CONTENTS

LIST OF FIGURES

LIST OF TABLES

CHAPTER 1

Introduction

In four decades, genomics has transformed the biological sciences and has penetrated well beyond them. The marriage of DNA sequencing techniques and computational infrastructures built to handle, store and analyse ever-increasing quantities of data has contributed to significant developments in:

- Our understanding of human history through our relationship to Neanderthals, Denisovans and other hominids (Pääbo, 2014);
- Our appreciation of the extent and diversity of life previously undetected by biological methods (Riesenfeld et al., 2004; Venter et al., 2004);
- Forensic science, food tracing and nature conservation (Arenas et al., 2017);
- Our picture of the Tree of Life and the evolutionary relationships within it (O'Malley et al., 2010);
- The reclassification of diseases resulting in improved diagnosis, prognosis and treatment options (Keating et al., 2016);
- Enhancements in the efficacy of selective breeding in agriculture (Lowe & Bruce, 2019);
- The reshaping of the fundamental models and metaphors with which we think about how living things develop and function (Keller, 2000).

© The Author(s) 2023
M. García-Sancho, J. Lowe, *A History of Genomics across Species, Communities and Projects*, Medicine and Biomedical Sciences in Modern History, https://doi.org/10.1007/978-3-031-06130-1_1

1

DNA sequencing has gone from being a highly specialised practice, requiring considerable labour and skill, to being routinely applied in ordinary laboratory work while also being conducted at great scale, speed and accuracy in factory-style genome centres. In the late-1970s, manually sequencing the tiny genome of a bacteriophage (a virus that infects bacteria) was a monumental task, one that earned Frederick Sanger, who led the group undertaking it, a Nobel Prize (Brownlee, 2014; Hutchison, 2007). The determination of the whole human DNA sequence (commonly referred to as the Human Genome Project) took more than a decade, at a cost initially estimated at $3 billion. It started in the 1990s and concluded in 2003, expanding in speed and scale throughout.

Progress since then has been so dramatic that, more recently, well over fourteen million coronaviruses have been sequenced and shared via the Global Initiative on Sharing Avian Influenza Data.[1] Another example that illustrates how far genomics has come, is that the cost of sequencing a whole human genome was estimated to be about £7000 in 2020, multiple orders of magnitude below the original budget of the Human Genome Project (Schwarze et al., 2020).[2]

In 1999, four years before the Human Genome Project was officially concluded, the National Center for Biotechnology Information of the USA created a new database called RefSeq. The purpose of this database was to serve as a centralised repository that would gather the ongoing reference sequence of the human genome and those of other species completed or in progress. Those reference sequences were and still are curated and freely released to the research community. They serve as canonical representations of their designated species and are graded according to

---

[1] https://www.gisaid.org/ (last accessed 29th November 2022). The COVID-19 Genomics UK (COG-UK) Consortium alone has sequenced over two million SARS-CoV-2 viruses: https://www.cogconsortium.uk/ (last accessed 29th November 2022).

[2] Elsewhere, lower figures have been indicated (https://www.genome.gov/about-genomics/fact-sheets/Sequencing-Human-Genome-cost, last accessed 29th November 2022), though these may not include the full range of costs involved in all aspects of the sequencing process, including processing, storage and curation of the resulting data.

their level of comprehensiveness, representativeness and quality (Ostell, 2013, pp. 72–74; Tatusova et al., 2014, p. 135).[3]

The number of entries in RefSeq has grown exponentially, from complete sequences representing just over two thousand different species in 2003, to 125,116 in November 2022.[4] On top of this, RefSeq also curates and stores a higher number of partial sequences, as well as variants and other versions of complete reference genomes. Life scientists from every discipline all around the world can access the sequences and curatorial metadata. In processing each existing and upcoming entry, RefSeq curators attempt to achieve a balance between respecting the differences across the stored sequences while avoiding a Tower of Babel of different communities producing separate datasets that would require considerable efforts to integrate, use and compare outside their contexts of creation. Yet in fostering this universal—or at least commensurate—language, some of the distinctions between the individual reference genomes are flattened, and indeed lost.

In what follows, we make some of these distinctions visible again by looking at the history of the production of three reference genomes: those of the baker's and brewer's yeast *Saccharomyces cerevisiae* released in 1996 and published in 1997; *Homo sapiens*, published in 2001 as a working draft and in more definitive form in 2004; and the pig *Sus scrofa*, initially released in 2009 and published in 2012. Taken together, these three genomes embody overlapping trajectories of change and differentiation in the practices, goals, organisation and status of genomics research. While yeast is both a model organism in basic biomedical science and a tool for the brewing and biotechnology industries, pigs were mainly sequenced for

---

[3] RefSeq distinguishes "reference genomes", "representative genomes" and "variant genomes". Throughout, when we refer to reference genomes, we are referring to objects that are designated by RefSeq as "reference genomes" and "representative genomes". When the distinction between these becomes relevant in our narrative, we will specify which RefSeq category we are referring to. For RefSeq, "reference genomes" are "manually selected 'gold standard'" high-quality complete genomes. "Representative genomes" are designated standard genomes for a given species of organism, while "variant genomes" constitute "genome variations within the species" (Tatusova et al., 2014, p. 135).

[4] See https://www.ncbi.nlm.nih.gov/refseq/statistics/ (last accessed 29th November 2022).

agricultural purposes, but also to serve objectives of human medicine—for instance, helping organ transplantation. Sequencing *H. sapiens* became the most prominent area of genomics, one believed to have potentially invaluable clinical payoffs.

By examining the substantially different ways in which these endeavours were conducted across the three organisms, this book argues that producing a whole-genome reference sequence was not always the main—nor the universally accepted—objective of genomics, as the growing entries in RefSeq may suggest. What these now centrally curated reference sequences represented, and the uses to which they were put, also varied substantially across the communities that produced them, in spite of the commensuration work of RefSeq and cognate institutions and repositories.

The rest of this introductory chapter summarises the main features of genomics and how it historically emerged from the practices that have subsequently accompanied it and conferred its identity: mapping and sequencing DNA, and processing the resulting data with information technologies, including databases.[5] We then present the key concepts and analytical tools that we use throughout the book and outline how we develop them in the remaining seven chapters. We argue that popular and scholarly accounts have tended to excessively emphasise the Human Genome Project in the history of genomics, due to the perceived impact and high profile of this initiative. We refer to this Human Genome Project-centred history as the canonical, master narrative of genomics, and relate its structure to the hourglass model that prior historiography has applied to the study of heredity throughout the nineteenth and twentieth centuries. As in the case of the study of heredity (Barahona et al., 2010), the hourglass model aids the comprehension of the institutional and infrastructural landscape of genomics, while falling short in capturing its broader history. We escape the boundaries of the hourglass model by looking at non-human genomic endeavours and documenting the deep entanglement between the creation of reference genomes

---

[5] Our outline is necessarily brief and focused on the episodes and practices that we examine in more detail later in the book. For more comprehensive reviews, see Müller-Wille & Rheinberger, 2012, Chs. 7–8; Morange, 2020, Ch. 27 and our timeline below (Fig. 1.2).

and the communities that were involved in their production. We propose the term *genomicist* to capture the crucial role of communities in the construction of genomic data and materials, and highlight both inclusive and exclusive mechanisms in the formation and operation of those communities.[6]

1.1   Genomics, DNA Mapping and Sequencing

The sequencing of DNA is the determination of the order of the four 'bases' along each of the two complementary strands of nucleotides that wind around each other to produce the molecule's double-helical structure: adenine, thymine, cytosine and guanine, known by their initials—A, T, C and G. Sequencing is central to genomics. However, genomics involves far more than just this, and sequencing can be conducted outside of genomics research and for other biological molecules, such as RNA and proteins. Indeed, while the history of sequencing—of proteins, RNA and then DNA—can be traced back to the 1950s, 1960s and 1970s respectively, genomics proper is recognised to have arisen only in the 1980s (García-Sancho, 2010). Its antecedents were not only sequencing practices, but also the mapping of chromosomes (bodies containing DNA in the cell), and the development of information technologies to process the resulting map and sequence data.

Chromosome mapping dates back to the early twentieth century and is conducted in order to find certain landmarks in them, such as genes (de Chadarevian, 2020; Hogan, 2016; Rheinberger & Gaudillière, 2004).[7] It was known since the early days of mapping that genes constitute only a small portion of chromosomes; after the discovery of the structure of DNA in 1953, genes were increasingly identified with partial, specific segments of the nucleotide sequence within the chromosome. The third central practice of genomics, the processing of the resulting map and sequence information using databases and computational methods, started to be applied to DNA in the 1970s. Similar practices involving other biological and medical data, such as the elucidation of protein sequences or the three-

---

[6] Our central idea of entanglement between genomes and communities of genomicists expands arguments that we formulated elsewhere, such as the distinction between 'thin' and 'thick' sequencing (Lowe, 2018) and the existence of different ways of sequencing that affect the ontological status and affordances of the resulting sequence data (Leng et al., 2022).

[7] On the metaphor of the genome as a territory to be mapped, see Dreger (2000), Gaudillière and Rheinberger (2004) and Winther (2020), Ch. 8.

dimensional structures of proteins, can be traced back to the decades following World War II (Strasser, 2019, Ch. 3; de Chadarevian, 2002, Ch. 4).

What makes genomics distinct from sequencing and these other practices, when they are considered separately? While it is important to avoid the error of being too inclusive, there is also the risk that a strict and exclusive definition of genomics can project the way that genomics developed—or at least a particular trajectory of it—back on to the past. To put it bluntly, there is a danger of a winner's narrative: that those who succeeded in making their vision of genomics a reality—or who are currently in charge of the institutional manifestation of it—dictate the boundaries of the field and project them retrospectively (Suárez-Díaz, 2010).

Areas of scientific endeavour, particularly ones with disciplinary names and associated journals, databases, brick-and-mortar facilities and well-funded institutions, are social and sociological phenomena. This means that the demarcation and boundary work performed by influential social groups and networks shapes the reality of the field. But scientific fields, disciplines and other phenomena are not only social creations and objects in this top-down political sense. They are also comprised of configurations of methods, techniques, technologies, theories, models, research programmes and commitments, norms and the careers, interests and activities of less-prominent scientists. These are no less infused with the social, cultural and political, but they are elements that deny the exclusivity of elite political, cultural and social mechanisms to define what scientific endeavours like genomics are.

It is not our job to provide an exhaustive and authoritative definition of genomics that takes account of these considerations. We can note, however, and show throughout this book, that the historical configuration of genomics involved a multi-directional, often dialectic, interaction between elite actors, less influential bench biologists and computer experts, all of whom mobilised differing visions, methods and forms of organisation. Genomics necessarily involves some form of sequencing and/or mapping of the genome, wherein the products—in the form of data—are stored and analysed using computational (informatics) infrastructures. To constitute genomics, this must be associated with a more general effort to construct a systematic representation of the genome, either in whole or in part.

The term 'genome' long antedates the idea of 'genomics', being coined by the German botanist Hans Winkler in 1920 to denote "the haploid chromosome set" (as translated in Lederberg, 2001). The haploid set constitutes one of each pair of chromosomes; so for humans that have a total of 46 chromosomes made up of 23 pairs,[8] the haploid set constitutes 23 chromosomes. Scholars have noted that the term genome, and genomics itself, aims to capture something comprehensive, a totality (Rheinberger & Müller-Wille, 2017; Stevens, 2013). Does this mean that something can only be genomic if it aims at the *complete* mapping or sequencing of a genome? Not necessarily. On the basis of achieving total completeness or comprehensiveness, barely anything could constitute genomics. Additionally, what constitutes completeness or comprehensiveness is not fixed; as we see later in the book, but particularly in Chap. 7, the goal posts are always moving. One may say that, as long as there is a concerted effort being made towards that end, it is genomics. However, the indeterminacy of what constitutes the end-point means that there is no strict criterion for ruling any given endeavour either in or out. The idea of a process or journey towards a goal means that the line between 'true' genomics and mere sequencing and mapping is somewhat blurry. How close does one need to be to the ever-receding end-point to be doing genomics?

Instead, we prefer to recognise genomics through its *systematicity* and its treatment of the genome as the *substrate* of its efforts. By systematicity, we mean that there is some concerted—and often collective—effort to identify and establish relations between multiple objects in and across the genome. By substrate, we mean that the genome is the field of operations for this activity: that which is to be mapped and the map itself. This does not mean that the whole genome needs to be mapped—or sequenced— for an effort to be deemed genomic. We distinguish systematicity from comprehensiveness and argue that in the history of genomics—especially during the early days—there were a substantial number of systematic but not comprehensive efforts, in the form of concerted operations that only addressed certain regions of target genomes.

Our criteria do not imply that all research that tries to identify genes in the genome can be classed as genomics. If a molecular geneticist was able to identify a gene that they had good reason to believe was implicated in some process in the cell, sequence that gene and then study the way it is

---

[8] 22 pairs of non-sex chromosomes, and typically one pair of sex chromosomes, XX or XY, though numerous exceptions to these figures exist in humans, and the numbers and sets of distinct chromosomes differ in other organisms. The full complement of 46 chromosomes in humans is the *diploid* set. The meaning of 'genome' has, inevitably, shifted over time (Keller, 2011).

expressed—how it results in the production of a specific protein—this falls well short of being genomics in both aspects of our guideline. It only considers a single object in the genome. Even in cases where two or more genes were involved in the process of interest, if the research does not consider the relations between them in terms of them being objects in the genome it would still not fulfil our second, 'genome-as-a-substrate' criterion. If, instead, the researcher was using known products of genes relating to a biological process of interest in order to identify and map multiple DNA sequences across the genome—ideally in collaboration with other laboratories—they would have shifted towards a more genomic way of working. This is because the focus is now on the genome as a territory to be mapped, rather than just on individual genes. Indeed, as we show in the next chapters, this kind of activity and the communities that converged around it became key drivers of genomics research from the 1990s onwards.

The invention of DNA sequencing methods in the 1970s was crucial to the forging of genomics. One of the main pioneers was Frederick Sanger, who had previously worked to discern the sequence of amino acids—the fundamental building blocks of proteins—in insulin, for which he won the Nobel Prize in 1958. He then moved on to RNA, the intermediary molecules in the process by which stretches of DNA form the basis for the synthesis of proteins with specific amino acid compositions. While other researchers in the mid-1970s such as Allan Maxam and Walter Gilbert also developed DNA sequencing methods, the technique that Sanger and his team devised at the Medical Research Council's Laboratory of Molecular Biology in Cambridge (UK) became the dominant approach before the creation of newer methods in the twenty-first century (García-Sancho, 2012, Chs. 1–2).

Sanger's technique required extremely time-consuming and labour-intensive bench work, as well as considerable technical and interpretive skills. The refinement of manual methods alongside increasing automation of parts of the process—including the invention and ongoing improvement of automated sequencing machines from the mid-1980s—enabled more and more to be sequenced in less time (García-Sancho, 2012, Chs. 5–6).[9] As the 1980s proceeded, therefore, the quantities of DNA sequence data were rapidly expanding year-upon-year.

---

[9] For an explanation of how the manual and automated sequencing techniques work, see https://genomicsincontext.wordpress.com/dna-sequencing-and-its-history/dna-sequencing-from-manual-biochemistry-to-industrial-genomics/ (last accessed 29th November 2022).

Alongside this were developments in mapping genes and other markers on the chromosomes. Genetic mapping had been pioneered by Thomas Hunt Morgan and his colleagues in the 1910s, working with the fruit fly *Drosophila melanogaster*. As in most animals, *Drosophila's* chromosomes are paired in two sets within its cell nucleus. Morgan's team observed, tracked and recorded different variant traits—such as the eye colour or wing shape—in many thousands of these flies, which were systematically bred and assessed (Kohler, 1994). The traits were presumed to result from different mutant versions of genes occurring across the chromosomes.

Morgan and his team exploited two facets of genetics: linkage and recombination. Linkage means that certain genes are commonly inherited together, which in the fly experiments meant that the associated traits were linked across generations. Recombination, discovered by the Morgan laboratory in their explorations of genetic linkage, happens during the creation of the sex cells (gametes), a process called meiosis in which the pairs of chromosomes separate. In it, parts of one of a pair of chromosomes can swap places with the corresponding parts of the other member of the pair. This means that the linkage between genes can be broken.

Morgan's laboratory realised that they could use this to find out the relative positions of genes on the fly's chromosomes: the further apart genes were, the more likely it is that a recombination event would occur between them, breaking their linkage. The frequencies of co-occurrence of versions of particular genes could be used to ascertain their relative proximity and order on the chromosomes. An array of relatively simple traits inherited from parent to offspring fly—such as the aforementioned eye colour and wing shape—enabled the group to map the *Drosophila* chromosomes and to further discern chromosomal dynamics in doing so. These maps of estimated chromosomal positions started to be called genetic linkage maps (see the upper part of Fig. 1.1).[10]

It took several decades for this approach to be applied to humans. When it did, inter-generational studies of families experiencing disproportionate numbers of cases of particular medical conditions could be used to identify the kind of genetic basis underlying them and to perform some analyses to assess the linkage relationships (Comfort, 2012; Lindee, 2005).

---

[10] They are also often referred to as just genetic maps, or linkage maps. Yet, for clarity, we use the term genetic linkage map throughout the book.

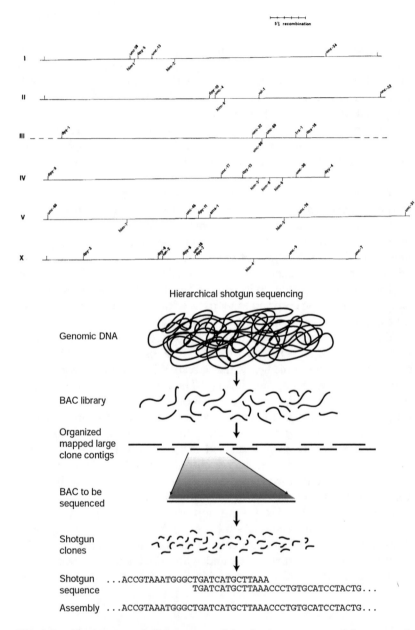

Fig. 1.1 Above, a genetic linkage map of the six chromosomes of the nematode worm *Caenorhabditis elegans*, elaborated by molecular biologists Sydney Brenner, Robert Horvitz and Jonathan Hodgkin in the 1970s, the decade that the chromosome workshops started. Below, a diagrammatic representation of how a physical

(continued overleaf)

This practice received a considerable boost when in the 1960s, molecular biologists began detecting polymorphic (many-variant) genetic markers that could be positioned on the chromosomal structures. These markers provided a greater number of landmarks for identification and analysis of variation beyond the small number of individuals suffering medical conditions or showing morphological traits that could be observed with the naked eye, and therefore mapped using the principles of genetic linkage. As we show in subsequent chapters, from 1973, human and medical geneticists periodically gathered in chromosome mapping workshops, with the first one held at Yale University. These workshops enabled attendees to systematically pool their mapping results—some of them obtained through molecular methods—and achieve an increasingly higher resolution in the location of genes and other markers of mainly medical interest.[11]

The first genetic linkage map encompassing the whole human genome obtained through molecular markers—Restriction Fragment Length Polymorphisms or RFLPs—was published in 1980. It deployed a type of protein (restriction enzymes) that cleaved the DNA molecule at specific sequence sites. When applied to DNA samples from multiple individuals, if their sequences diverged, the cleavage would produce different patterns of fragments. These different fragment patterns could be detected and used to map the sequence-specific genome regions where the restriction enzymes acted (Botstein et al., 1980). The same enzymes had been used from the mid-to-late 1970s as part of the recombinant DNA technologies, a suite of methods that enabled researchers to cleave and isolate specific fragments of the genome of one organism and transfer them into another.

Fig. 1.1 map is produced from a BAC library and assembled into a sequence—in this case, the reference sequence of the human genome. The physical map is the third illustration starting from the top ("Organized mapped large clone contigs") and the sequence is the bottom illustration ("Assembly"). Above image: Reproduced from Hodgkin, J, Horvitz, R, Brenner, S, Nondisjunction mutants of the nematode *Caenorhabditis elegans. Genetics*, 1979, *91*(1), 67–94: Fig. 1 on p. 70, by permission of Oxford University Press. Below image: Reprinted by permission from Springer Nature Customer Service Centre GmbH: Springer Nature, *Nature* (https://www.nature.com/), Initial sequencing and analysis of the human genome, International Human Genome Sequencing Consortium, 2001: Fig. 2 on p. 863

---

[11] The enduring impact of cytogenetics—the observation and analysis of chromosomes—on genetics and genomics has been observed by Soraya de Chadarevian (2020) and Andrew Hogan (2016) with particular reference to the medical context.

For instance, as a result, human genes synthesising insulin—a protein used for the treatment of diabetes—could be expressed in a controlled way in bacteria (Rasmussen, 2014; Yi, 2015).

These molecular methods propelled the creation of a different type of map in the 1980s. Rather than representing the approximate location of genes and markers on the chromosomes—as the genetic linkage maps did—this new map visualised a set of 'physical' DNA fragments ordered as overlapping lines across the genome (see the lower part of Fig. 1.1). In organisms with larger genomes, the construction of these physical maps required the prior generation of libraries to store and manage the thousands of fragments into which the DNA contained in the different chromosomes would be broken.

Producing a 'DNA library' or 'genome library' involves using restriction enzymes and other recombinant techniques to insert DNA from the organism to be mapped into the genome of another organism (Hutchison, 2007; Loenen et al., 2014). As well as functioning as warehouses of the DNA inserts, the host organisms can also be used to amplify the fragments to be mapped, multiplying their number. This is achieved through the reproductive cycle of the host organism, which results in the production of cloned copies of the original inserted DNA. The libraries can be screened as well, for instance by hybridisation: using the property of chemical complementarity by which, in a double-stranded DNA molecule, adenines always bond with thymines and cytosines with guanines. Building on this, a probe containing a specific sequence can be designed to detect and locate particular fragments to which it will hybridise: chemically bond, due to the complementarity of its bases.[12]

In the early days of sequencing, viruses or circular chromosomes called plasmids—present in bacteria such as *Escherichia coli*—were used as host organisms for libraries, but these were limited in storage capacity. In 1987, though, Yeast Artificial Chromosomes (YACs) were developed, offering considerably larger storage capacity. Later, in 1992, Bacterial Artificial Chromosome (BAC) libraries were created, with several quality-related advantages over YACs to compensate for their smaller capacity.

Ordering the inserted DNA fragments of these libraries in physical maps enabled researchers to isolate and access those fragments, which could be

---

[12] Another key object in the use of genome libraries and genomics research more generally is the primer. Primers are DNA fragments designed to specifically attach to a sequence and trigger the amplification of a target genome region using the enzyme DNA polymerase. This enables researchers to obtain multiple copies of a particular stretch of DNA they seek to sequence, detect or otherwise investigate.

used for sequencing purposes or any other sort of genetic experiment. The overlaps detected between the fragments also allowed their assembly into a reference sequence, as was done with the human and other genomes (see lower part of Fig. 1.1). A central argument of this book is that the way in which libraries were constructed, and mapping was combined with sequencing, crucially distinguished the production of the yeast, human and pig reference genomes, thus embodying different forms of organising genomics, and affecting the potentialities and limitations of the resulting sequence data.

The growing ability to map and sequence DNA presented a problem: what to do with the resulting data. In 1980, the first global database to gather DNA sequences was launched. This was the Nucleotide Sequence Data Library, sponsored by the European Molecular Biology Laboratory as a shared repository to which the life sciences community could both submit their sequencing results and access the data contributed by others (García-Sancho, 2012). In 1982, the US National Institutes of Health (NIH) created an equivalent repository—GenBank, on which RefSeq would later be built—and, two years later, the DNA Data Bank of Japan started its operation. During their early years, these repositories struggled to keep up with processing the increasing quantities of sequence data being produced, while simultaneously having to confront the problem that much of what was being produced was kept by the laboratories that performed the work and not shared with the wider community. In 1987, the three databases reached an agreement by which their entries would be mirrored and users would be able to access the same information regardless of the repository they queried. Their curators also started persuading journal editors to make submission to one of the databases compulsory ahead of the publication of new DNA sequences, something that became increasingly customary in the 1990s (Strasser, 2019, Chs. 5–6; Stevens, 2018).

That same year of 1987, the journal *Genomics* was founded. It was co-edited by prominent medical geneticists Victor McKusick and Frank Ruddle, who in the previous decade had played a leading role in organising the first chromosome mapping workshop at Yale University. The first editorial of *Genomics*, entitled "A new discipline, a new name, a new journal" stated that mapping and sequencing DNA should go "hand in hand" since both practices had the "same objective". McKusick and Ruddle regarded mapping and sequencing genes as "the way to go" and the resulting sequence data as the "ultimate map" or the "Rosetta Stone" from which "the complexities of gene expression in development" could be discerned and the "genetic mechanisms of disease interpreted". For the "newly developing discipline" of mapping and sequencing DNA, the co-editors "adopted the term GENOMICS" (McKusick & Ruddle, 1987, p. 1, capitals in the original; see also Kuska, 1998). In the late-1980s and especially the 1990s, *Genomics* established itself as a platform for the dissemination of mapping and sequencing results, along with other journals that reported on the progress of ongoing genomic research.

At this time, scientists and administrators began to consider the full mapping and sequencing of the genomes of different species. Already in the late-1970s, the tiny genomes of viruses had been sequenced, but the scale-up to even bacteria was daunting given the skills and time that the existing techniques required. From the mid-1980s onwards, however, serious proposals to map and sequence the human genome were presented and a number of national programmes began. As we show later in the book (Chap. 3), the most ambitious of these was the Human Genome Project (HGP), which started as a joint endeavour of the NIH and laboratories of the Department of Energy of the USA.

By 1990, an array of human and non-human genome projects were underway. Some, like that for the nematode worm *Caenorhabditis elegans* and the American side of yeast genome sequencing, were conceived as pilots for human genome sequencing, allowing methods and approaches to be tried and evaluated, then adapted and improved for the bigger task of tackling a larger genome. Others, like the European side of yeast genome sequencing (Chap. 2), and the mapping of the pig genome (Chap. 5), were driven by the research aims of particular communities of scientists working on the biology of those organisms. As we argue, it was in the specificities of the interactions between these communities and their target genomes where differences between the genome projects arose and distinct ways of practising and organising genomics were configured (for a timeline illustrating milestones in the history of genomics across these species and some select others, see Fig. 1.2).

Genomics came into the public spotlight with the ambitious plans to sequence the entire DNA of humans. These plans—and particularly their materialisation in the HGP—have, quite naturally, attracted considerable attention both in scholarly and non-scholarly literature. In the late 1990s, the US programme coalesced with other initiatives into a transnational effort to determine a reference sequence of the whole human genome. The label HGP was kept, but the meaning of this, in both the popular imagination and for the scientists and administrators involved, shifted from the national US project to designate a broader, multi-national endeavour (Fortun, 1999). The reference sequence was published between 2001 and 2004 by an International Human Genome Sequencing

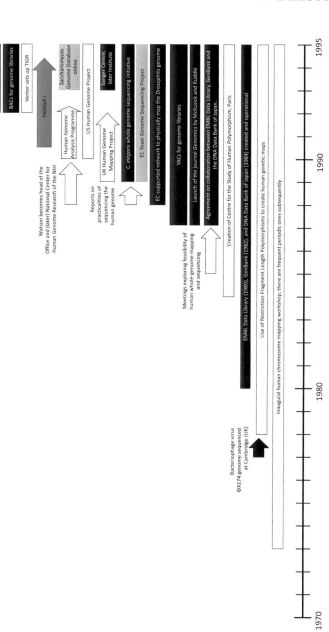

Fig. 1.2 Timeline representing historical milestones in DNA mapping and sequencing, as well as genomic research. White arrows refer to human genomics, light grey to yeast genomics and dark grey to pig genomics. Black arrows refer to technical or infrastructural developments. Elaborated by Jarmo de Vries from information compiled by James Lowe. For a larger version of this figure that can be zoomed in and out, see https://www.pure.ed.ac.uk/ws/portalfiles/portal/290406301/Fig_1_2_zoomable_final.pdf (last accessed 29th November 2022)

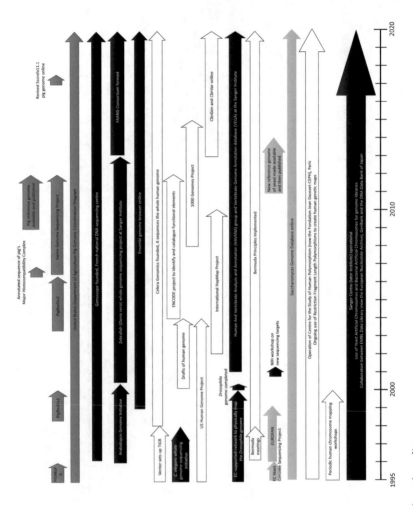

Fig. 1.2 (continued)

Consortium (IHGSC) formed by institutions from different countries, mainly the USA, UK, France, Germany, Japan and China (Chap. 4).[13] This was heralded as the entry of biology into the world of big science (Collins et al., 2003; Glasner, 2002; Hilgartner, 2013), a term characterising large-scale, coordinated scientific projects usually in the physical or engineering sciences, such as the World War II Manhattan Project, the Apollo space programme, or the creation and operation of CERN, the European centre for nuclear research (Barnes & Dupré, 2008, p. 43; Lenoir & Hayes, 2000).[14]

A central thesis of this book is that the excessive emphasis on the determination of the human reference sequence has led the history of genomics to be presented in a somewhat narrow fashion. By focusing on genomic work concerning non-human species—namely yeast and pig—and outside the HGP framework, we aim to capture a more richly-textured trajectory in which genomics forked, diversified and permeated in different ways across many areas of the life sciences and the world beyond them. We do this, in part, by unpacking the history of certain aspects of genomics that have come to be conceived of in a teleological manner: that they were created or happened in a certain way because that is how genomics would inevitably develop. These include the multiple possible ways in which genomes can be sequenced—with the HGP representing one strategy among many—and the diverse nature and utility of the reference sequences that are available today in the RefSeq database.

---

[13] As noted above, the name Human Genome Project and the acronym HGP are commonly used to refer to both the specific US programme and the later international initiative. In the remainder of this introductory chapter, our usage of HGP aligns to the latter sense: a coordinated effort that led to the production of the human reference sequence. Later in the book, and particularly in Chap. 3, we distinguish between the US human genome programme and later developments, designating the former as 'US-HGP' and differentiating it from the effort led by the International Human Genome Sequencing Consortium (the IHGSC endeavour).

[14] Some scholars of the life sciences query the novelty of the big science designation, drawing upon historical examples of large-scale coordinated endeavours that very much predated the HGP—and indeed the Manhattan Project—such as eighteenth-century voyages of discovery, surveys of the natural world, concerted ecological research programmes and networks of collection and information exchange—for example, associated with great museums, botanic gardens or around figures such as Charles Darwin (Aronova et al., 2010; Capshew & Rader, 1992; Strasser, 2019; Vermeulen, 2013). Others, while recognising that genomics does indeed constitute something new, highlight key differences between the way 'Bigness' manifests in the life sciences, in comparison to the physical or engineering sciences. The reasons for collaborating, and for forming networks and/or centralised facilities or resources, differ across the sciences, and even within the life sciences (Vermeulen, 2016; Vermeulen et al., 2013).

Based on the idea that the human reference sequence is often conceived of in a totemic manner, we now draw analogies between an HGP-centred history of genomics and the hourglass metaphor that some scholars have used to model and interrogate the history of heredity (Barahona et al., 2010). In this hourglass representation, there are two periods featuring heterogeneous activities conducted by a wide array of actors, one before and one after a bottleneck which is narrower in both content and participation. In the case of genomics, the neck of that hourglass corresponds to the later stages of the HGP (1996–2003), an initiative that has shaped the institutional landscape and infrastructures for mapping and sequencing endeavours well beyond itself. In what follows, we look beyond that narrow neck, and past an hourglass-based view of genomics more generally. We do this by paying attention to the needs and objectives of some often overlooked communities of researchers and the interactions they have with their target genomes, of both human and non-human species.

1.2   MOVING AWAY FROM A HUMAN GENOME PROJECT-CENTRED HISTORY OF GENOMICS

Since its inception, genomics has been an area with a significant concentration of humanities and social science scholarship. In 1988, a programme to examine the 'Ethical, Legal and Social Implications' (ELSI) of genomics was announced by James Watson, co-discoverer of the double helical structure of DNA and then head of the NIH Office for Human Genome Research. ELSI was formally launched in 1990 and awarded no less than 5% of the budget that the NIH would devote to human genomics. Other programmes encompassing 'Ethical, Legal and Social Aspects' were also launched in the early years of genomics. The one sponsored by the European Commission began as a small element of the second Framework Programme for Research and Innovation, running from 1987 to 1991. Projects and collaborations aiming to analyse the socio-ethical dimensions of genomics were particularly strong in the USA, UK, Netherlands, Germany and Canada.[15]

Sociological and ethical studies of human genomics have been particularly prominent, reflecting the societal concerns about the implications of the new technologies and the use of sequence data (e.g. see Gannett, 2019). These investigations have taken advantage of the possibility to pursue ethnographic approaches, examining the decision-making,

---

[15] See, for instance, Kevles & Hood, 1992; Sloan, 2000; Glasner & Rothman, 1998; Atkinson et al., 2007.

organisation and re-configuration of this new science as it happened (Hilgartner, 2017; Stevens, 2013). Histories have also been published, initially by people close to those involved, for example, Robert Cook-Deegan's *The Gene Wars* (1994; see also Gaudillière & Rheinberger, 2004). Philosophical accounts have explored the re-interpretations of the role of genes and genetics in the development of organisms in the light of the findings of genome projects (Keller, 2000; Moss, 2003). This includes aspects such as the smaller than expected number of human genes, the definition and identification of 'functional elements' (for example in the ENCODE—Encyclopedia of DNA Elements—project) and the so-called 'missing heritability' problem (e.g. Griffiths & Stotz, 2013; Guttinger & Dupré, 2016).

The existing historiography of genomics has been dominated by a particular phase of the HGP: that between the internationalisation, and radical scaling and speeding up of the project in the mid-to-late 1990s and the 'completion' of the reference sequence in the early 2000s. This was indeed the phase in which the vast majority of the data was produced. It was made especially salient by the story of a 'race' between the IHGSC, funded by an array of public bodies and charities, and the competing corporate effort led by Celera Genomics and its charismatic and controversial head, Craig Venter (Davies, 2001).[16]

This phase was one in which an extraordinary concentration of sequencing capacities was effected in a small number of institutions, with large and increasing numbers of sequencing machines, and ever-developing pipelines to produce, assemble and assess sequence data. Pipelines are series of successive software tools and algorithms configured to refine and validate inputs from sequencing to enable the resulting data to undergo further processing and be integrated into data infrastructures. In those pipelines, the sequences are assembled, with the parts growing smaller in number and larger in size, and more connected to each other (Fig. 1.1, bottom illustration). Many smaller laboratories and centres that had been involved in the earlier stages of human genome mapping were progressively sidelined from the effort. The advent of the reference genome heralded an era that became commonly known as 'post-genomic', reinforcing the equation of genomics with the HGP. 'Post-genomics' constituted an emergence from the narrow tunnel of the human reference sequence.

---

[16] The metaphor of a race has been criticised by Bartlett (2008), and the framing of competition between private and public sector projects—and values—has been qualified by other scholars (e.g., Fortun, 2006; García-Sancho, Leng, et al., 2022; Maxson Jones et al., 2018).

The canonical history of genomics—with its emphasis on the HGP—can be portrayed as an hourglass. In its upper part, there were a number of collective efforts to map the human genome and sequence those of other 'pilot' organisms such as yeast and the worm *C. elegans*. These efforts involved heterogeneous collections of institutions, some specialising in genomics, and others concerned with particular aspects of biology, such as anthropology, evolution, cell biochemistry or medical genetics. The later stages of the HGP from 1996 to 2003 constitute the narrow neck, tapered in because of the smaller number of institutions involved, the singularity of the aims of the programmes, and the radical abstraction of the potential genomic variation that was being captured in a single, consensus reference sequence. Then, in the lower part of the hourglass, there is an opening out to the world of post-genomics (Fig. 1.3, left).

This hourglass model refers to both the scope of genomics and the historical trajectory that the HGP-centred narrative conveys. According to this narrative, the pre- and post-genomic stages were wider in their range of activities and institutional variety, with the HGP resembling the hourglass neck through its focus on the production of a reference sequence at specialist genome centres. This narrative projects a winner's history in which the HGP is an obligatory passage point through which the sand in the hourglass flows: it is both the triumphant culmination of the pre-genomics stage and the opening to the post-genomic world.

The metaphor of an hourglass has also been used to productive effect when considering the history of the scientific study of heredity. In the second half of the nineteenth century, this research deployed a broad conception of heredity. In this, the roles of environment and inter-generational processes operating at different levels were explored and used to explain observed hereditary phenomena across a range of contexts. The advent of genetics as a discipline narrowed this sense of heredity, and also restricted the range of potential causal factors investigated and appealed to from the early 1900s onwards. This funnel effect, which was strengthened with the establishment of DNA as the genetic material, is what historians identify with the neck in the hourglass representing the study of heredity (Fig. 1.3, right). Then, later in the twentieth-century and into the twenty-first, the concept of heredity has once again been opened up and linked with examinations of organismal development, epigenetics, evolution and interactions with the environment, to produce new configurations such as evolutionary developmental biology. These remove the partitions between a version of heredity understood in terms of the inter-generational transmission of

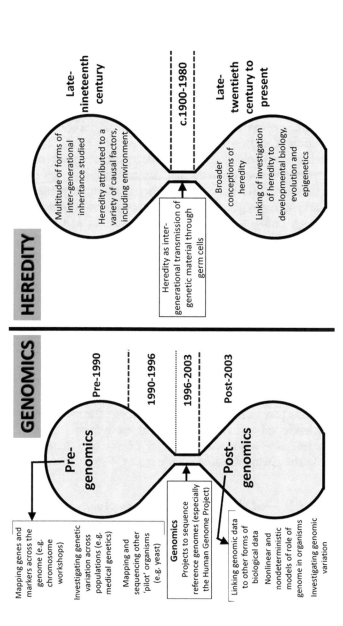

Fig. 1.3 An illustration of the two hourglass models we describe. The hourglass on the left represents the canonical history of genomics, as centred on the Human Genome Project. The hourglass on the right depicts the history of the scientific treatment of heredity over the nineteenth and twentieth centuries. In both cases, the hourglass models portray a change over time from a variety of practices, approaches and organisational forms (the upper part of the hourglass) to a narrower development (the neck of the hourglass) and then a return to a more diverse configuration (the lower part of the hourglass). Figure elaborated by both authors. For a larger version that can be zoomed in and out, see https://www.pure.ed.ac.uk/ws/portalfiles/portal/290406890/Fig_1_3_increased_final.pdf (last accessed 29th November 2022)

genetic material and other objects of biological research. We are now very much in the wider, lower part of the hourglass (Barahona et al., 2010).

While recognising the general utility of this metaphor, in making it explicit, its proponents have specifically interrogated the potential value and limitations of the hourglass model in the historiography of heredity. Could the hourglass be a "historiographical artifact" resulting from "historical research centered on a few actors and fields, most of them located in the American and British scenarios" (Barahona et al., 2010, p. 7)? Indeed, heredity was implicated in a wide range of endeavours beyond the mainstream genetics research that has traditionally been the focus of historical (and social scientific and philosophical) inquiry: medicine, agriculture, anthropology, genealogy, natural history and taxonomy, physiology, embryology and evolution. However, a cautious and critical use of the hourglass model has enabled its proponent historians to advance knowledge on these endeavours without neglecting the role and influence of the narrow neck representing genetics research.[17]

It is in this heuristic way that we intend to approach the hourglass model in the history of genomics. As we show later in the book, the effects of the HGP in the history of genomics are visible and self-evident. Key current institutions and infrastructures, such as RefSeq, were the products of its momentous impetus. The infrastructures, processes and materials produced through the HGP also shaped contemporary and subsequent genome initiatives, such as the sequencing of the yeast and pig genomes, respectively. In the USA, the NIH made the yeast initiative part of its national human genome programme: it was a pilot project through which technologies were developed and tested during the early-to-mid 1990s, thus preceding the intensive sequencing phase of *H. sapiens* (Chap. 2). Later on, in 2003, the Swine Genome Sequencing Consortium was

---

[17]Additionally, the hourglass model enables its proponents to unveil and scrutinise the tension between the desire to draw long-term lineages on the one hand and historicise and contextualise work in particular eras and domains on the other. One may want to trace the ways in which aspects of the upper part of the hourglass still survived and manifested in the neck, and were related to new developments in the lower part. But this should not come at the cost of equating twenty-first century interest in concepts of epigenetics with analogous examples of the ways that scientists connected organismal development and evolution in the late-nineteenth century (Barahona et al., 2010). This problem is not exclusive to historians: some scientists seek to draw historical parallels between their own interests and the ideas and practices of their predecessors (Scott Gilbert and Brian K. Hall are excellent examples of modern biologists interested in nineteenth century organismal development, see 1991 and 2009 respectively).

formed. It made use of the infrastructures and processes developed at the Sanger Institute, a leading member of the IHGSC (Chap. 5). It was leading members of the IHGSC that advocated for the subsequent transition to a 'post-genomic' era. When depicting this transition, its advocates often implicitly deployed an hourglass metaphor, with the HGP featuring in the narrow neck (Fig. 1.4).

Yet, however influential, the organisational model of the HGP, with its emphasis on concentration and maximised rates of production, was just one among other forms of genomics that historically emerged throughout the 1980s and 1990s: we argue that it was an unusual and rather exceptional one (Chap. 3). The other configurations demonstrate that the history of genomics is more complex and richly textured than the master narrative of the HGP and its representation in hourglass form may suggest. In order to appreciate this multifaceted history and its multiple genealogies, we need to look beyond the HGP and examine genome projects in human and non-human species that occurred before, during and after it. Another crucial way of moving beyond the restrictions of the hourglass model is placing the communities that produced the genomes—rather than the sequence end products—at the centre of our history.

1.3   'Thick Sequencing', Communities and 'Genomicists'

This book and the long-term historical narrative it encompasses enables us to probe, expand and develop a number of conceptual tools. While we use some of them for the first time here, we had originally proposed others elsewhere. Among the latter, we extend our distinction between *thin* and *thick* sequencing from its original context in making sense of pig genomics (Lowe, 2018), out to the history of genomics more generally. Thin sequencing is the compilation of the string of DNA nucleotides in order, while thick sequencing comprises all the processes, materials and organisational configurations that make the products of genomics—including the 'thin' sequence, but not limited to it—usable by a variety of potential actors. Thin sequencing is a feature of the narrowest point of the neck of the hourglass: it is the determination of the order of bases, whether manually or in a more automated way. This is not necessarily a simple task, as it requires the interpretation of recorded signals that are not always unequivocal. To understand the nature of genomics, however, and how its resulting outcomes can be taken up by different users in distinct ways, examining this part of the process alone is insufficient.

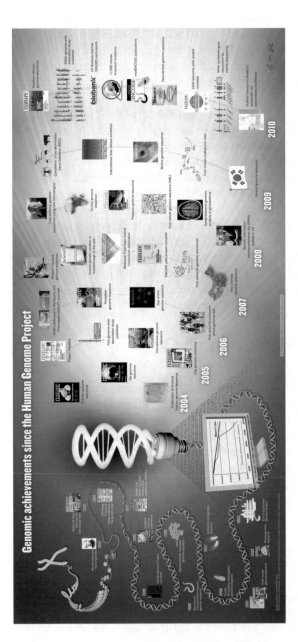

Fig. 1.4 A depiction of events preceding and succeeding the Human Genome Project (HGP) that illustrated an article co-authored by Eric Green and Marc Guyer in 2011. Green and Guyer were key scientific and administrative figures during the development of the HGP. After its conclusion, they were appointed director and deputy director, respectively, of the National Human Genome Research Institute of the NIH and tasked with planning what was by then called the 'post-genomic' era. In the illustration, the HGP is portrayed as a bulb powered by prior scientific achievements and illuminating subsequent milestones. The structure of events resulting from this past 'powering' the future places the HGP in a position that is analogous to the pinch-point of an hourglass. Reprinted by permission from Springer Nature Customer Service Centre GmbH: Springer Nature, *Nature* (https://www.nature.com/), Charting a course for genomic medicine from base pairs to bedside, Green and Guyer, 2011: Figure 1 on p. 205. A high-resolution version of this image is available in the open-access version of the article, which can be found online at: https://www.nature.com/articles/nature09764 (last accessed 29th November 2022). We thank Catherine Heeney for drawing our attention to the image

Capturing the thickness of sequencing means examining the obtaining and selection of DNA, its storage in DNA libraries, its mapping, the choice to sequence DNA fragments (clones) in YAC, BAC or other types of library, the extent of the coverage of the genome, and the selection of particular areas for more or less rigorous sequencing.[18] The sequences so generated then need to be assembled and annotated. All of these steps require decisions about what is to be abstracted from the variation that the different individual genomes exhibit in nature and what variation is to be represented in the final result. There are more stable aspects of this process, such as common pieces of software, sequencing and informatics pipelines, quality and validation standards, but the products also depend on the decisions and choices made in the whole thick sequencing process (Lowe, 2018). It is the thickening of our historical approach to sequencing—by focusing on practices such as library construction, mapping and annotation—that enables us to probe the hourglass representation and examine processes, trajectories and lineages beyond the narrow (thin) neck.

Through a thick sequencing framework, the differences between sequencing endeavours across species and how this affects the outcomes of genomics research—including reference genomes—become more manifest. One of the ways in which we capture these differences is by exploring the participation—or lack thereof—of particular communities of scientists in the production of reference genomes. These communities can be identified by coalescence around a particular object, such as a species, and/or a biological unit of it such as a cell. Additionally, or alternatively, they can be oriented around one or several biological processes such as heredity in the case of genetics, evolution or particular molecular mechanisms. These alliances are usually cemented and reinforced by common disciplinary membership and training, and participation in modes of scholarly communication and interaction such as a particular set of journals and conferences. These communities typically share "epistemic cultures" (Knorr-Cetina, 1999), and the extent of collaborative relations will be denser within members of a given community than between members of different communities.

---

[18] Coverage relates to the depth of sequencing: how many times on average that any given nucleotide in the sequence has been determined. 2X coverage means that, on average, a nucleotide will have two data points, 5X, five, and so on. Higher numbers would be more likely to iron out any random errors, resulting in a higher quality sequence.

There is no hard-and-fast rule for drawing the boundaries of particular communities, and weaker supra-communities or more specific sub-communities can also be identified. The notion of a community has long interested historians of science and scholars working in Science and Technology Studies (e.g. Shapin & Thackray, 1974). From the early days of both fields, a considerable amount of literature has explored the factors that lead scientists to group into communities and the dynamics of those groupings, from growth to stability, amalgamation, fragmentation or disappearance. Various mechanisms that glue communities together have been highlighted, among them common styles of thought or ways of knowing (Harwood, 1993; Pickstone, 2000), shared moral economies or working worlds (Agar, 2020; Kohler, 1994; Strasser, 2011) and particularly intense collaborative relationships (Vermeulen et al., 2013).

When we deploy the notion of community in this book, we refer to particular sets of individuals, laboratories and associated research practices converging around the description of a genome. Many of these consciously self-identify with communities, acting in concert to launch programmes and initiatives, and featuring specific conferences and venues of publication in common. Yet these communities are not homogeneous, and they may not exhibit the same characteristics or level of resolution. For instance, the community of yeast researchers we discuss (Chap. 2) is more heterogeneous than the medical geneticists we also survey (Chaps. 3 and 4). The pig genome community that we introduce (Chap. 5) is and was much smaller than both of these, but is in many respects broader, featuring different kinds of disciplinary backgrounds and researchers who have worked on other species, in addition to the pig. But, as we show, it was no less coherent a community for all that and acted as a community in shaping the genomics of their chosen species in a decisive and consequential manner. Genomics, and the object of a genome, can only be understood in relation to particular communities that it shapes as well as being shaped by, and wider social and technical configurations that it also impacts.[19]

---

[19] The notion of "working worlds", introduced by historian of science Jon Agar, helps us understand this entanglement between genomes and communities. Working worlds are spheres of activity that pose and frame particular problems, which scientists tackle by constructing and working with abstract representations (Agar, 2012, 2020). In this book, we consider the working worlds of medical geneticists that engage with real patients and the clinic, livestock geneticists that orient towards the needs of selective breeding for agriculture, and the development and use of one of a handful of model organisms in the biological sciences: yeast. What the reference genomes arising from these working worlds represent—and the problems that they are meant to address—varies significantly.

Our notion of communities builds on scholarship that considers the genome a rhetorical and practical space, as much as a material object (Szymanski et al., 2019). In this space, pre-existing scientific groupings can converge or fragment. Those, like the yeast biologists, who are more successful in defining and shaping the genome in their own terms, are in turn further unified by their orientation around the object of the genome. Human and medical geneticists, by contrast, formed a genome community that differed from the one assembled by the participants in the HGP.[20] This rhetorical and pragmatic definition enabled us elsewhere to highlight different characteristics of genomics research depending on the communities involved with a given genome: a strict separation of producers and users in the case of the human reference genome (García-Sancho, Leng, et al., 2022), different degrees of proximity and distance between yeast sequencing and particular research goals (García-Sancho, Lowe et al., 2022) and processes of bricolage or reuse of tools and resources that were deployed in the generation of the pig reference genome (Lowe, Leng, et al., 2022).

One conclusion arising from this community framework is that genomics can be regarded as a set of tools that enable groups of scientists to do different things and achieve different objectives with their target genomes (Lowe, García-Sancho, et al., 2022). Throughout the remaining seven chapters of this book, we propose the notion of yeast, human and pig genomicists as (often collective) subjects that make the history of genomics. In this process of construction, the genomicists mould their target genomes according to their necessities. They thus shape what these genomes represent and what they can do with them, sometimes quite consciously and deliberately.

This focus on communities of genomicists allows us to discern greater diversity and complexity in the history of genomics. In what follows, we show that yeast, human and pig genomicists have exhibited different mechanisms of inclusion and exclusion of particular sets of scientists and institutions. These have shaped each community differently and changed their compositions—and sometimes their roles—over time. The

---

[20] In this way, the genome is analogous to the "epistemic space" that was opened up for heredity and its scientific investigation in the mid-nineteenth century, giving rise to the historical trajectory that has been analysed through the hourglass model (see above). While narrower than the space of heredity, the genome shares heredity's "depend[ence] on a vast configuration of distributed technologies and institutions connected by a system of exchange" (Müller-Wille & Rheinberger, 2007, p. 25).

genomicists working on *S. cerevisiae* were relatively stable before, during and after the production of their reference genome, while in *H. sapiens* the leading genomicists of the early days were replaced by a different community based at specialist genome centres. For *S. scrofa*, the range of genomicists expanded, due to the convergence of a longstanding community of pig geneticists with practitioners from one of these specialist genome centres. These different trajectories further show that the history of genomics cannot be reduced to a single framework or periodisation.

Previous historiography has narrowly focused on a few, homogeneous genomicists: the participants in the HGP, recipients of the grants to determine the human reference genome and heads of the new institutions of genomics research: the genome sequencing centres. By looking at other less visible genomicists—those working on non-human organisms and beyond the HGP framework—we emphasise their agency as historical subjects and their capacity to pursue their own goals rather than following a teleological, pre-defined pathway. It is in the specificity of those goals and their agency in pursuing them where the interactions between the genomes and their communities occur and we identify trajectories and lineages that diverge from the canonical history of genomics. In other words, when a heterogeneous and inclusive array of genomicists is considered, genomics becomes something other than a static, retrospectively constructed field: it becomes a science (and history) in the making.

1.4   Outline of Chapters and Structure of Our Argument

The book is divided into three parts, comprising two chapters each. Taken collectively, these three parts de-centre the historiography of genomics: from a focus on *H. sapiens*; from an emphasis on the HGP; and, finally, from excessive attention to the determination of DNA sequences themselves (what we defined above as 'thin sequencing').[21] We achieve this by

---

[21] On other de-centring exercises in the historiography of science, see Andrew Cunningham and Perry Williams's work on the early-modern period (1993). They argue that what is now considered to be modern science did not emerge out of a single, sudden and epic event such as the so-called Scientific Revolution. Instead, there were a series of more gradual transformations that, over the sixteenth and seventeenth centuries, led to forms of knowledge-production more in line with our current understanding of science. A similar argument can be made with the HGP: however revolutionary and epic this event is presented, it does not in itself fully capture the emergence of genomics research.

exploring genomic endeavours around yeast, human and pig—including their reference genome projects—that started in the mid-1980s and concluded towards the late-2010s.[22] The sources that have enabled us to reconstruct these endeavours are oral histories, published literature—including scientific, administrative and policy reports—and archival materials. For the oral histories, we approached individuals ranging from Nobel Prize-winning scientists to administrators, lower-profile researchers and those devising and running the infrastructures of genomics. Our archival sources include catalogued and uncatalogued collections, as well as grey literature (see Appendix A and Appendix B at the end of the book for a complete list). We have also found extant and archived web pages to be useful in reconstructing parts of the history of genomics that had a lower public profile and lack an extensive secondary literature concerning them.

Part I of the book addresses what we call the distributed model of genomics. It starts with an account of the determination of the reference sequence of yeast: a non-human genome project that ended in 1996, just before the scaling-up of the HGP. The yeast effort enables us to show a greater variety of institutions and ways of organising mapping and sequencing practices than the ones behind the production of the human reference genome. Chapter 2 documents how institutional and organisational diversity was especially manifest in the European Commission-funded Yeast Genome Sequencing Project, which was not intended to serve as a pilot for the HGP, as the NIH *S. cerevisiae* genome programme was.

Similarly, a focus on the collective and systematic mapping work that preceded the large-scale sequencing characteristic of the latter stages of the HGP reveals a variety of heterogeneous human genome programmes. As we argue in Chap. 3, the HGP was but one among those many programmes: its focus on the rapid, industrial production of a reference sequence of the whole human genome was a particular—and rather singular—characteristic that distinguished the HGP from the others. The other, non-HGP programmes were more collective and inclusive of existing

---

[22] Our choice of these three species is necessarily selective, but as noted above encompasses different kinds of organisms used in distinct domains. Likewise, we have had to be selective in the choice of genomic projects and geographical scope concerning these species. Our focus on international initiatives—particularly those supported by the European Commission—has allowed us to provide an overview of the history of genomics that involves many different countries. In spite of this, further research on other species and geographical settings—most pressingly, Asia—would be valuable to complement and develop the arguments and perspectives that we raise in this book.

communities of medical geneticists. In order to accelerate the production of the reference sequence, the IHGSC that conducted the later stages of the HGP sidelined a large proportion of human and medical genetics institutions from its operation, starting in 1996. Yet these human and medical genetics communities continued their genome efforts, thus forming trajectories that the canonical winner's history of genomics overlooks.

Part II compares the production of the human reference genome with those of other species, especially the pig *S. scrofa*. Chapter 4 presents a main participant in the production of the human reference sequence: the Sanger Institute. Chapter 5 shows how this institution also played a major role in the subsequent sequencing of the pig genome that started in 2006, three years after the HGP was deemed concluded. At a first glance, the pig genome thus seems to be strongly modelled on the HGP. Yet, the broader history of pig genomics allows us to qualify that impression. If we take into account the early pig genome mapping work, started at the same time that the HGP was in the 1990s, we see that the scientific communities working on the agricultural genetics and immunogenetics of *S. scrofa* were intensely involved then and, unlike human and medical geneticists, continued to be. Indeed, institutions working on the genetics of pig immune response and traits relevant for selective breeding processes were important drivers and participants in the Swine Genome Sequencing Consortium that organised, managed and coordinated the reference genome work.

Taken together, Chaps. 4 and 5 continue the de-centring exercise that we started in Part I. In this case, the de-centring is not only due to our consideration of non-human species (pigs, as well as yeast) but also to our addressing of longer-term trajectories: considering genome mapping, as well as sequencing. We look at the sources of the DNA libraries from which the reference sequences were obtained and show that in both cases they were derived from a narrow pool of a few humans and pigs. Yet in the case of *S. scrofa*, the engagement of the early mapping communities in the sequencing operation eased the connection of the resulting reference genome with more general immunogenetic goals and the development of data and tools to aid the improvement of agriculturally-relevant breeds. These were the problems that motivated the mapping activity of pig genomicists before their involvement in whole-genome sequencing.

Part III comprises Chaps. 6 and 7. In it, we address a number of features that have been commonly attributed to post-genomics, such as connection of genomic data to other forms of biological data, and an attention to variation and diversity. We examine the annotation of reference genomes and other functional and systematic studies of sequence data. By the former, we mean the elucidation of the effects of particular genes and other genetic elements in the organism. By the latter, we mean the determination of patterns of variation within a given species or between species to inform, among other endeavours, evolutionary biology. We argue that our 'thick sequencing' approach—addressing the long-term processes by which DNA data become reference genomes—enables us to show that these practices have been deeply entangled throughout the whole history of genomics rather than necessarily following the completion of the HGP or any other reference sequence project.

Furthermore, in the case of the pig, the close involvement of the communities of immunogeneticists and agriculturally-oriented geneticists from the early days of genome mapping transformed annotation practices at the Sanger Institute into more collective and distributed endeavours. This paved the way to collaboration between two different communities of genomicists, one centred around the Sanger Institute and the other derived from the wider pig genetics community involved in mapping practices.

In our concluding Chap. 8, we explore the implications of our study beyond the realms of the history, philosophy and sociology of science. One of the preoccupations of science policymakers and funders in the wake of the HGP has been the notion of a 'translational gap' between the availability of masses of genome data and the exploitation of them, for example in effective new treatments or diagnostic tests in the clinic: 'from bench to bedside', as the slogan goes. We argue that this translational gap is an artifact of the particular configuration and history of the HGP: its model of concentrated production and the rigid division it implied between the producers of the reference sequence and the communities that would later use it in biomedical and clinical research. Other genomic endeavours that deployed more inclusive strategies show more immediacy and connection between the compilation of the data and its mobilisation towards particular goals. Our historical investigation thus illuminates ways of reducing the temporal, cognitive and conceptual distance between genomic data and user communities.

Dissatisfaction with reference genomes has given rise to new initiatives to represent genomic variation and to connect genomes to other forms of biological data and processes. As we show throughout, these qualms are based on trying to attribute particular functions to reference genomes and to make them carry weight that they were not designed or conceived for.

Our book highlights that many of these problems stem from the contingent and historically-driven processes of reference genome construction. Without a historical reconstruction, these processes and their consequences on the resulting reference genomes are flattened and rendered invisible.

References

Agar, J. (2012). *Science in the twentieth century and beyond*. Polity Press.

Agar, J. (2020). What is science for? The Lighthill report on artificial intelligence reinterpreted. *The British Journal for the History of Science, 53*(3), 289–310.

Arenas, M., Pereira, F., Oliveira, M., Pinto, N., Lopes, A. M., Gomes, V., et al. (2017). Forensic genetics and genomics: Much more than just a human affair. *PLoS Genetics, 13*(9), e1006960.

Aronova, E., Baker, K. S., & Oreskes, N. (2010). Big science and big data in biology: From the international geophysical year through the international biological program to the Long Term Ecological Research (LTER) Network, 1957–Present. *Historical Studies in the Natural Sciences, 40*(2), 183–224.

Atkinson, P., Glasner, P., & Greenslade, H. (2007). *New genetics, new identities*. Routledge.

Barahona, A., Suárez-Díaz, E., & Rheinberger, H.-J. (2010). *The hereditary hourglass. Genetics and epigenetics, 1868–2000*. Max Planck Institute for the History of Science Preprint 392. Retrieved December 4, 2022, from https://www.mpiwg-berlin.mpg.de/sites/default/files/Preprints/P392.pdf

Barnes, B., & Dupré, J. (2008). *Genomes and what to make of them*. The University of Chicago Press.

Bartlett, A. (2008). *Accomplishing sequencing the human genome*. PhD dissertation, Cardiff University.

Botstein, D., White, R. L., Skolnick, M., & Davis, R. W. (1980). Construction of a genetic linkage map in man using restriction fragment length polymorphisms. *American Journal of Human Genetics, 32*(3), 314–331.

Brownlee, G. G. (2014). *Fred Sanger – Double Nobel Laureate: A biography*. Cambridge University Press.

Capshew, J. H., & Rader, K. A. (1992). Big science: Price to the present. *Osiris, 7*, 2–25.

Collins, F. S., Morgan, M., & Patrinos, A. (2003). The Human Genome Project: Lessons from large-scale biology. *Science, 300*, 286–290.

Comfort, N. (2012). *The science of human perfection: How genes became the heart of American medicine*. Yale University Press.

Cook-Deegan, R. (1994). *The gene wars: Science, politics, and the human genome*. W. W. Norton and Company.

Cunningham, A., & Williams, P. (1993). De-centring the 'big picture': The Origins of Modern Science and the modern origins of science. *The British Journal for the History of Science, 26*, 407–432.

Davies, K. (2001). *The sequence: Inside the race for the human genome.* Weidenfeld & Nicolson.

de Chadarevian, S. (2002). *Designs for life: Molecular biology after World War II.* Cambridge University Press.

de Chadarevian, S. (2020). *Heredity under the microscope: Chromosomes and the study of the human genome.* The University of Chicago Press.

Dreger, A. D. (2000). Metaphors of morality in the Human Genome Project. In P. Sloan (Ed.), *Controlling our destinies: Historical, philosophical, ethical, and theological perspectives on the Human Genome Project* (pp. 155–184). University of Notre Dame Press.

Fortun, M. (1999). Projecting speed genomics. In M. Fortun & E. Mendelsohn (Eds.), *The practices of human genetics* (pp. 25–48). Kluwer Academic.

Fortun, M. (2006). Celera Genomics: The race for the human genome sequence. In A. Clarke & F. Ticehurst (Eds.), *Living with the genome: Ethical and social aspects of human genetics* (pp. 27–32). Palgrave Macmillan.

Gannett, L. (2019). The Human Genome Project. In Zalta, E. N. (Ed.), *The Stanford Encyclopedia of Philosophy* (Winter 2019 Edition). Retrieved December 4, 2022, from https://plato.stanford.edu/archives/win2019/entries/human-genome

García-Sancho, M. (2010). A new insight into Sanger's development of sequencing: From proteins to DNA, 1943–1977. *Journal of the History of Biology, 43*(2), 265–323.

García-Sancho, M. (2012). *Biology, computing and the history of molecular sequencing: From proteins to DNA, 1945–2000.* Palgrave Macmillan.

García-Sancho, M., Leng, R., Viry, G., Wong, M., Vermeulen, N., & Lowe, J. W. E. (2022). The Human Genome Project as a singular episode in the history of genomics. *Historical Studies in the Natural Sciences, 52*(3), 320–360.

García-Sancho, M., Lowe, J. W. E., Viry, G., Leng, R., Wong, M., & Vermeulen, N. (2022). Yeast sequencing: 'Network' genomics and institutional bridges. *Historical Studies in the Natural Sciences, 52*(3), 361–400.

Gaudillière, J.-P., & Rheinberger, H.-J. (2004). *From molecular genetics to genomics: The mapping cultures of twentieth-century genetics.* Routledge.

Gilbert, S. F. (1991). *A conceptual history of modern embryology.* Springer.

Glasner, P. (2002). Beyond the genome: Reconstituting the new genetics. *New Genetics and Society, 21*, 267–277.

Glasner, P., & Rothman, H. (Eds.). (1998). *Genetic imaginations: Ethical, legal and social issues in human genome research.* Routledge.

Green, E., Guyer, M., & National Human Genome Research Institute. (2011). Charting a course for genomic medicine from base pairs to bedside. *Nature, 470*, 204–213.

Griffiths, P., & Stotz, K. (2013). *Genetics and philosophy: An introduction.* Cambridge University Press.

Guttinger, S., & Dupré, J. (2016). Genomics and postgenomics. In Zalta, E. N. (Ed.), *The Stanford Encyclopedia of Philosophy* (Winter 2016 Edition). Retrieved December 4, 2022, from https://plato.stanford.edu/archives/win2016/entries/genomics

Hall, B. K. (2009). Tapping many sources: The adventitious roots of evo-devo in the nineteenth century. In M. D. Laubichler & J. Maienschein (Eds.), *From embryology to evo-devo: A history of developmental evolution* (pp. 467–498). The MIT Press.

Harwood, J. (1993). *Styles of scientific thought: The German genetics community, 1900–1933.* The University of Chicago Press.

Hilgartner, S. (2013). Constituting large-scale biology: Building a regime of governance in the early years of the Human Genome Project. *BioSocieties, 8,* 397–416.

Hilgartner, S. (2017). *Reordering life: Knowledge and control in the genomics revolution.* The MIT Press.

Hogan, A. J. (2016). *Life histories of genetic disease: Patterns and prevention in postwar medical genetics.* Johns Hopkins University Press.

Hutchison, C. A., III. (2007). DNA sequencing: Bench to bedside and beyond. *Nucleic Acids Research, 35*(18), 6227–6237.

International Human Genome Sequencing Consortium. (2001). Initial sequencing and analysis of the human genome. *Nature, 409,* 860–921.

Keating, P., Cambrosio, A., & Nelson, N. C. (2016). "Triple negative breast cancer": Translational research and the (re)assembling of diseases in post-genomic medicine. *Studies in History and Philosophy of Biological and Biomedical Sciences, 59,* 20–34.

Keller, E. F. (2000). *The century of the gene.* Harvard University Press.

Keller, E. F. (2011). Genes, genomes, and genomics. *Biological Theory, 6,* 132–140.

Kevles, D., & Hood, L. (Eds.). (1992). *The code of codes: Scientific and social issues in the Human Genome Project.* Harvard University Press.

Knorr-Cetina, K. (1999). *Epistemic cultures: How the sciences make knowledge.* Harvard University Press.

Kohler, R. (1994). *Lords of the fly: Drosophila genetics and the experimental life.* The University of Chicago Press.

Kuska, B. (1998). Beer, Bethesda, and biology: How "Genomics" came into being. *JNCI: Journal of the National Cancer Institute, 90*(2), 93.

Lederberg, J. (2001). 'Ome Sweet' Omics – A genealogical treasury of words. *The Scientist,* (April 2001).

Leng, R., Viry, G., García-Sancho, M., Lowe, J., Wong, M., & Vermeulen, N. (2022). The sequences and the sequencers: What can a mixed-methods approach reveal about the history of genomics? *Historical Studies in the Natural Sciences, 52*(3), 277–319.

Lenoir, T., & Hayes, M. (2000). The Manhattan Project for biomedicine. In P. R. Sloan (Ed.), *Controlling our destinies: Historical, philosophical, ethical, and theological perspectives on the Human Genome Project* (pp. 29–62). University of Notre Dame Press.

Lindee, M. S. (2005). *Moments of truth in genetic medicine.* Johns Hopkins University Press.

Loenen, W. A., Dryden, D. T., Raleigh, E. A., Wilson, G. G., & Murray, N. E. (2014). Highlights of the DNA cutters: A short history of the restriction enzymes. *Nucleic Acids Research, 42*(1), 3–19.

Lowe, J. W. E. (2018). Sequencing through thick and thin: Historiographical and philosophical implications. *Studies in History and Philosophy of Biological and Biomedical Sciences, 72*, 10–27.

Lowe, J. W. E., & Bruce, A. (2019). Genetics without genes? The centrality of genetic markers in livestock genetics and genomics. *History and Philosophy of the Life Sciences, 41*, 50.

Lowe, J. W. E., García-Sancho, M., Leng, R., Wong, M., Vermeulen, N., & Viry, G. (2022). Across and within networks: Thickening the history of genomics. *Historical Studies in the Natural Sciences, 52*(3), 443–475.

Lowe, J. W. E., Leng, R., Viry, G., Wong, M., Vermeulen, N., & García-Sancho, M. (2022). The bricolage of pig genomics. *Historical Studies in the Natural Sciences, 52*(3), 401–442.

Maxson Jones, K., Ankeny, R. A., & Cook-Deegan, R. (2018). The Bermuda triangle: The pragmatics, policies, and principles for data sharing in the history of the Human Genome Project. *Journal of the History of Biology, 51*(4), 693–805.

McKusick, V. A., & Ruddle, F. H. (1987). A new discipline, a new name, a new journal. *Genomics, 1*, 1–2.

Morange, M. (2020). *The black box of biology: A history of the molecular revolution.* Harvard University Press.

Moss, L. (2003). *What genes can't do.* The MIT Press.

Müller-Wille, S., & Rheinberger, H.-J. (2007). Heredity–The formation of an epistemic space. In S. Müller-Wille & H.-J. Rheinberger (Eds.), *Heredity produced: At the crossroads of biology, politics, and culture, 1500–1870* (pp. 3–34). The MIT Press.

Müller-Wille, S., & Rheinberger, H.-J. (2012). *A cultural history of heredity.* The University of Chicago Press.

O'Malley, M. A., Martin, W., & Dupré, J. (2010). The tree of life: Introduction to an evolutionary debate. *Biology & Philosophy, 25*, 441–453.

Ostell, J. (2013). What's in a Genome at NCBI? In *The NCBI handbook 2nd edition.* National Center for Biotechnology Information (US).

Pääbo, S. (2014). *Neanderthal man: In search of lost genomes.* Basic Books.

Pickstone, J. V. (2000). *Ways of knowing: A new history of science, technology and medicine.* Manchester University Press.

Rasmussen, N. (2014). *Gene jockeys: Life science and the rise of biotech enterprise.* Johns Hopkins University Press.

Rheinberger, H.-J., & Gaudillière, J.-P. (2004). *Classical genetic research and its legacy: The mapping cultures of twentieth-century genetics.* Routledge.

Rheinberger, H.-J., & Müller-Wille, S. (Trans. A. Bostanci). (2017). *The gene: From genetics to postgenomics.* The University of Chicago Press.

Riesenfeld, C. S., Schloss, P. D., & Handelsman, J. (2004). Metagenomics: Genomic analysis of microbial communities. *Annual Review of Genetics, 38,* 525–552.

Schwarze, K., Buchanan, J., Fermont, J. M., Dreau, H., Tilley, M. W., Taylor, J. M., et al. (2020). The complete costs of genome sequencing: A microcosting study in cancer and rare diseases from a single center in the United Kingdom. *Genetics in Medicine, 22,* 85–94.

Shapin, S., & Thackray, A. (1974). Prosopography as a research tool in history of science: The British scientific community 1700–1900. *History of Science, 12*(1), 1–28.

Sloan, P. R. (2000). *Controlling our destinies: Historical, philosophical, ethical, and theological perspectives on the Human Genome Project.* University of Notre Dame Press.

Stevens, H. (2013). *Life out of sequence: A data-driven history of bioinformatics.* The University of Chicago Press.

Stevens, H. (2018). Globalizing genomics: The origins of the International Nucleotide Sequence Database Collaboration. *Journal of the History of Biology, 51,* 657–691.

Strasser, B. J. (2011). The experimenter's museum: GenBank, natural history, and the moral economies of biomedicine. *Isis, 102*(1), 60–96.

Strasser, B. J. (2019). *Collecting experiments: Making big data biology.* The University of Chicago Press.

Suárez-Díaz, E. (2010). Making room for new faces: Evolution, genomics and the growth of bioinformatics. *History and Philosophy of the Life Sciences, 32*(1), 65–90.

Szymanski, E., Vermeulen, N., & Wong, M. (2019). Yeast: One cell, one reference sequence, many genomes? *New Genetics and Society, 38,* 430–450.

Tatusova, T., Ciufo, S., Fedorov, B., O'Neill, K., Tolstoy, I., & Zaslavsky, L (2014). About prokaryotic genome processing and tools. In *The NCBI handbook 2nd edition.* National Center for Biotechnology Information (US).

Venter, J. C., Remington, K., Heidelberg, J. F., Halpern, A. L., Rusch, D., Eisen, J. A., et al. (2004). Environmental genome shotgun sequencing of the Sargasso Sea. *Science, 304*(5667), 66–74.

Vermeulen, N. (2013). From Darwin to the census of marine life: Marine biology as big science. *PLoS ONE, 8*(1), e54284.

Vermeulen, N. (2016). Big Biology. *NTM Zeitschrift für Geschichte der Wissenschaften, Technik und Medizin, 24*, 195–223.

Vermeulen, N., Parker, J. N., & Penders, B. (2013). Understanding life together: A brief history of collaboration in biology. *Endeavour, 37*(3), 162–171.

Winther, R. G. (2020). *When maps become the world*. The University of Chicago Press.

Yi, D. (2015). *The recombinant university: Genetic engineering and the emergence of Stanford biotechnology*. The University of Chicago Press.

PART I

The Diversity of Genomics

CHAPTER 2

Distributed and Concentrated Strategies in the Sequencing of the Yeast Genome

John Sulston, the scientist who led the British contribution to the human reference genome sequence, considers that what is now called the Human Genome Project really got started with simpler organisms in 1989. That year, he passed a point of no return in his career that led him to see the sequencing of whole genomes through the scaling-up of technologies and scientific teams as the only way forward. This moment, which Sulston compares to a "prison door" shutting behind him, occurred during the seventh international meeting on the nematode worm *Caenorhabditis elegans*, held at Cold Spring Harbor Laboratory (CSHL) on the south coast of Long Island Sound, in New York state. There, Sulston and his associates Alan Coulson and Robert Waterston unveiled a physical map of ordered DNA fragments that encompassed the whole genome of *C. elegans*. This worm, of about one millimetre length, had become both Sulston's obsession and a widespread model organism in genetics research over the preceding 20 years. In 1983, after tracing all the divisions of *C. elegans* cells during embryonic and post-embryonic development, Sulston had embarked on assembling the map of its genome, first with Coulson and later in collaboration with Waterston. When James Watson, director of CSHL and Nobel Prize winner for his co-discovery of the structure of DNA, saw the map at the meeting, he exclaimed: "you can't see it without wanting to sequence it, can you?" (Sulston & Ferry, 2002, pp. 13–14).

© The Author(s) 2023
M. García-Sancho, J. Lowe, *A History of Genomics across Species, Communities and Projects*, Medicine and Biomedical Sciences in Modern History, https://doi.org/10.1007/978-3-031-06130-1_2

Watson was by then combining his long-term directorship of CSHL with a new appointment as associate director for Human Genome Research at the US National Institutes of Health (NIH). His remark propelled a frantic series of meetings in which Sulston, Coulson and Waterston committed to sequence 3% of the worm's genome, the largest portion of DNA that had been tackled to date. Watson offered to support the operation through the Office for Human Genome Research, which the NIH had established in 1988. Following favourable review of the detailed proposals, the NIH funded the whole sequencing enterprise in the USA—led by Waterston's team at Washington University in St Louis (WU)—and one-third of Sulston's work at the Laboratory of Molecular Biology (LMB) in Cambridge (UK). The rest of the funding was provided by the UK Government through its national Human Genome Mapping Project (Chap. 3). This international initiative started in 1989, just months after the CSHL meeting. Three years later, in 1992, three of the 100 million nucleotides of the worm genome had been completed. The sequencing effort was presented as a "pilot system" to test the technologies and feasibility of addressing the human genome, as well as interpreting the resulting data (Sulston et al., 1992, p. 37). The human genome comprises 3 billion DNA nucleotides, and so is about 30 times larger than that of *C. elegans.*

What is less known—and absent from Sulston's account—is that *C. elegans* was one of the drivers in the sequencing of another genome: that of the baker and brewers' yeast, *Saccharomyces cerevisiae*. Prior to that fateful 1989 meeting, Waterston had started using Yeast Artificial Chromosomes (YACs) in the physical mapping of the worm. This tool, developed in the 1980s using recombinant DNA techniques, allowed the insertion of foreign genetic material into yeast cell cultures. By using the replication mechanisms of yeast—a single-celled fungus—researchers could multiply (clone) the foreign inserts and obtain enough DNA of the organism they were working with. Waterston inserted the worm DNA fragments that he wanted to map into YACs and then multiplied them, producing a library of *C. elegans* fragments stored in *S. cerevisiae* cells. The fragments could then be isolated and their position within the worm genome determined. Given that this procedure yielded large amounts of yeast as well as *C. elegans* DNA in the cell cultures, Waterston included a project in his NIH grant to sequence *S. cerevisiae* on top of the worm, to distinguish the DNA of the two species. The project, which ran in parallel with the *C. elegans* sequencing effort, enabled WU to join an incipient multi-national and multi-institutional initiative to complete the entire yeast genome.

This chapter shows how Waterston's two-species effort signalled the emergence of a handful of groups that absorbed unprecedented amounts of funding for comprehensively mapping and sequencing whole genomes. Despite these comprehensive efforts starting with microscopic organisms such as yeast and *C. elegans*, the intention of their sponsors from the outset was to concentrate resources and capacities that these groups could later deploy for the human genome. Apart from WU, the NIH channelled its funding towards another yeast sequencing group at Stanford University (Szymanski et al., 2019, p. 36). In 1993, when the yeast sequencing effort was proceeding apace towards completion, the NIH transformed these two groups into the Genome Sequencing Center at WU and the Genome Technology Center at Stanford. That same year, Sulston left the LMB to become the founding director of the Sanger Institute, an institution that would comprehensively map and sequence the *C. elegans*, yeast and human genomes with funding from the UK Government and, especially, the British biomedical charity Wellcome Trust (Chap. 4).[1] In the sequencing of *S. cerevisiae*, WU, Stanford and the Sanger Institute cooperated and competed with smaller institutions from Canada, Japan and the USA, as well as a transnational consortium of laboratories sponsored by the European Commission (EC).

The chapter compares the strategy of concentrating funding and resources in specific groups that subsequently became genome centres, with the approach that the EC undertook in the sequencing of yeast. Unlike the NIH and the Wellcome Trust, the EC avoided channelling funding into just one or a few teams and preferred instead to distribute its support across a wider range of laboratories based in multiple European

---

[1] The Sanger Institute was originally called Sanger Centre and kept this name until 2001, five years after the completion of the yeast genome. The Stanford Genome Technology Center was named Stanford DNA Sequencing and Technology Center during the *S. cerevisiae* effort and until 2000. The Genome Sequencing Center at Washington University became known as the Genome Center at Washington University in 2009, the Genome Institute at Washington University in 2011 and the McDonnell Genome Institute at Washington University in 2015. To avoid confusion, we uniformly use Sanger Institute, Stanford Genome Technology Center and Genome Sequencing Center at WU throughout, unless we more concisely designate the latter two as 'Stanford' and 'WU', respectively.

countries. This partly followed from the agenda of what was then called the European Community—from 1993 the European Union—and the opportunities that a networked yeast genome project presented for fostering political and economic integration among member-states (Parolini, 2018). In contrast with its counterparts in the UK and the USA, the EC did not regard yeast sequencing as a springboard to tackle the larger human genome: it was, rather, a means to encourage and cement cross-European scientific and industrial collaboration, based on the potentialities of yeast as a model and industrial organism. This led the yeast sequencing consortium to be dominated by academic and corporate laboratories that were already investigating *S. cerevisiae* as a biotechnological object, a brewing instrument, or a model organism for genetics and cell biological research. The consortium also included a group of companies— some of them start-ups arising out of universities and publicly-funded research institutes—that provided sequencing services for *S. cerevisiae* and other genome projects sponsored by the EC.

The coordinated action of these laboratories created a distributed approach to sequencing that, as we show below, was re-implemented in subsequent EC projects and shared by the majority of national human genome programmes that emerged from the late-1980s onwards (Chap. 3).[2] We argue that this approach diverges from the canonical history of genomics—and its hourglass representation (Chap. 1)—in that a heterogeneous array of institutions exhibited diverse ways of sequencing DNA and modes of interacting with each other and external bodies. Crucially, these institutions persisted in their operation, without being replaced by a more homogeneous landscape of genome centres. However, the concentrated strategy that the NIH and Wellcome Trust pursued, along with changes in EC policies, increasingly reduced the visibility and scope of this distributed mode of genomics into the 2000s.

---

[2] Elsewhere, we have used the term "network genomics" to designate this approach and explored the implications of this way of sequencing for the historiography of genomics and biotechnology (García-Sancho, Lowe, et al., 2022). Our preference for *distributed* here seeks to emphasise the contrast between the EC strategy and the concentration of resources into Sulston and Waterston's groups, as well as that at Stanford.

2.1 OUT OF *C. ELEGANS* SEQUENCING

By 1989, the year of the crucial CSHL meeting, Sulston, Coulson and Waterston had established themselves as key drivers of the *C. elegans* community. As historians have documented, this tiny worm had become a widespread model organism for genetics research in the 1970s (Ankeny, 1997), and Sulston and Coulson's 'fingerprinting' mapping techniques had subsequently emerged as an obligatory passage point for the investigation of *C. elegans* genes. From the mid-1980s onwards, an increasing number of laboratories sent samples of *C. elegans* DNA to the LMB that they had previously identified as corresponding to—or located nearby—genes involved in behavioural, developmental or any other biological functions in the worm. Sulston and Coulson would position the DNA samples within their ongoing physical map and report the results back to the laboratories who had sent them (de Chadarevian, 2004; García-Sancho, 2012b).

Knowledge of the chromosome or chromosomal region in which their samples were located enabled the laboratories to progress their research, allowing them to detect and isolate other DNA fragments as part of the genes they were pursuing. Sulston and Coulson, on their side, could refine their maps by adding the samples they received and increasing the overall number of ordered fragments (Fig. 2.1). The samples were initially delivered to Cambridge as cosmid clones: colonies of bacteria that had propagated from a single one in which the genetics laboratories had inserted the DNA they wanted to be mapped. Upon fingerprinting analysis, Sulston and Coulson looked for overlaps between the sequence of the *C. elegans* DNA contained in the sample and others from cosmids they had already mapped. Overlapping sequences suggested that the corresponding DNA fragments were contiguous in the worm's genome. After Waterston joined the team in 1985, he screened for additional sequence matches with the larger YACs that he compiled in St Louis.

This strategy of completing the *C. elegans* map while providing a service to other laboratories shifted significantly when Sulston, Coulson and Waterston embarked upon the sequencing of the worm's genome. Following their conversation with Watson at CSHL, the three scientists started advocating an approach in which their laboratories at WU and LMB would sequence the whole worm genome on their own initiative, rather than relying on requests and sample deliveries from *C. elegans* geneticists. This was the same approach that Watson sought to implement

Fig. 2.1 Left, Alan Coulson beside the physical map of *C. elegans*, pinned on the wall of the central theatre of Cold Spring Harbor Laboratory in 1989. James Watson looked at the map and proposed that he, John Sulston and Robert Waterston sequence the genome of this nematode worm. Right, the outcome of Sulston and Coulson's fingerprinting technique: an autoradiograph picture of a gel on which the DNA fragments to be mapped had been run, one in each column, along with marker fragments that were used for reference. Each black spot on the picture corresponds to a sub-fragment into which the original fragments had been fractionated. Since the enzymes used to fractionate the fragments cut at specific sequence sites, a matching pattern of spots in two or more columns—or across different autoradiographs—meant that the fragments overlapped in the genome. Left image courtesy of Barry Honda and retrieved from Jenny Shaw (2014) "Alan Coulson's science of collaboration", blog post produced for the Wellcome Library and available at: https://wayback.archive-it.org/16107/20210313022805/ http://blog.wellcomelibrary.org/2014/07/alan-coulsons-science-of-collaboration/ (last accessed 7th December 2022). Right image: reproduced from Sulston, S, Mallett, F, Staden, R, Durbin, R, Horsnell, T, & Coulson, A, Software for genome mapping by fingerprinting techniques. *Bioinformatics*, 1988, 4(1), 125–132: Fig. 1 on p. 126, by permission of Oxford University Press

for the mapping and sequencing of the human genome, the task he had been set in his new position at the NIH Office. As we show in the next chapter, in 1989 he was finalising an agreement with the US Department of Energy by which this institution and the NIH would contribute three billion dollars towards the completion of the human genome map and sequence between 1990 and 2005 (Chap. 3). This programme—unprecedentedly large in the molecular life sciences in its level of funding and 15-year time horizon—enabled the *C. elegans* mappers to undertake a comprehensive, whole-genome sequencing operation; one that would not be conditioned by external requests of sequence data.

In Cambridge, Sulston and Coulson approached the Medical Research Council (MRC) as a potential funder of the two-thirds of the worm's sequencing project that the NIH did not provide for. The MRC is the agency of the UK Government that funds and oversees biomedical research, including the operation of the LMB. In 1989, the same year as the CSHL meeting, the UK Treasury had granted the MRC an extra 11 million pounds to run a national programme to map and sequence the human genome. Yet the UK programme, called the Human Genome Mapping Project, only had guaranteed funding for 3 years compared to the 15 years of its US counterpart and was not committed to whole-genome mapping and sequencing. This led Sulston and Coulson to propose, in their funding application to the MRC, a phased approach that would start by focusing on targeted regions of the worm's genome, and then develop the efficiency and output of the sequencing techniques, culminating in a "factory style operation". The end goal was to move beyond the three-year support framework, and comprehensively sequence DNA fragments from mapped cosmid clones and YACs encompassing the entire *C. elegans* genome, so it would be completed "in a time not longer than 10 years".[3]

Waterston's parallel proposal to the NIH took the intended efficiency and comprehensiveness of such a factory model further. At the same time as the *C. elegans* operation, his department at WU had hosted another mapping initiative aimed at the yeast *S. cerevisiae* and led by Maynard Olson (Szymanski et al., 2019, pp. 435ff). Olson had provided yeast DNA to Waterston and other colleagues for the construction of YACs. The availability of this source of DNA, together with the use of YACs in the mapping of *C. elegans*, led Waterston to include a side project—called Project 2—to sequence *S. cerevisiae* as well as the worm in his formal

---

[3] John Sulston and Alan Coulson (1989): "Mapping and sequencing the genome of *Caenorhabditis elegans*", application to the UK Medical Research Council's Human Genome Mapping Project. Obtained through the Medical Research Council and available at the Wellcome Library, London, Papers and Correspondence of Sir John Sulston, file number PP/SUL/A/2/1/3. Later in this book (Chap. 4), we provide a more detailed analysis of Sulston and Coulson's proposal.

funding application to the NIH. From a *C. elegans* sequencing perspective, this project sought to distinguish between yeast and worm DNA in the YACs. Yet by using Olson's map, the yeast sequence data could also be assembled and stored, rather than just being discarded as contamination.

By the time Waterston was writing his proposal, Olson had moved from St Louis to a new Department of Molecular Biotechnology at the University of Washington in Seattle. Instead, yeast geneticist Mark Johnston joined Project 2 and more generally became responsible for yeast sequencing at WU after 1992. Olson's former technician at WU, Linda Riles, started working on the NIH grant and providing clones, including in YACs, whose sequencing would be collaboratively overseen by Johnston and Waterston.[4] Johnston inherited from Olson the spirit of distributing clones with the mapped yeast DNA fragments to other laboratories that were starting sequencing projects. Among these were the members of the EC consortium and two independent groups of Canadian and Japanese institutions led by McGill University and RIKEN (**Ri**kagaku **Ken**kyūjo, the Institute of Physical and Chemical Research), respectively.

Waterston's proposal was funded as part of the NIH contribution to the human genome through the National Center for Human Genome Research. This institution had succeeded the Office for Human Genome Research in 1989 and also sponsored one-third of Sulston and Coulson's endeavour (Fig. 2.2). Waterston's funding was for five years. In the case of yeast, his grant aimed to complete the whole of chromosome VIII of this organism "by mid-1994" and sequence "parts" of other chromosomes "totalling 2.5" million nucleotides. It was envisaged that, along with other efforts "in progress or planned worldwide", the entire *S. cerevisiae* genome would be determined "by the end of 1995".[5] The WU team completed the sequence of chromosome VIII by the target year of 1994 (Fig. 2.3) and led the determination of another full chromosome—XII—that was published in 1997, co-authored with the EC consortium. The WU group also contributed to chromosomes IV and XVI in collaboration with the Sanger Institute, Stanford University, the EC consortium, and McGill and

---

[4] Mark Johnston, interview with both authors via Skype, September 2020.

[5] Robert Waterston and Mark Johnston (1989) "Project 2." In R. Waterston, "Sequencing the *C. elegans* genome," p. 175. The NIH application, excluding Project 2, is available at the Wellcome Library in London, Papers and Correspondence of Sir John Sulston, file number PP/SUL/A/2/1/5. The "Project 2" part of the application was obtained through Mark Johnston and is courtesy of Robert Waterston.

Fig. 2.2 The cover of a 1992 meeting programme at Cold Spring Harbor featuring John Sulston and Robert Waterston as *C. elegans* worm cyclists competing with other organisms in a race to finish their genomes. The race is refereed by James Watson, pictured beside the 'Start' sign with his hand raised. Courtesy of Cold Spring Harbor Laboratory Archives. The image was introduced to us by Marina Schutz, who analysed it at the workshop "Cooperation and Competition in the Life Sciences", held in November 2019 at the Ludwig-Maximilian University of Munich. It is also available at: Papers and Correspondence of Sir John Sulston, Wellcome Library, London (UK), file number PP/SUL/A/6/16

Concordia Universities in Canada, as well as the University of Toronto Hospital for Sick Children.

Another element in common between the NIH-led sequencing of yeast and *C. elegans* was that, in both cases, the effort was coordinated between two groups. Yet in the case of *S. cerevisiae*, the WU group partnered with a US institution—Stanford University—rather than a British one. While the partnership between WU and the LMB derived from Waterston and Sulston's collaboration in the prior mapping of *C. elegans*, with yeast there

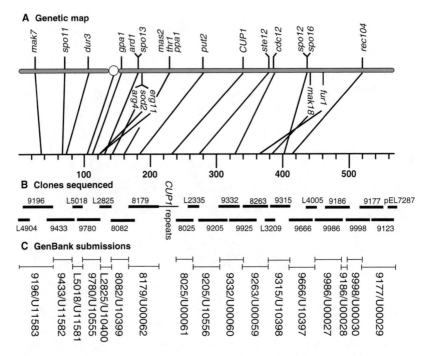

Fig. 2.3   Genetic linkage and physical maps that Mark Johnston used to sequence chromosome VIII of the yeast *S. cerevisiae*. In the genetic linkage map—upper part of the figure, marked as 'A'—the vertical lines represent relative positions on the chromosome and the numbers denote relative distances. In the physical map—middle of the figure, marked as 'B'—the horizontal lines represent partially over-lapping DNA fragments and the alphanumeric codes denote the samples from which those fragments were obtained, such as a cosmid library or another form of cloning system. The lower part of the figure, marked as 'C' corresponds to the data submitted and stored in GenBank, the global DNA sequence database that was established in the USA in 1982. The whole yeast chromosomal sequence was frac-tionated into various, non-overlapping manageable chunks that are represented as horizontal lines with their submission references listed below them. Johnston et al. (1994: 2080, Figure 1). Reprinted with permission from the American Association for the Advancement of Science

was not a strong inter-personal or inter-institutional connection beyond Johnston having completed a postdoctoral fellowship at Stanford before moving to St Louis. Johnston's research at Stanford had explored the genetics of glucose metabolism in *S. cerevisiae*. His mentor was Ronald Davis, a contributor to the development of the first recombinant DNA techniques at Stanford's Department of Biochemistry during the mid-to-late 1970s (Yi, 2015, Ch. 2). These techniques, which allow the transfer of genetic material from one organism to another, were used by Johnston to isolate and study the expression of the *GAL* genes responsible for the synthesis of proteins that process sugars in the yeast cell.

In 1990, the reputed human and microbial geneticist David Botstein moved to Stanford. Botstein had a history of collaboration with this institution, since in 1980 he had co-authored a seminal article with Davis and other colleagues proposing to map the human genome using the restriction enzymes that form the basis of the recombination procedure (the Restriction Fragment Length Polymorphism—RFLP—approach; Botstein et al., 1980; Chap. 1). In the recombination procedure, restriction enzymes cut DNA at specific sequence sites and therefore allow cleavage, transfer and insertion of genetic material from one organism to another where it can be expressed and studied under controlled conditions. Botstein was later appointed vice-president of Genentech, the biotechnology company that Stanford had created to develop and market commercial products derived from the recombinant technologies, such as human insulin expressed in bacteria (Rasmussen, 2014, Ch. 2; Yi, 2015, Chs. 4–5). By the time he moved to Stanford, Botstein had become an enthusiastic advocate of using yeast as both a model to investigate gene function and a system to express recombinant DNA molecules (Botstein & Fink, 1988).[6]

Botstein and Davis were regarded by Watson as representatives of a younger generation of molecular biologists, one that had scientifically and commercially exploited the elucidation of the double helix and other milestones around the structure and function of DNA. This perceived continuity led Watson and others who had been prominent in shaping molecular biology to mobilise a success story around biotechnology and recombinant DNA as a way of retaining their influence and authority. Botstein and Davis were tasked with co-leading a second NIH-funded yeast sequencing

---

[6] Botstein's advocacy built on a strong tradition of using yeast as a model organism for genetics and cell biochemical research. This tradition dated back to the mid-twentieth century and had witnessed important contributions from the neighbouring University of California Berkeley and WU-based researchers, before the start of the sequencing programme (Langer, 2016; Szymanski et al., 2019, pp. 434ff).

group (Szymanski et al., 2019, p. 436). Their Stanford team published the whole sequence of chromosome V in 1997 and was involved with WU in the collaborative sequencing of chromosomes IV and XVI of *S. cerevisiae*. It also hosted the *Saccharomyces* Genome Database, a centralised repository that was founded in 1993 and released to the community one year later (Chap. 7).

As the yeast and worm genome endeavours progressed, the St Louis sequencing teams were incorporated into the larger WU Genome Sequencing Center, and Davis and Botstein's group into the Stanford Genome Technology Center. Both centres extended their remit to contribute to the NIH arm of the (by then fully inaugurated) Human Genome Project of the USA: the one at WU via large-scale sequencing and the one at Stanford through the development of advanced instrumentation (Chap. 3). In 1993, the same year in which these US centres were founded, Sulston became director of the Sanger Institute and created specific divisions for the sequencing of the human and yeast genomes, as well as *C. elegans* (Chap. 4). A common feature of these three centres was their unprecedented level of funding—considerably higher than the groups from which they derived—and the conditioning of this support on an absolute prioritisation of the genome efforts, rather than them using the sequence data for research into human, yeast or *C. elegans* biology. This was the fundamental difference between them and the laboratories involved in the yeast sequencing project sponsored by the EC.

2.2   The Distributed Strategy
of the European Commission

The EC initiative was called the Yeast Genome Sequencing Project (YGSP) and started in 1989. In common with the St Louis effort—and later with Stanford and the Sanger Institute—it had the objective of producing a full reference sequence of *S. cerevisiae* that would encompass its whole genome and function as a representation of this type of yeast. Yet the strategy for achieving this common goal markedly differed between the EC, and the US and British projects. While Stanford, WU and the Sanger Institute comprehensively sequenced entire yeast chromosomes with a view to applying the expertise that they cultivated (and the technologies that they refined) through this to the human genome, the EC pursued the YGSP with a distributed approach.

This distributed approach aligned with the broader political and economic agenda of the EC, the executive arm of the European Community. It resulted in the EC funding being spread across a larger number and variety of both public and private institutions, contrary to the concentration of resources in the WU and Stanford teams, and later the Sanger Institute. Furthermore, the sequencing work undertaken by the YGSP participants was more immediately motivated by research objectives that went beyond producing a reference sequence of *S. cerevisiae*. Within the YGSP consortium, academic institutions conducted genetics, biochemical and cell biological research with the sequence data, while the companies and corporate laboratories used their experience on the project to foster their commercial activities and expand their customer bases.

The YGSP was part of the EC's Second Framework Programme for Research and Technological Development. By the time this programme was launched in 1987, the EC—and the European Community more generally—were pushing for stronger political and economic integration among their member-states. This process had gained new impetus with the adoption of the Single European Act—which also came into effect in 1987—and led to the development of policies that would crystallise in the establishment of the European Single Market. Research and development were regarded as key means towards economic convergence, with the EC conceived as the institution that would drive transnational cooperation in science and technology. The funds from the framework programmes were channelled through more specific schemes and, right from the start, biotechnology was identified as an emergent and key area to align scientific and economic agendas across Europe (Bud, 1993, Chs. 7 and 9; see also Bud, 2010, p. S18; Cantley, 1995). The Biotechnology Action Programme (BAP; 1985–1989) and later the Biotechnology Research for Innovation, Development and Growth in Europe (BRIDGE programme; 1990–1993) supported the first years of the YGSP.

Drawing on this political architecture, the sequencing of the yeast genome represented an opportunity to build international scientific networks and contribute to economic development within the increasingly integrated European market. The brewing, pharmaceutical and biotechnology industries in the European Community were, from the outset, considered players in both the sequencing effort—the consortium included corporate participants such as Carlsberg Laboratory and Pharmacia Biotech—and the Yeast Industrial Platform, a group of member-state companies to which the EC provided privileged access to the sequence data (Parolini, 2018).

As project coordinator, André Goffeau was responsible for embedding these principles into the day-to-day running of the YGSP. His dual status of yeast geneticist at the Catholic University of Louvain and civil servant at Directorate-General XII—the department of the EC that oversees research, science and technology—enabled him to press for a yeast sequencing operation as part of the transnational projects included in the framework programmes. With Watson, Goffeau shared a tradition of working with standardised organisms and the vision of a reference sequence as a valuable resource for his research community. From the 1950s onwards, Watson and other molecular biologists had adopted model organisms such as phage viruses or *C. elegans* as "exemplars" to study different life properties. Similarly, yeast geneticists, biochemists and cell biologists had bred and disseminated a specific strain of *S. cerevisiae*—named S288C—to conduct their experiments and other explorations of yeast biology, including genome mapping (Strasser & de Chadarevian, 2011; Szymanski et al., 2019, pp. 434ff).[7]

Yet the strategies that Watson and Goffeau imprinted into the NIH and EC yeast genome programmes differed substantially. The latter, in line with EC policy, regarded sequencing as a collective effort to be distributed among the research community, rather than assigned to a reduced number of groups. This resulted in a motley grouping of yeast biologists conducting the various aspects of the sequencing process in the YGSP (Fig. 2.4), a

---

[7] The convergence here of the creation of standards in biological research through model organisms, and the elucidation of maps and reference sequences in genomics, is telling. We revisit this parallel throughout the book. The standardisation and distribution of uniform strains and populations is a key aspect of model organism biology, to try to ensure that experimental work in different places is conducted on as similar objects as possible. As experimental systems (Rheinberger, 1997), model organisms are more appropriately assessed in terms of their productivity and generativity, rather than the extent to which they, or particular biological processes they manifest, represent either their own species, or a wider class of species (Ankeny & Leonelli, 2011; Leonelli & Ankeny, 2013). Model organisms are thus a class of experimental organism for which there are considerable resources and tools available for comprehensive biological characterisation: these constitute a formidable basis for further investigation and experimental intervention. Different model organisms are often used for different purposes, for instance *Drosophila melanogaster* for genetics, *C. elegans* for neural development, and zebrafish (*Danio rerio*) for developmental biology (Ankeny & Leonelli, 2020). The communities that orient around such organisms therefore exhibit distinct historical trajectories.

Fig. 2.4 Members of the consortium of the European Commission, along with representatives from Washington University, Stanford University, McGill University and the Sanger Institute, at the final conference of the Yeast Genome Sequencing Project, held in Trieste (25th–28th September 1996). They are wearing specially-designed shirts illustrating the parts of the yeast genome they worked on. André Goffeau is the third on the left and the picture features the following chromosome coordinators: Hervé Tettelin (far-left, chromosome VII); Mark Johnston (second-from-left, chromosome VIII and co-coordinator of chromosomes IV, XII and XVI); Bernard Dujon (behind Goffeau, chromosomes XI and XV); Horst Feldmann (middle-back, chromosome II); Ron Davis (third-from-right, chromosome V and co-coordinator of chromosomes IV and XVI); Howard Bussey (second-from-right, chromosome I and co-coordinator of chromosome XVI at McGill University); Bart Barrell (far-right, chromosomes IX and XIII and co-coordinator of chromosomes IV and XVI at the Sanger Institute). Agnès Thierry (middle-front), a member of Dujon's group, is also included in the picture. Photograph courtesy of Karl Kleine

much broader community than the specialist sequencing personnel based in the NIH-funded teams and, later, the genome centres. The distinct communities that coalesced around each programme became the very different kinds of *genomicists* that sent yeast sequencing on disparate trajectories: in the USA as an antecedent to the Human Genome Project; and as a more species-specific initiative in Europe that was closer to the research require-ments of academic and industrial laboratories working with *S. cerevisiae*.

Because of this, the membership of the EC consortium was always het-erogeneous and included a large number of institutions. The YGSP started with a three-year pilot phase aimed at chromosome III of *S. cerevisiae* and led by Stephen Oliver, a microbiologist from what was then called the University of Manchester Institute of Science and Technology (UMIST).[8] This phase trialled the distributed model and used mapped yeast clones—from the AB972 sub-strain of S288C—distributed by Olson and Riles at WU. The sequence of that chromosome was published in 1992 and involved 35 institutions from eleven different European countries, plus Kobe University in Japan and the New Jersey Medical School in the USA, which were external members of the EC consortium. The pilot phase re-affirmed in outline the basic distributed model of sequencing, though indicated that scale-ups in sequencing output would still be needed in the main phase of the project. An unexpected empirical outcome was the rev-elation of scores of genes—a majority of those found on the chromo-some—that were previously unknown to yeast biologists and would require extensive functional analysis to be characterised (Oliver et al., 1992; Vassarotti & Goffeau, 1992).

The full genome of the species was completed in 1996 and marked by a special issue of *Nature* that was published one year later, entitled *The yeast genome directory*. It reported on chromosomes led by the EC effort that had not yet appeared in the literature, as well as those led by Stanford, WU, McGill University and the Sanger Institute: nine out of the sixteen chromosomes of *S. cerevisiae* in all. By then, membership of the EC con-sortium had grown to 82 institutions. This consortium coordinated ten of the sixteen chromosomes of yeast—three of them in collaboration with American institutions and the Sanger Institute. The nucleotides deter-mined by the EC consortium represented almost 55% of the overall sequence produced (Parolini, 2018, Figure 1; see also Appendix A at the end of Parolini's article for a full list of EC laboratories) (Table 2.1).

Apart from Goffeau and Oliver's home institutions, another important nucleus of the EC programme was Munich. This city, which like Louvain and Manchester is furnished with a longstanding brewing tradition, had

---

[8] UMIST was a university focused on technical, scientific and engineering degrees indepen-dent from the more general Victoria University of Manchester. Both institutions amalgam-ated into the University of Manchester in 2004, a name that has not been changed since then (Wilson, 2008).

Table 2.1 An outline of the distribution of work among the different institutions involved in completing the *S. cerevisiae* genome, indicating the size of each chromosome, institutions participating in the sequencing work, scientist(s) coordinating the operation and year in which the sequence was published

Chromosome	Coordinator(s)	Additional laboratories involved	Sequence size, in base pairs	Year published
I	H. Bussey (McGill University, Canada)	1 from the European consortium (external member: New Jersey Medical School, USA), 1 from the USA and 1 from Canada	230,218	1995
II	H. Feldmann, (Ludwig-Maximilian University of Munich, Germany)	26 from the European consortium	813,184	1994
III	S. Oliver (UMIST, UK)	36 from the European consortium (including 2 external members: New Jersey Medical School, USA and Kobe University, Japan)	316,620	1992
IV	C. Jacq (Ecole Normale Supérieure, France) B. Barrell (Sanger Institute, UK) M. Johnston (Washington University, USA) R. Davis (Stanford University, USA)	24 from the European consortium	1,531,933	1997
V	R. Davis (Stanford University, USA)	N/A	576,874	1997
VI	Y. Murakami (RIKEN, Institute of Physical and Chemical Research, Japan)	4 from Japan	270,161	1995
VII	H. Tettelin (Catholic University of Louvain, Belgium)	32 from the European consortium, 1 from the USA	1,090,940	1997

(continued)

Table 2.1 (continued)

Chromosome	Coordinator(s)	Additional laboratories involved	Sequence size, in base pairs	Year published
VIII	M. Johnston (Washington University, USA)	1 from the European consortium	562,643	1994
IX	B. Barrell (Sanger Institute, UK)	1 from Europe	439,888	1997
X	F. Galibert (University of Rennes, France)	18 from the European consortium	745,751	1996
XI	B. Dujon (Pasteur Institute, France)	28 from the European consortium	666,816	1994
XII	J. Hoheisel (German Cancer Research Center, Germany) M. Johnston (Washington University, USA)	18 from the European consortium	1,078,177	1997
XIII	B. Barrell (Sanger Institute, UK)	N/A	924,431	1997
XIV	P. Philippsen (University of Basel, Switzerland)	19 from the European consortium.	784,333	1996
XV	B. Dujon (Pasteur Institute, France)	28 from the European consortium.	1,091,291	1997
XVI	A. Goffeau (Catholic University of Louvain, Belgium) H. Bussey (McGill University, Canada) R. Davis (Stanford University, USA) B. Barrell (Sanger Institute, UK) M. Johnston (Washington University, USA)	13 from the European consortium, 2 from Canada	948,066	1997

Elaborated by Jarmo de Vries using data from Engel et al. (2014, especially Table 1), *The yeast genome directory* (Goffeau et al., 1997), other chromosome sequence publications and the *Saccharomyces* Genome Database Community Wiki: https://wiki.yeastgenome.org/index.php/Systematic_Sequencing_Table (last accessed 7th December 2022)

developed an institutional architecture around biomedical research that resulted in different types of laboratories playing crucial and complementary roles in the YGSP. Due to the impracticality of individually reviewing all of the 82 institutions involved in this project here, we focus on the contributors from Munich as representative of the institutional diversity within the EC consortium. By looking at the connections that these Munich-based laboratories deployed both within and across the YGSP, we also reveal how the EC consortium members interacted between themselves and with the institutions—genome centres and smaller teams—that coordinated the sequencing of other yeast chromosomes.[9]

One leading figure from Munich in the YGSP was Horst Feldmann. A chemist by training, Feldmann was a principal investigator at the Institute of Physiological Chemistry, a research institution affiliated to the Ludwig-Maximilian University of Munich (LMU, the largest university in this city) and the Klinikum of the LMU, a teaching hospital. Since 1968, Feldmann had led his own group using yeast as a model to investigate the genes that control the activity of transfer RNA (tRNA). tRNA molecules are involved in the process of translation by which proteins are synthesised in the cell. They interact with messenger RNA (mRNA) molecules produced through the expression of DNA sequence, with each kind of tRNA bringing a particular amino acid, thereby adding to the growing chain that will make up the protein, as specified by the original DNA that was expressed.

Feldmann had learned sequencing methods during a short visit to the LMB, the institution where Sulston and Coulson were based (Feldmann, 2008, p. 291). The laboratory that Feldmann visited was led by Frederick Sanger, the inventor of the first protein, RNA and DNA sequencing techniques and the scientist after whom the Sanger Institute was named (García-Sancho, 2010). At the time of Feldmann's visit, the *C. elegans*

---

[9] Our Munich-centred institutional typology complements an investigation of other EC consortium members that we have published elsewhere. In that publication, we showed how some of the YGSP institutions were more proximate to the final users of the sequence, while others were more distal. Likewise, some of the participants in the EC consortium conducted sequencing work that was directed to specific research goals, while others practiced undirected sequencing (García-Sancho, Lowe, et al., 2022). Here, we depict the institutional heterogeneity within the YGSP through analysing the differing ways in which the Munich-based institutions organised their day-to-day involvement in yeast sequencing.

mapping project had not yet begun and Coulson was starting his career as a technician in Sanger's group. Feldmann's mentor in Cambridge was another Sanger technician, Bart Barrell, who would later run the yeast sequencing operation at the Sanger Institute (Chap. 4).

By the time the YGSP started in 1989, Feldmann's research was focused on specific genomic regions involved in the distribution of tRNA genes on the yeast chromosomes. These regions, called Ty elements, are found in multiple copies in the genome, as they can create new copies of themselves and jump from one area of the chromosome into others; they are structurally and functionally related to retrotransposons and retroviruses.[10]

Feldmann saw the emerging yeast genome initiative as an opportunity to exploit his sequencing expertise and, at the same time, use both the EC funding and sequence data to further his investigation of Ty elements. He joined the consortium from the start and, apart from being involved in the chromosome III pilot effort, led the completion of chromosome II in 1994, coordinating other consortium members. During this time, Feldmann's team combined the YGSP work with comparison of Ty sequences and analysis of the implications of their conservation: the preservation of particular sequences across different evolutionary lineages— including those featuring yeast species—deriving from a common ancestor (Feldmann, 2008; Stucka et al., 1992; Fig. 2.5). Their detailed knowledge of the Ty regions proved essential in the subsequent completion of other yeast chromosomes, where Feldmann and his collaborators offered crucial intelligence on repetitive patterns in the sequence data to other institutions with less specific expertise.[11]

---

[10] Rolf Stucka, interview with Miguel García-Sancho, Adolf Butenandt Institute, Munich, November 2019. Retrotransposons and retroviruses reverse the normal flow of genetic information through an enzyme that produces DNA sequences from RNA and incorporates those sequences into the genome.

[11] We thank Mark Johnston for indicating Feldmann's role to us, in an interview via Skype in September 2020. This kind of contribution was also offered by other members of the EC consortium such as Edward Louis, whose expertise on *S. cerevisiae* telomeres enabled the Sanger Institute and WU to tackle these repetitive regions at the ends of chromosomes. Like Feldmann, Louis was based in a biomedical institution—the Institute of Molecular Medicine—that is affiliated to a university and clinical setting: the University of Oxford and John Radcliffe Hospital (García-Sancho, Lowe, et al., 2022, note 44). Apart from yeast, the Institute of Molecular Medicine played an important role in the UK's national Human Genome Mapping Project (Chap. 3). Repetitive sequences are a key indicator in some genetic diseases and, therefore, a focus of medical genetics research.

Another prolific contributor of yeast sequences—and participant in the chromosome II effort with Feldmann's group—was Genzentrum, an institution co-managed by LMU and the Max Planck Institute for Biochemistry (MPIB, also based in Munich). Genzentrum had been established in 1984 as part of a network of centres with which the German Federal Government sought to foster research using the fledging recombinant DNA techniques: there were other gene centres in Berlin, Cologne and Heidelberg. The Munich Genzentrum was headed by Ernst-Ludwig Winnacker, one of the pioneers of molecular biology in Germany, and it incorporated a set of shared facilities in which scientists from LMU and MPIB could access recombinant DNA and sequencing techniques. On top of this, Genzentrum was equipped with a suite of laboratories in which early career researchers could start their trajectory towards becoming principal investigators.[12]

One of these early career researchers, Horst Domdey, became heavily involved in the sequencing of yeast. Like Feldmann, he combined YGSP work with research on the genetics of *S. cerevisiae*; in this case, the transcription of DNA into mRNA (Fig. 2.5). Yet given the availability of advanced technological facilities at Genzentrum, Domdey also embarked on sequencing work supported by other EC schemes, namely the Human Genome Analysis Programme (see below).[13]

Unlike Feldmann's laboratory, in which manual sequencing methods were used, Genzentrum was equipped with state-of-the-art automated instruments that were starting to be marketed by the companies DuPont and Applied Biosystems (García-Sancho, 2012a, Chs. 5–6). The funding from the EC genome programmes provided crucial support to the day-to-day running of these sequencing machines, which were used by other Munich-based researchers working on immunology and animal

[12] E.L. Winnacker and other authors (undated): *Laboratory of Molecular Biology and Biochemistry: München / Martinsried*, Hoechst Archives, Frankfurt, file number H0049176. We thank Magnus Altschäfl at LMU for generously providing access to this archival record.

[13] Horst Domdey, interview with Miguel García-Sancho, Genzentrum, Munich, November 2019. See also: "Horst Domdey: Gene expression in yeast." In E.L. Winnacker and other authors (undated): *Laboratory of Molecular Biology and Biochemistry: München / Martinsried*, Hoechst Archives, Frankfurt, file number H0049176, p. 8; Obermaier, Gassenhuber, et al. (1995); Obermaier, Stachowitz, et al. (1995).

Fig. 2.5 Top-left, Horst Feldmann's group at Ludwig-Maximilian University of Munich in the late-1980s, including Gertrud Mannhaupt (far-left), Rolf Stucka (behind the line) and Christa Schwarzlose (second-from-right), all of them involved in the sequencing of chromosome II of *S. cerevisiae*. Other group

(*continued overleaf*)

genetics.[14] In contrast with the Sanger Institute and the genome centres at Stanford and WU, the EC funds did not fully cover the sequencing operations at Genzentrum. Due to this, the facilities needed to combine sequencing work for the EC programmes with a service role supporting research grants undertaken by Genzentrum's early-career investigators, as well as other scientists at LMU and MPIB.

In 1994, Genzentrum created a start-up company, MediGene, to both commercialise medical products derived from their research and conduct on-demand sequencing work. Brigitte Obermaier, who led the genome analysis team in charge of YGSP assignments at Domdey's group, became head of sequencing services at this company. MediGene was one among a set of mainly German firms created out of academic research that conducted contract sequencing work for other institutions or concerted genome projects (García-Sancho, Lowe, et al., 2022; Zeller, 2001). They were especially active during the YGSP and other subsequent EC genome programmes: thirteen different companies offering sequencing services were involved as co-authors in *The yeast genome directory*. MediGene participated in the

Fig. 2.5 members depicted are Hans Lochmüller, Susanne Mitzel and Robert Krieg (second, third and fourth-from-left), as well as Uschi Obermeier (far-right). Bottom-left, Horst Domdey in the mid-to-late 1980s and, right, a laboratory during the early years of Genzentrum. *Sources*: Top-left picture, Feldmann (2008, p. 300), copyright (2008), with permission from Elsevier; bottom-left and right pictures: E. L. Winnacker and other authors (undated): Laboratory of Molecular Biology and Biochemistry: München / Martinsried, pp. 8 and 4. Reproduced with permission from Genzentrum and available at Hoechst Archives, Frankfurt, file number H0049176

[14] One of these researchers was Hans Georg Zachau, who had mentored Feldmann and moved with him to Munich from the Institute of Genetics in Cologne. While Feldmann had focused his LMU laboratory on the investigation of yeast, Zachau worked on human genes controlling the synthesis of immunoglobulins (Feldmann, 2008, pp. 285ff). Gottfried Brem, another early career researcher at Genzentrum, conducted a substantial amount of sequencing of pig DNA, despite neither LMU nor MPIB participating in any EC-sponsored swine genome programme (Chap. 5). Pig-related products are important in the economy of Bavaria, the state of which Munich is the capital: "Gottfried Brem: Department of Molecular Animal Breeding." In E.L. Winnacker and other authors (undated): *Laboratory of Molecular Biology and Biochemistry: München / Martinsried*, Hoechst Archives, Frankfurt, file number H0049176, p. 7.

completion of two different chromosomes of *S. cerevisiae*, with these sequencing operations being the company's most profitable line of business.[15]

MediGene, along with Genzentrum and Feldmann's group, were representative of what we have called elsewhere the network model of genomics: a versatile and heterogeneous array of institutions that exhibited different motivations to produce DNA sequence data (García-Sancho, Lowe, et al., 2022). This heterogeneity gave the EC consortium members flexibility, and encouraged an ability to adjust to changing circumstances. Within the consortium, scientists and institutions could produce large amounts of sequence data for various users—e.g. the customers of MediGene—or behave more like a traditional life sciences laboratory and use the sequences they determined for specific research purposes: e.g. Domdey and Feldmann's groups. Even within the same laboratory, the sequences were often contributed to the YGSP after being used for more immediate research work. By contrast, the genome centres deriving from WU, Stanford and Sulston's group only practised one model of DNA sequencing that was more distal to the final user and led to the production of large amounts of data without advanced concrete knowledge of what its purpose and destination would be.

At the core of the EC network, another Munich-based institution—the Martinsried Institute for Protein Sequences (MIPS)—compiled the various sequencing results and assembled them into full chromosomes and later a reference genome. MIPS also played a crucial role in assessing the quality of the sequences and inferring biological features from the DNA data. This institution was located in the same campus as Genzentrum, the MPIB and other biomedical laboratories of LMU in Martinsried, a suburb in the south-west of Munich.

MIPS had originated in the late 1980s as a unit of the MPIB Department of Protein Chemistry. This department was particularly strong in the determination of protein sequences, whose techniques had preceded the emergence of DNA sequencing. Pehr Victor Edman, a key figure in the development and automation of protein techniques during the 1950s and 1960s (García-Sancho, 2010, pp. 284ff), had moved to Munich in 1972 and finished his career there.

The first objective of MIPS was to harmonise and unify the different protein sequence databases operating in Europe, the USA and Japan. Yet its director, Werner Mewes, saw the sequencing of yeast as an opportunity to extend to DNA MIPS's expertise in data handling, analysis and

---

[15] Brigitte Obermaier, interview with Miguel García-Sancho, IZB Building, Munich, November 2019. Horst Domdey, interview with Miguel García-Sancho, Genzentrum, Munich, November 2019.

standardisation.[16] Unexpectedly, MIPS became the institution chosen by Goffeau to channel the sequencing results produced by the YGSP. He and other EC administrators engineered a funding system in which all the institutions involved in the *S. cerevisiae* consortium were incentivised to swiftly produce and submit their sequences to MIPS. MIPS scientists checked the accuracy and quality of the sequences before assembling them into chromosomes and, eventually, a whole genome. They also made the sequences public after a period of 6 to 12 months in which, according to the terms of the EC contract, the sequencing laboratories were entitled to exclusive exploitation of the data (Joly & Mangematin, 1998).

The embargo period was another difference between the EC consortium and the genome centres, which made their yeast sequence data immediately available.[17] This further reflects the contrasting philosophies underlying the distributed and concentrated strategies and what the adoption of one or the other required in terms of support and organisation. The concentration of funding in WU, Stanford and the Sanger Institute meant that these institutions could exclusively focus on producing the sequence without any other financial needs that would require some kind of diversification. Their designation as genome centres emphasised this exclusivity of sequence production and differentiated them from other institutions supported by the NIH, the MRC and the Wellcome Trust. By contrast, the distribution of the YGSP budget among a much larger number of institutions resulted in these institutions having to combine the EC support with other sources of funding. One way of achieving this was using the sequence data that each laboratory produced as a springboard that would ease the award of either other contracts—especially in the case of the sequencing companies—or research grants exploring different aspects of yeast biology. As a yeast biologist himself, Goffeau recognised this necessity and protected the competitive advantage of the sequencing institutions through the exclusive exploitation window. At the same time this model was being implemented, other scientists and EC administrators attempted to export the distributed strategy to the sequencing of other organisms.

---

[16] "Arbeitsgruppe Datenbank für Proteinsequenzen (MIPS). Leiter: Hans Werner Mewes" and "Datenbank für Proteinsequenzen (MIPS). Leiter: Hans Werner Mewes." Both *in Max-Planck Gesellschaft Jahrbuch [Yearbook of the Max Planck Institutes]*, volumes 1992 (p. 138) and 1994 (pp. 128–129) respectively. Karl Kleine, telephone interview with both authors, October 2019.

[17] As we show in Chap. 3, the early-to-mid 1990s witnessed a heated debate about how best to disseminate and exploit DNA sequence data. Although the NIH advocated for free and immediate release of yeast sequences, it initially considered patent protection in the case of *Homo sapiens*.

2.3   DISTRIBUTED SEQUENCING AND LARGER GENOMES

As the 1990s progressed, the EC extended its sequencing programmes to the genomes of other organisms of interest to science and industry, such as the fruit fly *Drosophila melanogaster*, the bacterium *Bacillus subtilis* and the plant *Arabidopsis thaliana*. All three of these species were, by then, model organisms and *B. subtilis* had been used extensively as a biotechnological cell factory for the production of multiple chemicals. The consortia that the EC created for those sequencing efforts shared the features of the one it had established for the YGSP: the membership was as inclusive as possible and sought to foster cooperation among member-states. As with the YGSP, the full sequencing of those organisms involved cooperation and competition between the EC-sponsored laboratories and other institutions, mainly in the USA and Japan.

The EC project to sequence the genome of *A. thaliana* began in 1993, with eighteen institutions concentrating on two chromosomes: 4 and 5. Like *S. cerevisiae*, *A. thaliana* has an economically-sized genome comprising five chromosomes. It has, though, over ten times the number of nucleotides as yeast. This European effort was later joined by separate initiatives in France, Japan and the USA.[18] These became formally coordinated in 1996 with the creation of the Arabidopsis Genome Initiative, which was completed in 2000. The active participation of researchers specialising in *Arabidopsis* biology within the EC consortium enabled them to become crucial actors in the sequencing of genome regions that were difficult to tackle with existing large-scale sequencing methods, a role that had been played in the YGSP by Feldmann's team and other groups with expert knowledge of yeast biology and genetics.[19]

---

[18] https://cordis.europa.eu/project/id/BIO2930075 (last accessed 7th December 2022). Internet sources have been particularly important for our reconstruction of *Arabidopsis* genomics. Primary sources such as reports and project websites from the mid-1990s onwards were made available—and are still available—online, reflecting the public accountability and dissemination policies of the EC and the community of *Arabidopsis* researchers themselves. While some secondary literature exists on the history of *Arabidopsis* genomics (most notably by Sabina Leonelli, e.g., 2007), it was necessary to return to the online primary sources to confirm certain details and to situate *Arabidopsis* genomics within our overall narrative.

[19] See note 11, above, and 'The Multinational Coordinated *Arabidopsis thaliana* Genome Research Project—Progress Report: Year Six.' Available at: https://www.nsf.gov/pubs/1997/nsf97131/nsf97131.htm (last accessed 7th December 2022).

Another similarity with the YGSP was that the *Arabidopsis* genome was parcelled out, both to the different parts of the international collaboration and to the individual participating laboratories in the European network. In Europe, MIPS once again played a leading role as an informatics centre, reviewing the quality of the sequences submitted by the EC-sponsored laboratories, assembling parts of the genome and analysing it.[20] Additionally, many of the same sequencing companies that had contributed significant portions of the European sequence in the YGSP came on board in the main phase of the EC *Arabidopsis* project in 1996, following the conclusion of the pilot begun in 1993.[21] Even within the European distributed model, however, a move towards concentration of larger sequencing capacities into a few institutions was evident.

Arabidopsis has been described as the "botanical *Drosophila*", and the completed genome sequence—the first for a plant—augmented its more general status as a model organism for plant biology. Arising out of the project to sequence the reference genome, The Arabidopsis Information Resource became an altogether more all-encompassing data infrastructure with ambitions far beyond being a mere repository of DNA sequence and other molecular data. Its promoters conceived it as a means of providing a common basis for representing the species as a whole through the integration of multiple different kinds of data and knowledge, and from this to serve as a platform for the exploration of less well-catalogued species (Leonelli, 2007).

Genome sequencing of *D. melanogaster* was performed by the European *Drosophila* Genome Project, a consortium comprising ten laboratories that the EC began to fund in 1997.[22] This built on previous EC-supported efforts to physically map the *Drosophila* genome from 1988. As well as smaller centres conducting the sequencing, the Sanger Institute performed an analogous role to MIPS in the yeast and *Arabidopsis* projects, assessing and assembling the sequence submitted to it by the participating laboratories.[23] Like the other projects mentioned above, Europe joined forces with

[20] https://www.arabidopsis.org/weedsworld/Vol1/sequencing.html (last accessed 7th December 2022); https://www.nsf.gov/pubs/1997/nsf97131/nsf97131.htm (last accessed 7th December 2022); http://arabidopsisresearch.org/images/publications/mascreports/2002_MASCreport.pdf (last accessed 7th December 2022).

[21] https://cordis.europa.eu/project/id/BIO4960338 (last accessed 7th December 2022).

[22] https://web.archive.org/web/19990208004153/http:/edgp.ebi.ac.uk/ (last accessed 7th December 2022).

[23] https://cordis.europa.eu/project/id/BIO4960506 (last accessed 7th December 2022).

American institutions. These included the publicly-funded Berkeley *Drosophila* Genome Project, another NIH-sponsored genome centre at Baylor College of Medicine and FlyBase, a genetics and genomics database with antecedents dating back to the 1930s, but that was made available on the internet in 1992.

An important commercial player involved was Celera Genomics. This company derived from prior partly-charitable and partly-corporate sequencing efforts led by Craig Venter (Adams et al., 2000; Drysdale & Bayraktaroglu, 2000).[24] Celera was founded in 1998 and began contributing to the *Drosophila* project shortly after. It performed the bulk of the sequencing that resulted in the completion of the reference genome in 2000 (Dove, 2000). Seeking as it did to become a leader in the nascent field of bioinformation, the company used the fly initiative as a testbed for its advanced sequencing and informatics pipelines. As we see in subsequent chapters, Celera's involvement in genomics was far more extensive than its most famous role in the sequencing of the human genome and concomitant controversy about commercially protecting and restricting access to the resulting data (Chap. 4). One of the key innovations it generated through the *Drosophila* sequencing project was in the annotation of the genome.

We discuss annotation in depth later in the book (Chap. 6), but it is worth noting here the community-based annotation approach that the *Drosophila* genomicists pioneered: the so-called 'jamboree'. The jamboree was an 11-day event held at Celera headquarters in Maryland in November 1999. Over 40 researchers from across the world—mainly from the publicly-funded *Drosophila* projects—converged there. The aim was to join together the biological expertise of the *Drosophila* community with computer scientists to identify the genes and other key landmarks in the masses of sequence data produced by Celera (Pennisi, 2000). This would give the fly researchers access to the sequence data and Celera an idea of how they could add value to the data they were producing, a key consideration for their commercial strategy.

From this jamboree event, a model of community annotation was created that informed Celera's later analysis of the human genome in

---

[24] https://web.archive.org/web/19990208004153/http:/edgp.ebi.ac.uk/ (last accessed 7th December 2022).

conjunction with medical geneticists (García-Sancho, Leng, et al., 2022) and that served as inspiration for the annotation of the pig genome (Chaps. 5 and 6). This model was, again, based on combining the expertise of a large-scale sequencing institution—Celera—and smaller laboratories with specialist knowledge in the target genomes, many of whom were funded through the distributed model of the EC. As with the yeast and *Arabidopsis* genome projects, the intelligence provided by the specialist laboratories proved crucial to adding value to Celera's sequences and for interpreting specific, biologically-relevant regions of the resulting reference genomes.

In spite of these contributions, the EC's distributed strategy became increasingly challenged with the emergence of concerted efforts aimed at larger and more complex genomes. With the Human Genome Project in the USA underway since 1990, European scientists and EC administrators debated the effectiveness of collaborative consortia and a networked sequencing operation for more ambitious targets. Some defended the advantages of involving the community in the production of the sequence, as had been the case in the YGSP and contemporary initiatives. Others, however, highlighted the difficulties of recruiting a sufficient number of laboratories to sequence exponentially larger genomes and, especially, coordinating their activity and outputs.[25]

Adopting a more concentrated model was complicated by the EC's political need to support multiple institutions across the continent and the scarcity of large-enough institutions that could become sequencing centres for all of the European Union. The Sanger Institute, established in 1993, was a sizeable genome centre based in Europe, but it was already supported in its day-to-day operations by two UK bodies—the MRC and the Wellcome Trust—which gave it some independence from the EC approach. Another potential candidate, the European Molecular Biology Laboratory in Heidelberg (EMBL), never wanted to become a genome centre despite being equipped with advanced technology and expertise, including a centralised database to store DNA sequences

---

[25] An example of these debates is in the transcript of a discussion among scientists and administrators involved in the YGSP at the end of the programme: Programme of the Final European Conference of the Yeast Genome Sequencing Network, Trieste, September 25th–28th 1996, pp. 24, 31, 80–84. Personal papers of Karl Kleine, obtained 22nd November 2019.

(García-Sancho, 2011).[26] This lack of orientation towards large-scale genomics resulted in the EMBL participating as a standard sequencing laboratory in the EC consortia and MIPS being selected over this institution as the informatics and data assembly coordinator of the genome projects.

One way in which the EC oriented its operation to larger genomes was through programmes that did not seek to sequence them in full. Examples of this were the Pig Gene Mapping Project (PiGMaP, see Chap. 5), and the Human Genome Analysis Programme (HGAP). The latter started in 1990 and involved the creation of a consortium of human and medical genetics laboratories from different member-states. Rather than determining the full sequence of the human genome—which is almost 300 times larger than the yeast genome—the HGAP laboratories sought to refine existing linkage and physical maps. This involved narrowing down the location of genes or gene markers connected to the research interests of the groups forming the consortium. For this, the consortium members used a technique called complementary DNA (cDNA) sequencing that, rather than tackling the whole human genome, yielded only the parts involved in the synthesis of mRNA, a key intermediary in the production of proteins in the cell. The HGAP laboratories divided the genome into different regions—normally connected to the genes and proteins they were working on—and formed groups devoted to mapping and cDNA sequencing (Chap. 3).

Despite only targeting specific genome regions, these groups required enhanced sequencing capacities. This resulted in Genzentrum in Munich and other institutions with advanced instrumentation—often consisting of centralised, shared equipment—becoming especially active in HGAP. MediGene, Genzentrum's start-up, was split into two companies, with Obermaier's sequencing services arm becoming an independent brand called MediGenomix. In 1998, MediGenomix joined with other German sequencing firms involved in the YGSP—AGOWA, GATC and QIAGEN—as well as Biomax Informatics, to form the Gene Alliance, a

---

[26] The EMBL had been created in the 1970s and was collectively funded by various European governments. One of its objectives was supporting promising early-to-mid career molecular biologists through visiting fellowships (Krige, 2002). The awardees wanted to make the most of their time conducting research at Heidelberg, and were therefore seldom inclined to devote a substantial portion of it to the routine, comprehensive sequencing of organisms with larger genomes (Albayrak, 2015).

consortium that aimed to pool their sequencing capacities and capabilities to secure contracts for genome sequencing projects.

The Alliance was formed against the backdrop of the greater intensification and centralisation of whole-genome sequencing efforts propelled, to a large extent, by the emergence of Celera in 1998. Sequencing was becoming larger-scale and higher-throughput. The corporate alliance model was an attempt to keep up in terms of capacity when acting jointly, while enabling individual firms to retain their independence and specific expertise. In addition to their individual capacities, by coming together the Alliance could purchase new high-throughput sequencing machines (capillary sequencers) at a lower unit cost than was available for a smaller order.[27]

A change in the model and direction of funding at the EC level made it more difficult for individual companies like these to operate alone. Three changes in the years around the turn of the millennium were particularly significant. One was a general shift in research policy, namely a lessened emphasis on establishing wide networks of collaborators, something that had opened the door to smaller laboratories and companies. The drive towards European integration was diverted instead towards the building up of large and well-resourced "centres of excellence" that could enable Europe to compete at a global level across a variety of scientific fields.[28] The second change was more specific to genomics, with a shift towards funding functional genomics research (Chap. 7), rather than whole-genome mapping and sequencing (Desaintes, 2008; Gannon, 2000). This removed a potential market from the small sequencing companies that had been able to invest and grow through projects such as the YGSP and *Arabidopsis* genome sequencing. Finally, the mode of funding also changed, with the removal of the payment-per-nucleotide system that had characterised the earlier genome programmes. Participants were instead paid for labour and materials, favouring non-profit research institutes and universities.[29]

However, the private sequencing companies based chiefly in Germany had built up a customer base beyond EC-funded projects. Through the Gene Alliance, they were able to build on their experience of working with

[27] Brigitte Obermaier, interview with Miguel García-Sancho, Munich, November 2019. Brigitte Obermaier, telephone interview with James Lowe, June 2021.

[28] This was tied to the agenda of the then European Commissioner for research, Philippe Busquin (who served 1999–2004), to create a European Research Area that would ease researcher mobility, thus enabling European scientists to move to these "centres of excellence" rather than head to the USA (EMBO Reports, 2000).

[29] Thomas Pohl, telephone interview with both authors, September 2019.

other laboratories and private firms needing sequencing services. In addition to forming a collaboration in 1999 with Genome Pharmaceuticals Corporation around agrogenomics and pharmacogenomics—thus coupling the Alliance's sequencing capacity to the Corporation's functional genomics expertise—they were contracted to sequence the whole genomes of two species: the bacterium *Chlamydia pneumoniae* and the fungus *Aspergillus niger*.[30]

The project to sequence *C. pneumoniae* derived from an agreement between the Alliance and the German pharmaceutical company Byk Gulden (a subsidiary of the chemical firm Altana) in 1998. *C. pneumoniae* is a bacterium that causes pneumonia and has been implicated in other diseases such as atherosclerosis. The full sequence of its genome was completed in nine months.[31] The Alliance operated along similar lines to the yeast and *Arabidopsis* sequencing projects, a model that the companies were familiar with. Various tasks such as library preparation were divided out among the companies, the sequencing itself was parcelled out to the four non-bioinformatics firms, while Biomax had a role analogous to MIPS in the EC-sponsored projects. A key advantage of the Alliance was that the allotting of work could benefit from the different expertise and business models of its members, with tasks also arranged to enable parallel projects to be managed by individual companies and the Alliance as a whole, alongside their more regular operations. Obermaier of MediGenomix has described the Alliance's model as "interactive", with monitoring and evaluation of the ongoing sequencing conducted by Biomax, leading to the identification of regions of lower quality and coverage requiring additional work.[32]

In 2000, the Alliance was contracted by the research arm of the large Dutch company DSM (which operates in multiple fields, including nutrition and healthcare products) to sequence the genome of *Aspergillus niger*, a fungus and species of mould that the company used to produce enzymes and organic acids. The bioinformatics capacity of the consortium was just as crucial as its sequencing output to securing the contract, with the ability of Biomax to fully annotate this genome being a key attraction

---

[30] https://web.archive.org/web/20040106234422fw_/http:/www.gene-alliance. com/1_4_news.htm (last accessed 7th December 2022).

[31] https://www.biomax.com/lib/press-releases/1999_altana_ga_e.pdf (last accessed 7th December 2022).

[32] Brigitte Obermaier, telephone interview with James Lowe, June 2021.

for DSM.[33] QIAGEN took the lead for the Alliance, and the annotated genome sequence was announced in 2007 (Pel et al., 2007).

By then, the Gene Alliance itself was no longer taking on new business. They found that they were increasingly unable to compete on price with the ever-larger centres in east Asia, North America and Europe, to win whole-genome sequencing contracts. Additionally, the increasing availability of next-generation sequencing machines fostered a shift in the business models of the companies. They enabled them to develop new services and markets based on quick, often overnight, sequencing for research in academia and industry. Without the centripetal force of large contracts through the Alliance, the trajectories of the companies diverged. They had always had to differentiate from each other, as they remained competitors in the same market and region. This was reinforced when the countervailing tendencies that were keeping them aligned in a network diminished.[34] With the demise of the Alliance, one of the last vestiges of the European distributed strategy of genomics waned.

The models of the Gene Alliance and HGAP co-existed with the genome centres that proliferated in the USA, UK and other countries during the 1990s. Apart from being single institutions rather than groups or consortia, these genome centres were the executive arm of initiatives that sought to produce a reference sequence of the whole human genome. The publication of that reference sequence between 2001 and 2004 led to the genome centres being identified with a single, coherent and unified 'Human Genome Project'. Since they were not part of this whole reference sequence effort, the HGAP and work of the Gene Alliance were largely forgotten and excluded from the success narrative of genomics.

Our shift from human to non-human genome projects and addressing of the distributed programmes that the EC sponsored throughout the 1990s has enabled us to present other historical configurations of genomics, beyond the success narrative. As the experience of the HGAP shows, this wider diversity of institutions, approaches and genomicists also applied to human genomics. In the next chapter, we further explore this broader history of human genomics by showing that most of the national human

[33] https://web.archive.org/web/20040106234422fw_/http:/www.gene-alliance.com/1_4_news.htm (last accessed 7th December 2022). Due to the lack of secondary literature on the Gene Alliance, archived web pages and interviews with those involved have been valuable resources in reconstructing its history.

[34] Brigitte Obermaier, telephone interview with James Lowe, June 2021.

genome projects that proliferated worldwide in the 1990s adopted the distributed strategy of the EC and the Gene Alliance during their early years. We also document the factors through which the concentrated model became identified in the public imagination with a single Human Genome Project, which funnelled the diversity of approaches to human genomics into just one.

REFERENCES

Adams, M. D., Celniker, S. E., Holt, R. A., Evans, C. A., Gocayne, J. D., Amanatides, P. G., et al. (2000). The genome sequence of *Drosophila melanogaster*. *Science, 287*(5461), 2185–2195.

Albayrak, G. (2015). *The emergence of European approach to the Human Genome Project*. Gazi University, Ankara (Dissertation, Jean Monnet Scholarship).

Ankeny, R. (1997). *The conqueror worm: An historical and philosophical examination of the use of the nematode Caenorhabditis elegans as a model organism*. PhD dissertation, University of Pittsburgh.

Ankeny, R., & Leonelli, S. (2011). What's so special about model organisms? *Studies in History and Philosophy of Science Part A, 42*(2), 313–323.

Ankeny, R., & Leonelli, S. (2020). *Model organisms*. Cambridge University Press.

Botstein, D., & Fink, G. R. (1988). Yeast: An experimental organism for modern biology. *Science, 240*(4858), 1439–1443.

Botstein, D., White, R. L., Skolnick, M., & Davis, R. W. (1980). Construction of a genetic linkage map in man using restriction fragment length polymorphisms. *American Journal of Human Genetics, 32*(3), 314.

Bud, R. (1993). *The uses of life: A history of biotechnology*. Cambridge University Press.

Bud, R. (2010). From applied microbiology to biotechnology: Science, medicine and industrial renewal. *Notes and Records of the Royal Society, 64*(suppl_1), S17–S29.

Cantley, M. (1995). The regulation of modern biotechnology: A historical and European perspective. In H. J. Rehm & G. Reed (Eds.), *Biotechnology set* (pp. 505–681). Wiley-VCH.

de Chadarevian, S. (2004). Mapping the worm's genome: Tools, networks, patronage. In J. P. Gaudilliere & H. J. Rheinberger (Eds.), *From molecular genetics to genomics: The mapping cultures of twentieth century genetics* (pp. 113–128). Routledge.

Desaintes, C. (2008). Research on animal model organisms funded by the European Commission's framework programmes. *Disease Models & Mechanisms, 1*(4–5), 209–212.

Dove, A. (2000). *Drosophila* sequence "done". *Nature Biotechnology, 18*(4), 365–365.

Drysdale, R., & Bayraktaroglu, L. (2000). The *Drosophila* genome: So that's what it looks like! *Yeast, 1*, 724067.

EMBO Reports. (2000). A European research identity. *EMBO reports, 1*(2), 96–99.

Engel, S. R., Dietrich, F. S., Fisk, D. G., Binkley, G., Balakrishnan, R., Costanzo, M. C., et al. (2014). The reference genome sequence of *Saccharomyces cerevisiae*: Then and now. *Genes|Genomes|Genetics, 4*(3), 389–398.

Feldmann, H. (2008). A life with yeast molecular biology. *Comprehensive Biochemistry, 46*, 275–333.

Gannon, F. (2000). The discovery of functional genomics. *EMBO Reports, 1*(5), 386–387.

García-Sancho, M. (2010). A new insight into Sanger's development of sequencing: From proteins to DNA, 1943–1977. *Journal of the History of Biology, 43*(2), 265–323.

García-Sancho, M. (2011). From metaphor to practices: The introduction of "Information Engineers" into the first DNA sequence database. *History and Philosophy of the Life Sciences, 33*(1), 71–104.

García-Sancho, M. (2012a). *Biology, computing and the history of molecular sequencing: From proteins to DNA, 1945–2000.* Palgrave Macmillan.

García-Sancho, M. (2012b). From the genetic to the computer program: The historicity of 'data' and 'computation' in the investigations on the nematode worm *C. elegans* (1963–1998). *Studies in History and Philosophy of Biological and Biomedical Sciences, 43*(1), 16–28.

García-Sancho, M., Leng, R., Viry, G., Wong, M., Vermeulen, N., & Lowe, J. (2022). The Human Genome Project as a singular episode in the history of genomics. *Historical Studies in the Natural Sciences, 52*(3), 320–360.

García-Sancho, M., Lowe, J., Viry, G., Leng, R., Wong, M., & Vermeulen, N. (2022). Yeast sequencing: 'Network' genomics and institutional bridges. *Historical Studies in the Natural Sciences, 52*(3), 361–400.

Goffeau, A., Aert, R., Agostini-Carbone, M., Ahmed, A., Aigle, M., Alberghina, L., et al. (1997). The yeast genome directory. *Nature, 387*(6632).

Johnston, M., Andrews, S., Brinkman, R., Cooper, J., Ding, H., Dover, J., et al. (1994). Complete nucleotide sequence of *Saccharomyces cerevisiae* chromosome VIII. *Science, 265*(5181), 2077–2082.

Joly, P. B., & Mangematin, V. (1998). How long is co-operation in genomics sustainable? In P. Wheale, R. Von Schomberg, & P. Glasner (Eds.), *The social management of genetic engineering* (pp. 77–90). Ashgate.

Krige, J. (2002). The birth of EMBO and the difficult road to EMBL. *Studies in History and Philosophy of Biological and Biomedical Sciences, 33*(3), 547–564.

Langer, E. M. (2016). *Molecular ferment: The rise and proliferation of yeast model organism* research. PhD dissertation, University of California San Francisco, San Francisco, CA.

Leonelli, S. (2007). Growing weed, producing knowledge: An epistemic history of *Arabidopsis thaliana. History and Philosophy of the Life Sciences, 29*(2), 193–223.

Leonelli, S., & Ankeny, R. (2013). What makes a model organism? *Endeavour, 37*(4), 209–212.

Obermaier, B., Gassenhuber, J., Piravandi, E., & Domdey, H. (1995). II. Yeast sequencing reports. Sequence analysis of a 78·6 kb segment of the left end of *Saccharomyces cerevisiae* chromosome II. *Yeast, 11*(11), 1103–1112.

Obermaier, B., Stachowitz, S., Blum, B., Domdey, H., Schwager, F., & Arnold, G. (1995). The partial sequence analysis of human cDNAs. In M. Hallen & A. Klepsch (Eds.), *Human Genome Analysis Programme* (pp. 29–39). IOS Press.

Oliver, S. G., van der Aart, Q. J. M., Agostoni-Carbone, M. L., Aigle, M., Alberghina, L., Alexandraki, D., et al. (1992). The complete DNA sequence of yeast chromosome III. *Nature, 357*(6373), 38–46.

Parolini, G. (2018). Building human and industrial capacity in European biotechnology: The Yeast Genome Sequencing Project (1989–1996). *Technical University of Berlin preprint series.* Retrieved December 7, 2022, https://depositonce.tu-berlin.de//handle/11303/7470

Pel, H. J., de Winde, J. H., Archer, D. B., Dyer, P. S., Hofmann, G., Schaap, P. J., et al. (2007). Genome sequencing and analysis of the versatile cell factory *Aspergillus niger* CBS 513.88. *Nature Biotechnology, 25*(2), 221–231.

Pennisi, E. (2000). Ideas fly at gene-finding Jamboree. *Science, 287*(5461), 2182–2184.

Rasmussen, N. (2014). *Gene Jockeys: Life science and the rise of biotech enterprise.* The Johns Hopkins University Press.

Rheinberger, H.-J. (1997). *Toward a history of epistemic things: Synthesizing proteins in the test tube.* Stanford University Press.

Strasser, B. J., & de Chadarevian, S. (2011). *The comparative and the exemplary: Revisiting the early history of molecular biology* (pp. 317–336). *History of Science.*

Stucka, R., Schwarzlose, C., Lochmüller, H., Häcker, U., & Feldmann, H. (1992). Molecular analysis of the yeast Ty4 element: Homology with Ty1, copia, and plant retrotransposons. *Gene, 122*(1), 119–128.

Sulston, J., Du, Z., Thomas, K., Wilson, R., Hillier, L., Staden, R., et al. (1992). The *C. elegans* genome sequencing project: A beginning. *Nature, 356*(6364), 37–41.

Sulston, J., & Ferry, G. (2002). *The common thread: A story of science, politics, ethics and the human genome.* Bantam Publishers.

Szymanski, E., Vermeulen, N., & Wong, M. (2019). Yeast: One cell, one reference sequence, many genomes? *New Genetics and Society, 38*(4), 430–450.

Vassarotti, A., & Goffeau, A. (1992). Sequencing the yeast genome: The European effort. *Trends in Biotechnology, 10,* 15–18.

Wilson, D. (2008). *Reconfiguring biological sciences in the late twentieth century: A study of the University of Manchester.* Manchester University Press.

Yi, D. (2015). *The recombinant university: Genetic engineering and the emergence of Stanford biotechnology.* University of Chicago Press.

Zeller, C. (2001). Clustering biotech: A recipe for success? Spatial patterns of growth of biotechnology in Munich, Rhineland and Hamburg. *Small Business Economics, 17,* 123–141.

CHAPTER 3

The Human Genome Project(s)

One of the consequences of excluding a great deal of the collaborative genome efforts that proliferated in the 1980s and 1990s from the success story of genomics has been the assumption that human genomics corresponded to a single initiative or entity. This assumption portrays the Human Genome Project as one international endeavour that started and ended at defined dates, presented a set of stable participants, and operated according to a predefined plan: the large-scale production of a reference sequence of the whole human genome. The narrative of a single human genome effort consolidated in June 2000, when a consortium of funders, sequencing centres and bioinformatics institutions from Europe, Asia and North America presented a first draft of the full sequence of *Homo sapiens* in a ceremony chaired by the US president, Bill Clinton, and attended remotely by UK Prime Minister, Tony Blair (Chap. 4). Before—and contemporaneous to—this announcement, a number of multinational genome initiatives to sequence yeast (*Saccharomyces cerevisiae*), the fruit fly (*Drosophila melanogaster*) and the thale cress (*Arabidopsis thaliana*) were unfolding with substantial leadership from the European Commission (Chap. 2).

The draft human genome sequence was published in the journal *Nature* in 2001. This article referred to the sequencing effort as the "Human Genome Project" and defined this project as an "international

M. García-Sancho, J. Lowe, *A History of Genomics across Species, Communities and Projects*, Medicine and Biomedical Sciences in Modern History, https://doi.org/10.1007/978-3-031-06130-1_3

collaboration" that had started in 1990 and was scheduled to conclude with the release of a more final sequence, which appeared in a follow-up publication, also in *Nature*, in 2004 (International Human Genome Sequencing Consortium, 2001, pp. 860 and 862; 2004). Since then, press coverage, popular literature and a substantial amount of academic scholarship have depicted a single, international Human Genome Project.[1] The depiction of the role of the European Commission (EC) as a funder and broker of genomic endeavours has tended to be restricted to yeast sequencing and presented as an antecedent to the Human Genome Project. As we discussed earlier in the book, this consideration of *S. cerevisiae* as a pilot or model platform for human sequencing aligns more with the US yeast genome effort than with the EC one. The EC, rather, selected yeast as an industrially-significant organism that would foster economic growth and scientific collaboration across its member-states (Chap. 2; see also Parolini, 2018).

In this chapter, we continue augmenting the historical landscape of genomics and de-centring it beyond the production of a human reference sequence. We start by arguing that instead of a monolithic Human Genome Project (with capitals H, G and P), a plethora of national and international human genome initiatives co-existed from the mid-1980s onwards with different rationales, spokespersons and funding regimes.[2] As late as 1996, the strategy of tackling the whole human genome via the concerted action of a handful of large-scale sequencing centres was not yet dominant. Contemporary historical accounts (e.g., Cook-Deegan, 1994,

---

[1] This terminology was already present in the 1990s and early 2000s, especially among scholars based in the USA and working on the socio-ethical implications of "the Human Genome Project" (Kevles & Hood, 1992; Sloan, 2000). As we show later in the chapter, this literature often conflated the international sequencing effort with the US national human genome project. Europe-based scholars have tended to be more nuanced and distinguish different initiatives and approaches to the human genome (Glasner & Rothman, 1998). Later sociological and historical investigations have continued to refer to the Human Genome Project, acknowledging the multiple genealogies behind the genesis of this term (Hilgartner, 2017; Stevens, 2013).

[2] For a survey of different human genome efforts written at the time they were developing, see McLaren (1991), Cook-Deegan (1994, Ch. 14) and the section "European contributions" from the Spring 1991 newsletter of the UK Human Genome Mapping Project: Nigel K. Spurr (ed.) *G-Nome News*, volume 6: 30-69, National Archives of the UK at Kew (London), Medical Research Council Series, file number FD7/2745. In this chapter, we focus on the national human genome projects of the USA and UK, and to a lesser extent on the EC's Human Genome Analysis Programme.

Part Three) document that only one national initiative unambiguously sought, from the onset, to produce a physical map and a reference sequence of the entire human genome: the joint programme of the USA's Department of Energy (DoE) and National Institutes of Health (NIH). Due to this, a widely accepted meaning of the capitalised phrase 'Human Genome Project' during most of the 1990s was just the US national effort, which itself adopted that name.

We designate the US national programme throughout this chapter as 'US-HGP'. It formally commenced in 1990, when some other national human genome projects were already underway, and had as a defining characteristic the concentration of NIH and DoE funding in a series of centres that specialised in various aspects of genomics, such as physical mapping, large-scale sequencing, bioinformatics or technology development (Hilgartner, 2017, Ch. 2 and pp. 91-110). Some of these centres already existed and were devoted to other types of research, such as medical genetics or, in the case of those supported by the DoE, the effects of radiation on DNA. Others were created *de novo* to comprehensively sequence the human genome and those of pilot organisms, such as yeast (Chap. 2). All of the centres were committed to the objective of full genome mapping and sequencing, a feature that distinguished the US-HGP from many other contemporary human genome programmes. As we highlight, a leading architect in the design of the new centres and advocate of their whole-genome approach was the Nobel Prize-winning molecular biologist—and co-discoverer of the double helical structure of DNA—James Watson, who led the NIH arm of the US-HGP until 1992.

Among national human genome programmes, the objective of mapping and sequencing the whole human genome was unique to the US-HGP, as was the prominence of a leader such as Watson. The main insights of this chapter stem from comparing the US programme with another, less well-known national initiative: the UK Human Genome Mapping Project (HGMP). Launched one year earlier, in 1989, and funded by the British Government through its Medical Research Council (MRC), the HGMP did not create large-scale sequencing centres. As with many other emergent human genome projects, its strategy aligned with the distributed, network approach that the EC was forging for the sequencing of yeast (Chap. 2).

The HGMP enabled the MRC to secure funds from the UK Treasury for a Directed Programme of grants specifically tailored to map and sequence human DNA. The recipients of those grants were laboratories in the fields of human genetics and, especially, medical genetics. Those recipients and the ways they aimed to tackle the human genome were key differences between the British programme and the US-HGP. Rather than promoting a new breed of whole-genome-oriented practitioners, as the DoE and the NIH were fostering, the HGMP funded and coordinated research groups that kept working on specific parts of the human genome. In other words, the communities of genomicists that constituted each programme differed. Although the beneficiaries of HGMP grants collectively produced map and sequence data across the genome, they retained their individual identity as specialists in diseases or biological phenomena affecting only certain genome regions. Conversely, the specialism of the DoE and NIH-funded genomicists increasingly tended towards the large-scale mapping and sequencing of the entire human genome.

The HGMP beneficiaries used their grants to develop mapping and sequencing methods aimed at positioning, within human chromosomes, DNA fragments encompassing genes or gene markers associated with diseases, or any other biologically or medically relevant characteristic. They were assisted by a resource centre that the HGMP established as both a technological hub and a repository of the genomic data produced by the laboratories in receipt of Directed Programme funding (Balmer, 1998; Glasner, 1996). Apart from providing technical support and advice to the HGMP-funded laboratories, the resource centre pooled their mapping results and compiled them in databases.[3] It also conducted partial sequencing of the mapped DNA fragments, particularly the regions corresponding to genes that were thought to be involved in the genetic diseases that the HGMP laboratories investigated. This work was developed in collaboration with gene-specific sequencing groups sponsored by the Human Genome Analysis Programme, an initiative that the EC launched in 1990 that followed the distributed model it had just implemented for the yeast sequencing project (Chap. 2).

The HGMP Resource Centre differed from the US-HGP genome centres in two key aspects: (1) it fulfilled a service role and conducted mapping and sequencing work at the request—and based on the results—of the Directed Programme-funded laboratories rather than comprehensively

---

[3] As we see later in the book, the Resource Centre provided mapping tools and assistance to other communities throughout the 1990s, such as those involved in the mapping of the genome of the pig, *Sus scrofa* (Chap. 5).

sequencing at its own initiative; and (2) the map and sequence data it compiled represented only the areas of interest of the contributing laboratories and was thus not intended to be a *complete* representation of the whole human genome. As we argue, out of the HGMP, HGAP and other groupings of human and medical geneticists—such as the chromosome mapping workshops—a community of genomicists emerged, one that was larger and more diverse than the one working at the genome centres conducting the US-HGP.

This contrast enables us to conclude that a key factor distinguishing the US-HGP from the HGMP, and more generally from the distributed approach promoted by the EC, was in the assemblage of the research communities and funding regimes underlying each of them. In the case of the US-HGP, this assemblage embodied Watson's vision, his circle of influence and the joint funding provision of the DoE and NIH. Watson was a founder of molecular biology and, from the late-1960s onwards, had been instrumental in structuring this community from his position of director of Cold Spring Harbor Laboratory (CSHL). One of the pillars in this structuring process had been fostering the shared belief in the mechanistic action of genes and the community's commitment to detailed investigations of model organisms. It was hoped that a full molecular description of those organisms would unveil the role of genes in a myriad of biological processes.[4]

As we showed in the previous chapter, the worm *Caenorhabditis elegans* had become one such model organism. It was at CSHL where, in 1989, Watson met John Sulston, Alan Coulson and Robert Waterston, and instigated the start of the worm's sequencing project. The designated host institution of the project in the USA, Washington University, subsequently inaugurated a genome centre and undertook the sequencing of two other organisms: yeast and *H. sapiens*. This intensive and scaled-up approach differentiated the genome centres from the distributed model that the EC was promoting in its sequencing programmes (Chap. 2). The status of *C. elegans* and yeast as model organisms was one of the reasons that led Watson to regard them as suitable pilot platforms to inform the mapping and sequencing of the human genome. He did not hesitate in adopting the same genome centre model when the US-HGP—a self-contained,

---

[4] Other crucial figures in spreading the influence of molecular biology and promoting its mechanistic view of gene action were Sydney Brenner (discussed below) and co-elucidator of the DNA double helical structure, Francis Crick (Aicardi, 2016).

national initiative as opposed to the multi-country programmes of the EC—sponsored the comprehensive human genome sequencing effort. Under Watson's leadership, the US-HGP was the vehicle for producing a reference sequence from which the connections between genes and biological properties—implicating evolution, health and disease—could later be drawn.

The HGMP was also promoted by a founding figure of molecular biology: the proponent of *C. elegans* as a model organism, Sydney Brenner. Yet the MRC, partly due to the size of the UK relative to the USA, lacked the resources to launch a whole-genome initiative on its own. This led Brenner and the MRC to look at the communities of human and medical geneticists as possible allies to execute the project. Unlike molecular biologists, these communities were interested in variation rather than comprehensive standard descriptions as an entry point into investigating gene function. Consequently, their motivation to tackle the human genome was not achieving a complete reference sequence, but using the reference sequence data as a scaffold to aid in the determination of variants associated with diseases or evolutionary traits. Identifying and investigating variation, as opposed to establishing a canonical reference sequence, was thus a driving force behind the organisation of the HGMP and its indifference to adopting whole-genome approaches. From the viewpoint of the many laboratories supported by the HGMP Directed Programme, focusing on specific genome regions that could be compared with either other organisms or between patients suffering a genetic condition and non-sufferers was far more useful than mapping and sequencing the entire human genome.[5]

In what follows, we show that the differences in the funding systems, organisational models, communities and genomicists involved in the HGMP and US-HGP assemblages point to a diverse landscape. This diversity is difficult to grasp from a perspective that narrowly focuses historical inquiry on the human reference sequence published in 2001 and 2004. What is now associated with a single, coherent and successful Human Genome Project represents just one route through complex

---

[5] Elsewhere, we have characterised these different approaches through the categories of horizontal and vertical sequencing. Whereas horizontal sequencing would involve producing a one-dimensional reference genome, the vertical strategy would explore sequence variation in one specific genome region across individuals or different species, with the aim of augmenting clinical or evolutionary knowledge (García-Sancho, Leng et al., 2022).

historical terrain. Our ability to identify the web of pathways that criss-cross this terrain enables us to extend our historical interrogation from yeast to *H. sapiens*. The multiplicity of both parallel and interwoven lineages in the development of the HGMP and US-HGP indicates that the historical landscape was as heterogeneous in human as in non-human genomics. By looking both within and beyond human genomics, we can highlight the factors that led to the increasing prominence of the human reference genome. This enables us to assess its significance in a fresh light, while at the same time preventing it from narrowing our vision.

3.1   THE EXCEPTION RATHER THAN THE RULE

In 1988, Watson supplemented his CSHL directorship with a new role as associate director of the freshly-established NIH Office for Human Genome Research. He had held the CSHL position since 1968—15 years after co-elucidating the DNA double helix and 6 years after receiving the Nobel Prize—and transformed this institution into the most influential forum of molecular biology. CSHL held annual symposia in which the invitees, considered to be the international elite of molecular biologists, would discuss pressing scientific challenges. The 1986 symposium had been devoted to the *Molecular Biology of Homo sapiens* and became one of the first settings in which the feasibility of mapping and sequencing the human genome was assessed. The enormous size of the human genome—three billion DNA nucleotides compared to the 12 million of yeast and 100 million of *C. elegans*—made the viability and utility of the enterprise a matter of debate within and outside the CSHL meeting. In his 1988 CSHL director's report, Watson expressed concerns about his increased responsibilities and the stress of commuting. He considered, however, that the remit of the new NIH Office—implementing a national human genome programme in the USA—represented a one-time "opportunity". From this new position, Watson could let his scientific life "encompass a path from double helix to the three billion steps of the human genome".[6]

Watson's commitment to the sequencing of the human genome was shared by scientists and administrators at the DoE. However, rather than completing the molecular description of DNA—from ascertaining the

---

[6]"Director's report" in *Annual Report 1988—Cold Spring Harbor Laboratory:* 1-24, quote from p. 5. We thank Robert Cook-Deegan at Arizona State University for generously providing access to this record.

double helix to laying out its nucleotide sequence—what the DoE human genome advocates sought was to build on a longstanding tradition of investigating the genetic effects of radiation. This line of research had started after World War II, following the dropping of the atomic bombs and their devastating medical effects on local populations in Hiroshima and Nagasaki (Lenoir & Hays, 2000; Lindee, 1994). It had led to the reorientation of some of the personnel and research programmes of DoE-funded laboratories from physics to the life sciences. An example of this was Los Alamos National Laboratory, which after playing a leading role in the wartime race to develop the atomic bomb—it was the home of the flagship Manhattan Project—devoted a growing proportion of its mathematics expertise and computing resources to solve biological and medical problems.

Due to this, Los Alamos was chosen as the institution that would host the first centralised DNA sequence database in the USA—GenBank—in 1982 (Strasser, 2019, Ch. 5). A few months prior to the 1986 symposium at CSHL, DoE representatives organised a workshop in Santa Fe and subsequently announced a pioneering programme called the Human Genome Initiative.[7] As a result of this, the biomedical lines of research at two other DoE-sponsored institutions, the Lawrence Berkeley and the Lawrence Livermore National Laboratories, were strengthened and largely channelled towards technology development and genome-wide mapping and sequencing of human DNA. That the DoE network of national laboratories was equipped with personnel and infrastructures to conduct big science endeavours was a competitive advantage that favoured their early leadership in the incipient human genome work in the USA.[8]

How the DoE initiative converged with the NIH effort has been amply described in the literature (Cook-Deegan, 1994, Part Three; Hilgartner,

[7] The CSHL and Santa Fe meetings had been preceded by a workshop convened in 1985 by Robert Sinsheimer, a molecular biologist who was then chancellor of the University of California at Santa Cruz. Its participants cited multi-million dollar grants that the University had been awarded in the areas of particle physics and space science to argue for the necessity of a similar NIH investment in a human genome programme. No significant NIH move occurred until three years later (Sinsheimer, 1989).

[8] While the DoE's advocacy and pursuit of human genomics was motivated by the tracking of heritable radiogenic genetic mutations, they also needed to find new purposes for their national laboratories in the light of arms reduction treaties. In genomics, key DoE figures saw the potential for a big-enough science that would take the place of weapons development and make use of expensively-assembled facilities. This prompted the acerbic comment by David Botstein that the DoE's plans constituted a program for unemployed bomb-makers (Cook-Deegan, 1994, pp. 96–100, quote from p. 98).

2017, pp. 91-110). In 1988, two reports issued by the US National Academy of Sciences and the Office of Technology Assessment recommended a single national initiative that would initially focus on physical mapping and improving the existing instrumentation to create a platform for sequencing the human genome in the longer term. This led the DoE and NIH to merge their endeavours into the US-HGP, a 15-year programme that was launched in 1990 with a three billion dollar budget that was contributed towards by both agencies, the former through the Office of Health and Environmental Research and the latter through the National Center for Human Genome Research, an expanded version of Watson's Office that was later renamed as the National Human Genome Research Institute (NHGRI).[9] The explicit goal of the US-HGP was to produce a physical map and reference sequence of the whole human genome by 2005.

What we want to stress concerning the history of the US-HGP is how this initiative, and other contemporary human genome projects, disrupted the funding and organisational regimes of biomedicine. This disruptive effect has already been noted by scholars who have investigated the impact of big science and data-intensive approaches on different areas of contemporary biological and medical research (Leonelli, 2016; Stevens, 2013; Vermeulen, 2016).[10] With regard to genomics, Stephen Hilgartner has argued that it propelled a new "knowledge-control regime" that was distinct from existing disciplines, such as molecular biology. This regime constituted new categories of "agents, spaces, objects and relationships", and

[9] The National Center for Human Genome Research was established in 1989 and renamed as the NHGRI in 1997. Also in 1997, the DoE changed the name of its office to Biological and Environmental Research. Given that in subsequent chapters we refer to events that occurred under the new names, we use NHGRI throughout the book for ease of reading. Watson was appointed director of the National Center for Human Genome Research and remained in this position until his resignation in 1992. Elke Jordan, who had acted as day-to-day director of the NIH Office—while Watson was part-time associate director—became deputy director of the National Center until 2002, overseeing its transition to NHGRI and the early years of that institution.

[10] The existence and impact of data-intensive endeavours is not exclusive to the twentieth and twenty-first centuries. Historians have documented how expeditions to Asia, Africa and the Americas in the early-modern period led to the introduction of large amounts of new knowledge in Europe, and a feeling of "information overload" that was crucial for the emergence of natural history (Müller-Wille & Charmantier, 2012; Rosenberg, 2003). Building on this, Bruno Strasser (2019) has forcefully argued that the perspective of obtaining new knowledge through the collection, compilation and comparison of specimens and data about them has always existed in the life sciences and interacted in different ways with more experimental approaches.

allocated to them "entitlements and burdens" that led to novel ways of conceiving and disseminating knowledge (Hilgartner, 2017, p. 9).

Hilgartner's empirical work has focused on the US-HGP as an exemplar of new players—the genome centres—and new rules for processing, storing and sharing the data they produced. Crucially, the emergence of the knowledge-control regime of genomics was neither immediate nor uniform. It occurred gradually throughout the 1990s, with more intensity in some parts of the world than in others. The rest of this chapter emphasises the gradualism of the transformation within the US-HGP, and how other human genome programmes adopted different knowledge-control regimes. Some of these alternatives to the US-HGP, we argue, never converged with what Watson and his DoE colleagues advanced.

A challenge that the NHGRI faced was in transforming the funding culture of the NIH into a system that would enable large-scale mapping and sequencing. Like many other biomedical funders, NIH managers and administrators were used to issuing competitive calls for proposals and awarding grants across relatively large numbers of laboratories, following peer review of their applications. This differed from the DoE model, which rather than running a responsive grant mechanism would distribute their budget among a narrower cohort of recipients: its network of national laboratories. The DoE funding system had allowed the creation of a number of genome centres that prioritised the production of map and sequence data via the development of high-throughput technologies and the deployment of industrial modes of production. These genome centres were based in some of the DoE laboratories and had begun operating during the preceding Human Genome Initiative. Although Watson could not exempt the NHGRI from the NIH grant-award system, he established different, specific criteria when distributing US-HGP funds with the aim of fostering a similar type of operation to the DoE one.

The main criterion for NHGRI grants was whether the applicants and their home institutions could contribute to the establishment of a solid base of whole-genome mapping and sequencing centres. With this, Watson sought to avoid what he labelled the "cottage industry" approach, which he attributed to the sequencing of microorganisms (Watson, 1990, p. 45). This approach consisted in the formation of large inclusive consortia and required the distribution of resources as widely as possible among the communities working on the organisms to be characterised. Watson's attribution of "cottage industry" was initially aimed at the sequencing of the bacterium *Escherichia coli*, but as the 1990s progressed, the EC's Yeast Genome Sequencing Project emerged as the most widely cited example of cottage industry genomics (e.g. Palca & Roberts, 1992, p. 957).

For Watson, the cottage industry approach presented several logistical problems when applied to larger genomes. Instead, the NHGRI sought to gradually form a small set of funding recipients with industrial mapping and sequencing capacities that were not necessarily interested in conducting research using the resulting data. This change of ethos, however, did not become fully implemented in the USA until the mid-to-late 1990s, partly due to the resistances it encountered among some quarters of the genetics community.[11]

During the early days of the US-HGP, the NHGRI administrator in charge of distributing genome mapping grants was Jane Peterson. She worked hard to persuade laboratories equipped with the appropriate technologies and expertise to broaden the genome areas they would tackle. Some of these laboratories featured long-established teams of medical geneticists that had historically focused on smaller regions of human chromosomes encompassing genes or genetic markers connected to diseases.[12] Examples of this were Victor McKusick and Frank Ruddle's groups, at Johns Hopkins University and Yale University respectively. These two scientists (Fig. 3.1) had pioneered the chromosome mapping workshops, forums at which geneticists from all over the world shared their mapping results.

Started in 1973 and continued annually or biennially until the release of the human reference genome, these workshops produced human genome maps with increasing numbers of genes and markers on them, and at improved resolution (Fig. 3.2).[13] They achieved this through the collation of multiple partial results: those reported by individual genetics

---

[11] See, for instance, Ayala (1987); Baltimore (1987). Some commentators, including reputed biomedical scientists, argued that the potential outputs of the US-HGP, in the form of a full human genome map and sequence, did not justify an expenditure that would curtail other areas of life science research.

[12] Jane Peterson, interview with Miguel García-Sancho, National Human Genome Research Institute (Bethesda, Maryland), November 2018.

[13] Until 1991, such meetings were called Human Gene Mapping Workshops. These were subsequently replaced by Single Chromosome Workshops under the auspices of the Human Genome Organisation (HUGO; see Chap. 4).

Fig. 3.1 Victor McKusick (left photograph, second seated from left) with fellow medical geneticist P. S. Gerald; and Frank Ruddle (right photograph, standing wearing a white shirt) surrounded by, among others, G. J. Darlington and R. S. Kucherlapati. They were all attending the first chromosome workshop, held at Yale University in 1973. Both pictures from: New Haven Conference (1974, pp. 209 and 211); copyright © 1974 Karger Publishers, Basel, Switzerland

groups working on a specific disease or set of diseases at given chromosomal locations. By collectively gathering and pooling these results, the workshops gradually covered broader areas of the chromosomes and populated them with an increased number of landmarks (Jones & Tansey, 2015). In 1987, building on the success and consolidation of this model, McKusick and Ruddle co-founded *Genomics,* a journal devoted to the publication of mapping results (Kuska, 1998; Powell et al., 2007, pp. 13ff). Yet in order to achieve the US-HGP goals, the NHGRI needed to fund institutions—rather than collectives—whose mapping went well beyond the contributions to the chromosome workshops or the results published in the articles of *Genomics.*

On the sequencing front, the NHGRI initially funded a small number of individual grants aimed at model organisms with relatively small genomes, such as *E. coli, C. elegans,* and *D. melanogaster,* as well as a number of yeast (*S. cerevisiae*) chromosomes (Chap. 2). Some, but not all, of these grants were among the first set of genome centre grants funded in 1990. Strategically, not only were those grants intended to contribute towards the completion of the sequences of their target organisms but, more importantly in the long term, to act as platforms for technology development and the creation of the infrastructures for the establishment of sequencing centres. In 1996, the NHGRI awarded a set of six grants as pilots for human genome sequencing; these projects had a minimum

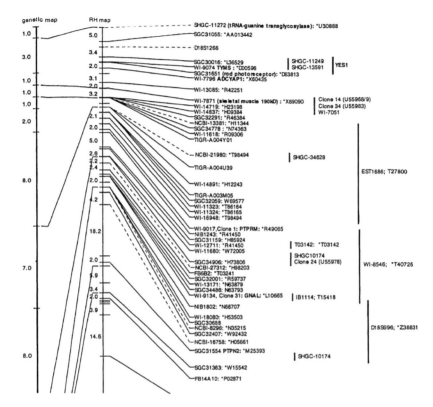

Fig. 3.2 Part of the genetic linkage map and physical map of human chromosome 18, as reported in the Fourth International Workshop devoted to its mapping, held in Boston (USA) in 1996. The genetic linkage map is displayed on the left of the picture and labelled as "Genetic map", with the physical "RH map" arrayed next to it (Radiation Hybrid—RH—maps are a form of physical map). From: Silverman et al. (1996), p. 119; copyright © 1996 Karger Publishers, Basel, Switzerland

target of sequencing one Megabase (one million nucleotides, or bases) of human DNA. With the information and experience gained in these pilots, the NHGRI scaled up its sequencing programme in 1999 with the funding of three Genome Sequencing Centers at Washington University, Baylor College of Medicine and the Whitehead Institute (of the Massachusetts Institute of Technology and Harvard University). At all three of these sites, the sequencing centres were outgrowths of previously funded genome centres and pilot sequencing projects.[14]

A defining characteristic of those sequencing centres was that their funding and organisation prioritised the completion of their target genomes over any other scientific or medical objective, including the mapping of genes or markers associated with diseases. This form of operation was difficult to deploy beyond the USA. For example, in most European countries, governments had neither the resources nor the motivation to create specific grants for large-scale genome mapping and sequencing at dedicated centres. Private and charitable funds, by contrast, had fewer constraints and could be more easily channelled to a particular enterprise or group, as opposed to having to support a wider scientific community. This was the case for the Wellcome Trust, a British charity that teamed up with the MRC in 1992 to create the Sanger Institute, an institution that substantially contributed to the completion of the yeast, *C. elegans,* human and pig genomes (Chaps. 4 and 5). Another example of a charitably-funded genome centre was Généthon, supported by AFM-Téléthon, the French Muscular Dystrophy Association. Established in 1990, this institution was devoted to comprehensive mapping and quickly became a world leader in the production of genetic and physical maps encompassing the

---

[14] Mark Guyer, director of the NHGRI's extramural (grant-funding) programme; personal communication with Miguel García-Sancho, National Human Genome Research Institute (Bethesda, Maryland), November 2018. On the history of the Center for Genome Research at the Whitehead Institute, which became part of the Broad Institute, see García-Sancho, Leng et al., 2022. Its leader, Eric Lander, had a vision of genome mapping and sequencing that differed from the traditional ways of working of medical geneticists. The group behind the Human Genome Sequencing Center at the Baylor College of Medicine, by contrast, was furnished with a strong tradition in medical genetics research. See: Jim Lupski, "Applications of sequencing in clinical genetics", presentation delivered at *The Evolution of Sequencing Technology: A Half-Century of Progress* meeting, organised by the Genentech Center and Cold Spring Harbor Laboratory Archives in Long Island, 16th–19th July 2015. Available at: http://library.cshl.edu/Meetings/sequencing/video-pages/Lupski.php (last accessed 14th December 2022). Most scientists in this group have had double affiliations, also belonging to the Department of Molecular Genetics of Baylor College. This enables them to perform large-scale sequencing at the Genome Center alongside medical genetics research at the Department of Molecular Genetics.

entire human genome.[15] Généthon combined whole-genome work at its own initiative with a service role, attending to mapping requests from the French medical genetics community. This service role differentiated it from the US sequencing centres (Jordan, 1993, pp. 131ff; Kaufmann, 2004).

In spite of their influence, Généthon and the Sanger Institute were exceptional cases outside the USA. The US-HGP was rather unique in its commitment to full human genome mapping and sequencing when compared to programmes introduced by other governments, especially in Europe.[16] Those other programmes did not distinguish the human genome work they sponsored as sharply from medical genetics research as the US-HGP did. For this reason, they refrained from focusing on producing a reference map and sequence of the full human genome and were closer to the distributed, networked organisation that the EC was implementing in its sequencing projects. This distributed form of organisation was more suitable for fostering communication and tailoring the genome work to the regions of interest of the local medical genetics communities. The British HGMP was one of the earliest examples of this way of approaching the human genome.

[15] As we have shown elsewhere, Généthon was the institution that submitted the largest volume of human DNA sequence data to public repositories prior to 1996, well above any other laboratory, including the Sanger Institute and the US-HGP genome centres (Garcia-Sancho, Leng et al., 2022, Table 1, p. 334). This sequencing, however, was conducted to enable mapping work rather than to comprehensively characterise the human genome. In 1996, the publicly-funded French atomic energy commission (Commissariat à l'énergie atomique; CEA) created Genoscope, a sequencing centre that contributed to the elucidation of the human reference genome, albeit to a lesser extent than its US and British counterparts (Ramillon, 2007). The French CEA was also involved in early mapping and sequencing work on the pig *S. scrofa*, as we see later in the book (Chap. 5).

[16] Some Asian programmes pursued whole human genome sequencing. During the 1980s, Japan invested heavily in the automation of sequencing techniques and deployed an ambitious human genome project (Fujimura, 2000; Yoshikawa, 1990; Cook-Deegan, 1994, Ch. 15). Yet the Japanese DNA sequencing machines were never as popular as those manufactured in the USA and Europe, and Japanese institutions performed below the British and US genome centres, despite being involved in the human reference sequence. China created high-throughput sequencing centres—namely the Beijing Genomics Institute—but only joined the production of the human reference genome in the late-1990s (Wang et al., 2021). In the Americas, Canada created human genome programmes that were more in line with the HGMP and the EC (Dusyk, 2007).

3.2   The UK Human Genome Mapping Project

In 1989, one year before the launch of the US-HGP, the British Government authorised the release of 11 million pounds to fund the HGMP, a three-year programme that would be managed by the MRC.[17] The key proponents of this initiative were Brenner, a senior scientist who had just left the Laboratory of Molecular Biology of Cambridge (LMB) after a successful 30-year tenure, and Walter Bodmer, a reputed geneticist who coordinated the research laboratories of the medical charity Imperial Cancer Research Fund (ICRF).[18] Keith Peters, a practising physician with ample experience in teaching and researching immunology at London's Hammersmith Hospital, had presented the HGMP proposal on Brenner and Bodmer's behalf to the Advisory Committee on Science and Technology (ACOST). This body directly reported to the UK Prime Minister—in this case Margaret Thatcher—on projects that were likely to generate impact and required rapid funding. It approved the HGMP on Peters' recommendation and transferred the funds in less than one year (Balmer, 1996).

The prime mover behind the HGMP was Brenner. He had moved to Cambridge (UK) in 1956 to begin his research career, having recently concluded his PhD. Watson had also moved to Cambridge at the same stage in his career and returned to the USA the same year Brenner arrived in the UK. Brenner became the main collaborator of physicist-turned-biologist Francis Crick, who had successfully worked out the structure of DNA with Watson. Up to the early-1960s, Crick, Brenner and Watson focused on what became known as the coding problem: how the order of the nucleotides comprising DNA affects the synthesis of specific proteins that are responsible for most of the structural and functional aspects of the living cell (de Chadarevian, 2002, Part II; see also Kay, 2000).

In 1962, the same year Watson and Crick were awarded the Nobel Prize, the LMB was founded as an MRC-supported institution that would host an increasingly influential group of biologists in Cambridge. Crick became the director of the LMB Division of Molecular Genetics and

---

[17] According to the UK Retail Price Index measure of inflation, the equivalent sum as of November 2022 would be about 26.7 million pounds. https://www.bankofengland.co.uk/monetary-policy/inflation/inflation-calculator (last accessed 14th December 2022).

[18] In 2002, ICRF merged with The Cancer Research Campaign to form Cancer Research UK. A substantial part of its laboratories have now been amalgamated into the Francis Crick Institute in London, see https://www.crick.ac.uk/about-us/our-history (last accessed 14th December 2022).

Brenner started a long-term line of research, adopting the nematode worm *C. elegans* as a model to investigate the genetics of development and behaviour. This enterprise sought a detailed description of the worm's neuron circuitry, as well as its development from embryo to adult, with the hope of finding the "programme" that connected brain activity and cell differentiation to particular *C. elegans* genes.[19] The project included crossing experiments in which Brenner attempted to produce mutant worms and identify specific genes associated with variation in properties such as size or mode of movement, as geneticists had done with the fruit fly *Drosophila* and other organisms. Brenner also recruited more junior associates that would carefully detail the fates of every single cell throughout the *C. elegans* life cycle—its cell lineages—and the position and synaptic connections of each neuron in its brain.[20] To this end, John Sulston joined the LMB in 1969 to chart the multiple divisions of cells during the worm's embryonic and post-embryonic development (de Chadarevian, 1998).

By the time Brenner first proposed to map the human genome, in 1986, the worm project was experiencing a profound transformation. The description of cell lineages and brain connectivity had been completed by the early-1980s and a project to construct a physical map of its genome had started under the leadership of Sulston and Alan Coulson (Fig. 3.3). Coulson was a research assistant who joined the team after working at another LMB division on the development of early DNA sequencing techniques. Brenner, however, was becoming increasingly sceptical about the possibility of matching the detailed information his team had gathered

---

[19] See Brenner (1973, p. 271; 1974, p. 71). This language of programmes, circuitry and information flows had been mobilised by cybernetics after World War II and imported into molecular biology by various researchers, among them François Jacob and Jacques Monod at the Pasteur Institute in Paris. During the early-1970s, Jacob and Monod published popular accounts that further spread the use of cybernetic vocabulary in biology (Kay, 2000; Rheinberger, 2006). The Pasteur Institute played a major role in the EC's yeast genome effort (Dujon, 2019) and Mark Johnston, one of the leading yeast sequencing scientists in the USA, started his career with the *S. cerevisiae* GAL system (Chap. 2), a closely related gene expression system to the one Jacob and Monod had explored back in the 1960s.

[20] On the deeper history of cell lineage research, reaching back to the late-nineteenth and early-twentieth century, see Guralnick (2002); Lowe (2016); and Maienschein (1978, 1990). Concerning cell lineage research on *C. elegans*, see de Chadarevian (1998) and Jiang (2013). The ability to trace the lineages of adult cells back to cell divisions earlier in development—and therefore the fates of those earlier cells and divisions—provides the basis for precise experimental intervention, for example being able to assess changes wrought on the process of development and resultant outcomes by a mutation in a gene or genes.

Fig. 3.3 Left, Sydney Brenner with co-discoverer of the double helical structure of DNA, Francis Crick, at the Laboratory of Molecular Biology of Cambridge in 1962. Right, John Sulston holding a section of the physical map of *C. elegans* around 1985 (pictures of the nematode worm are pinned to the wall behind him). Copyright of left image: Hans Boye/MRC Laboratory of Molecular Biology. Copyright of right image: MRC Laboratory of Molecular Biology. Both reproduced with permission

about cell divisions and synaptic transmissions in *C. elegans* to the genes Sulston and Coulson would identify in their map, given the complexity of developmental processes in multicellular organisms (Lewin, 1984).

Partly because of this, in the same year of his human genome map proposal, Brenner left the LMB and established a Molecular Genetics Unit that, despite being also supported by the MRC, was part of the School of Clinical Medicine of the University of Cambridge. In this Unit, Brenner continued some work on the genetics of *C. elegans* but left the physical map to Sulston and Coulson, who remained at the LMB. The other lines of research in Brenner's Unit were the development of genome mapping technologies and "certain aspects of gene evolution".[21]

---

[21] Anonymous (1986) "Extract of minutes of the Council meeting held on Thursday 17th July 1986—Molecular genetics: proposal from Dr S. Brenner (MRC Laboratory of Molecular Biology) for a new Unit under his direction (86/C616; file E243/130)", National Archives of the UK at Kew (London), Medical Research Council Series, file number FD12/1191, quote from unnumbered page. Brenner had also concluded, by 1986, his tenure as director of the LMB and was approaching retirement age. To some especially distinguished scientists reaching this career stage, the MRC offered to create a more personally-managed research unit for them.

Brenner's proposal was entitled "A physical map of the human genome" and it was submitted in November 1986 to the Cell Board, the body of the MRC that funded genetics research. In his case for support, he argued that it was by then "not clear" whether the resources needed for a "central facility" to sequence the entire human genome "would ever be made available". This led Brenner to advocate for the construction of a physical map not only as a "first necessary step towards the grander sequencing proposal, but also for the more immediate benefits" it could bring "to medical research and practice". Brenner's vision started with a laboratory that would "carry out" the mapping programme and "act as the reference centre for human genetics". A "central concept" of his strategy was to establish "cooperative links and not enter into competition with individual research projects". In this regard, Sulston and Coulson's ongoing physical map of *C. elegans* provided a "useful benchmark" for Brenner's intended human mapping enterprise.[22]

At the time of this proposal, Brenner was serving on the committee of the National Academy of Sciences that advised the US Government on the plausibility and best strategy for conducting a human genome project. By late 1986, the discussions were still nascent and the model of tackling the entire human genome at dedicated and comprehensive mapping and sequencing centres had not yet attained majority support. Nevertheless, this comprehensive and concentrated strategy was gaining momentum in the USA. The physical mapping exercise that Brenner envisaged for the UK and the reference laboratory that would execute it differed in many respects with what became the US-HGP.

First, and contrarily to Watson, who also served in the committee, Brenner did not support a whole-genome operation. For Brenner, the size of the human genome—30 times bigger than *C. elegans*—meant that a comprehensive mapping and sequencing initiative would yield a substantial volume of data that would not correspond to genes. Biomedical scientists were well aware that only a small fraction of human DNA constituted genic regions, i.e., those directly involved in the synthesis of proteins. By the mid-to-late 1980s and early-1990s, a large proportion of those scientists—especially within the human and medical genetics communities—regarded the remainder of the genome as 'junk DNA': repetitive sequences

---

[22] S. Brenner (1986) "A physical map of the human genome", National Archives of the UK at Kew (London), Medical Research Council Series, file number FD23/3441, quotes from pp. 1 and 2.

that were expected to be non-functional.[23] Based on this common wisdom, Brenner argued that mapping and sequencing the entire human genome was not a worthwhile enterprise (Brenner, 1990). He, however, maintained his commitment to detailed descriptions of organisms that, due to their simpler developmental processes, could be used to model the molecular basis of life properties.[24]

Secondly, the reference laboratory that would channel Brenner's genome project was conceived to operate at the behest of human and medical geneticists. This was largely due to the framing of his proposal against the background of the ongoing physical mapping of *C. elegans*. Since 1983, Sulston and Coulson had mapped ever-increasing areas of the worm's genome by fulfilling requests of laboratories working on specific *C. elegans* genes. This had been mutually beneficial and ensured that the mappers were regarded as important, foundational members of the *C. elegans* research community: Sulston and Coulson crucially contributed to the objectives of this community, while increasing the resolution of their physical map (García-Sancho, 2012). The genome centres that Watson established for the US-HGP lacked this community service role: they mapped and sequenced comprehensively, at their own initiative rather than addressing requests from other laboratories. Although the genomes of *C. elegans* and *S. cerevisiae* were part of the remit of these large-scale centres, the US-HGP approached the mapping and sequencing of both

---

[23] Biomedical research communities who were less focused on human genes and their role in disease were more mindful of the importance of non-genic regions. This was the case for developmental biologists, with whom Sulston and Coulson collaborated during the mapping of *C. elegans* and for whom some non-genic DNA exerted a key role in inhibiting or activating the mechanisms of protein synthesis (Chap. 4). Up to the mid-to-late 1990s, it was believed that the human genome was formed of around 100,000 genes. Following the publication of the first draft of the reference sequence (2001), this figure was reduced to 20,000 to 25,000 and the estimated percentage of protein-coding regions lowered to 1.5% of the DNA in the human genome. On the origins of the term 'junk DNA', see: https://judgestarling.tumblr.com/post/64504735261/the-origin-of-junk-dna-a-historical-whodunnit (last accessed 14th December 2022).

[24] One of the flagship projects of Brenner's newly-established Molecular Genetics Unit was the full sequencing of *Fugu*, a pufferfish whose genome is characterised by a high-density of genic regions and a lack of repetitive sequences. Using comparative approaches, Brenner believed that *Fugu's* genome would provide insights concerning the sequence and protein synthesis mechanisms of human genes with a fraction of the effort of tackling the human genome in its entirety (Venkatesh, 2019).

organisms as a means of easing the path to human genome work rather than engaging with the research necessities of worm and yeast biologists.[25]

Thirdly, and as a consequence of the above, Brenner's project sought to involve the existing human and medical genetics groups rather than creating a new community of genome centres and specialist genomicists. After receiving Brenner's proposal, the MRC sounded out the opinion of reputed scientists and institutions in search of arguments for approval or rejection, as well as possible sources of co-funding. One of Brenner's first allies was Bodmer, who belonged to a group of geneticists that in the 1960s and 1970s had pioneered the mapping of a region of the human genome called the Human Leukocyte Antigen system (HLA).[26] This region contains densely-packed and hypervariable genes implicated in the immune response to infection; the variability of many of these genes aided their mapping (Löwy, 1987; see also Heeney, 2021). In his role of director of research at ICRF, which he took up in 1979, Bodmer equipped the charity's laboratories with cutting-edge DNA mapping and sequencing technologies (Weston, 2014, esp. Chs. 2-4). Another supporter of Brenner's proposal was Peters, who in 1987 moved from Hammersmith Hospital to the University of Cambridge due to his appointment as Regius Professor of Physic and Dean of the School of Clinical Medicine. From this position, he oversaw the establishment of Brenner's Unit in the school and saw human genome mapping as an opportunity to connect genetics research with medical goals.[27]

Peters had also become life sciences adviser in ACOST and suggested this committee—directly reporting to the Prime Minister's Office—as a potential source through which the MRC could obtain the necessary

[25] The genome centre at Washington University was headed by Sulston and Coulson's *C. elegans* collaborator, Robert Waterston. As we show in the next chapter, this collaboration and Watson's intervention were crucial for the redefinition of the worm genome effort from *à la carte* mapping to comprehensive, large-scale sequencing, and for Sulston and Coulson's institutional migration to a UK-based genome centre (Chap. 4).

[26] Another scientist who was heavily involved in the mapping of the HLA region was French immunologist Jean Dausset. In 1984, following the award of the Nobel Prize, Dausset established the Centre for the Study of Human Polymorphism (Centre d'Etude du Polymorphisme Humain), from which Généthon was created. One of Dausset's associates, Daniel Cohen, led the human genome work at Généthon and collaborated with pig geneticists in the mapping of the equivalent swine region: the SLA (Chap. 5).

[27] Keith Peters, two-part interview with Miguel García-Sancho: in person (October 2013) and by telephone (December 2013). Peters had previously attempted to persuade Brenner to move to Hammersmith Hospital.

funding for the human mapping project. In 1988, he formally endorsed Brenner's proposal and presented it to an audience that included Thatcher and her chief scientific advisor. He emphasised his experience as a practising physician and argued that the resulting physical map would become "the central tool for basic and applied research in the medical sciences".[28] ACOST agreed to support the initiative, which was subsequently named as the HGMP. This support materialised in an extra 11 million pounds that the Treasury transferred to the MRC as an earmarked fund to be exclusively spent in a Directed Programme of grants and a Resource Centre for human genome mapping. The funding was for a three-year period (April 1989 to April 1992) subject to extension following a progress review.

From its inception, the HGMP sought to build an identity that distinguished it from other human genome projects, especially the one that was already set to start in the USA. The US National Academy of Sciences had issued its report a few months before Peters' presentation to ACOST and, by 1989, the NIH and DoE's agreement to join forces in the US-HGP was being ironed out. Given the extraordinary budget and timeframe of the US effort—three billion dollars over 15 years—an early concern for the HGMP was how to make a differentiated contribution with a fraction of the money and a much more limited time horizon.

Tony Vickers, the HGMP manager, argued in his first report to the MRC in 1991 that in the UK there was "no individual enthusiasm" for becoming involved "in mega-sequencing", a task that was "unlikely to yield rewards to compensate workers for the drudgery involved". The British biomedical community, however, had "substantial strengths" in "many fields of genetics" where human genome mapping offered "promise of immediate and substantial pay-off". This short-term pay-off had somehow been "left aside" by the US-HGP with its focus on comprehensive, large-scale work at genome centres that were distant from the communities that would use the map and sequence data. The HGMP sought to take advantage of the prompt exploitation of results by involving the human and medical genetics communities in the mapping exercise.[29]

---

[28] K. Peters (1988) "Mapping and sequencing the human genome", typescript of presentation to Margaret Thatcher with additional manuscript notes (courtesy of Keith Peters), quote from p. 1. See also National Archives of the UK at Kew (London), Medical Research Council Series, file FD23/3442, and Bodleian Library (Oxford), Papers and Correspondence of Sir Walter Bodmer, file MS. Bodmer 1304.

[29] T. Vickers (1991) "The UK Human Genome Mapping Project: project manager's report", p. 4 (courtesy of Tony Vickers).

Consequently, the research grants awarded by the Directed Programme supported groups that were either developing mapping and sequencing technologies, creating shared resources to aid in these operations, or focusing on chromosomal regions connected to various types of genetic conditions, among them disorders affecting blood (haemophilia), mental health (aneuploidy syndromes) and muscular mobility (myotonic dystrophy). Of the five institutions in receipt of the largest amount of funding (Fig. 3.4), four of them investigated different aspects of medical genetics: ICRF, the Human Genetics Unit of the University of Edinburgh, the Institute of Molecular Medicine at John Radcliffe Hospital in Oxford and Guy's Hospital in London.[30]

The outcomes of the Directed Programme grants were delivered to the Resource Centre. This institution was housed in the Clinical Research Centre, a unit that the MRC had established in 1970 at Northwick Park Hospital (in northwest London) to foster collaboration between biomedical research and clinical practice. The Resource Centre was organised into two divisions that were headed by a biological manager (Ross Sibson) and a computing manager (Martin Bishop). Their duties involved assisting HGMP awardees in various capacities, from conducting mapping and sequencing work on request, to providing punctual support through their advanced technology and expertise (Balmer, 1998; Glasner, 1996). To do this, Sibson and Bishop's teams liaised with the so-called "user community", addressed their feedback and ensured access to the shared resources. They also collated the map and sequence data coming from the grant-supported laboratories.[31]

---

[30] T. Vickers (1991) "The UK Human Genome Mapping Project: project manager's report", Appendix, pp. 30ff; and T. Vickers (1992) "MRC Review of the UK Human Genome Mapping Project: Project Manager's Report", Annex 1, pp. 100ff (both courtesy of Tony Vickers). The contributions of ICRF, Guy's Hospital and the Institute of Molecular Medicine to the HGMP are detailed below. The Human Genetics Unit in Edinburgh housed a group with a longstanding tradition of cytogenetic mapping, based on chromosomal images that allowed the detection of malformations and helped diagnose diseases at the city's Western General Hospital (de Chadarevian, 2020). The institution with the largest share of HGMP funding was the LMB in Cambridge, although more than half of this support was awarded to Sulston and Coulson's *C. elegans* sequencing project (see the caption to Fig. 3.4).

[31] Ross Sibson, interview with Miguel García-Sancho, Royal Liverpool University Hospital, March 2014, see also: "The concept of a user", in T. Vickers (1991) "The UK Human Genome Mapping Project: project manager's report", pp. 14-16 (courtesy of Tony Vickers). The Resource Centre hosted a sociological investigation of users' experiences that was conducted by Peter Glasner, Harry Rothman and Wan Ching Yee, all of them social scientists who were working on the HGMP. The study was supported by funds that the different human genome programmes devoted to ethical, legal and social aspects of genomics research (Glasner et al., 1998).

PLACE	INSTITUTION OR DEPARTMENT	NO	VALUE
			(£KS)
BELFAST	MEDICAL GENETICS	1	5
CAMBRIDGE	GENETICS	2	692
	MEDICINE	1	83
	MRC LMB	7	2,011
	MRC MOLECULAR GENETICS UNIT	2	710
	PATHOLOGY	8	580
CARDIFF	GENETICS	1	44
DUNDEE	BIOCHEMICAL MEDICINE	1	35
EDINBURGH	CENTRE FOR GENETICS RESEARCH	1	106
	MRC HUMAN GENETICS UNIT	12	1,465
	BIOCOMPUTING RESEARCH UNIT	1	76
	MOLECULAR BIOLOGY	1	19
GLASGOW	GENETICS	1	28
	MEDICAL GENETICS	1	8
HARWELL	MRC RADIOBIOLOGY UNIT	7	635
LEICESTER	GENETICS	2	166
LONDON	GUY'S	6	1,015
	MRC HBGU AND GALTON LAB UCL	9	641
	IMPERIAL CANCER RESEARCH FUND	8	1,565
	IMPERIAL COLLEGE	4	399
	INSTITUTE OF CANCER RESEARCH	1	2
	INSTITUTE OF CHILD HEALTH	2	22
	MRC LEUKAEMIA UNIT	1	104
	NIMR	1	79
	ST MARY'S	6	747
	UNIVERSITY COLLEGE/MIDDLESEX	1	58
MANCHESTER	CELL AND STRUCTURAL BIOLOGY	1	89
	MEDICAL GENETICS	1	182
OXFORD	BIOCHEMISTRY	6	741
	DYSON-PERRINS (CHEMISTRY)	1	77
	REGIONAL GENETICS CENTRE	1	65
	IMM AND SURGERY	8	1079
	PATHOLOGY	2	125
SALISBURY	WESSEX REGIONAL GENRTICS CENTRE	1	49
SOUTHAMPTON	BIOCHEMISTRY	1	69
(TOTALS)		108	14,349

Fig. 3.4 The level of grant support per institution that the UK Human Genome Mapping Project had awarded by 1992, in thousands of pounds. Of the overall £2,011,000 that the LMB (Laboratory of Molecular Biology) received, just

(*continued overleaf*)

By 1991, a probe bank and a library of Yeast Artificial Chromosomes (YACs) were being transferred from their originators—all of them Directed Programme awardees—to the Resource Centre. The probe bank had been compiled by Nigel Spurr, a researcher at Clare Hall Laboratories, one of the ICRF divisions that Bodmer had equipped with the latest mapping and sequencing instruments during the 1980s (Weston, 2014, Ch. 4). It consisted of a series of DNA fragments whose known sequence enabled screening and the detection of specific chromosomal locations. The YAC library was a collection of human DNA fragments inserted in yeast cells and kept under controlled conditions in cultures. It was used as a source for chromosome mapping and derived from a collaboration between David Bentley at Guy's Hospital in London and Kay Davies at John Radcliffe Hospital's Institute for Molecular Medicine. Both scientists were renowned for applying genetics research to medical problems—presented by the patients of their home hospitals—and were regular recipients of HGMP funding.[32]

Fig. 3.4 over half (£1,150,000) went to co-fund the start of Sulston and Coulson's *C. elegans* sequencing project. The *C. elegans* grant was an outlier in the funding policies of the HGMP and, as such, is further examined in Chap. 4. Source: "Table 1: Distribution of HGMP awards (numbers and volume) amongst centres" in T. Vickers (1992) "MRC Review of the UK Human Genome Mapping Project: Project Manager's Report", p. 13. Report courtesy of Tony Vickers; Table 1 reproduced by kind permission of the Medical Research Council, as part of UK Research and Innovation

[32] N. Spurr (1990) "UK DNA probe bank: how it will function", *G-Nome News: the newsletter from the UK Human Genome Mapping Project,* number 3—February 1990, pp. 8-9, available online at https://groups.google.com/g/bionet.molbio.genome-program/c/dLgdQ83qTWY/m/9eqUC4Ic3DMJ (last accessed 14th December 2022). D. Bentley and K. Davies (1990) "Yeast artificial chromosome resources and genome mapping", *G-Nome News: the newsletter from the UK Human Genome Mapping Project,* number 7—Summer 1991, pp. 17-20, National Archives of the UK in Kew (London), Medical Research Council Series, file number FD7/2745. Only a few years later, in 1993, Bentley was chosen by Sulston to lead the mapping and sequencing of whole human chromosomes at the Sanger Institute, a UK-based genome centre that the MRC and the Wellcome Trust established (Chap. 4).

On top of housing and managing these shared tools, the Resource Centre started an in-house sequencing programme using complementary DNA (cDNA) methods. These methods allowed researchers to sequence only the DNA that is transcribed to produce messenger RNA, a vital step in protein synthesis. They therefore enabled the capturing of protein-coding genes in the DNA. The HGMP Directed Programme Committee decided, in 1990, that Sibson's division would apply this technique to "tissue"-specific and "developmental stage"-specific DNA, as well as the mapped fragments that the Resource Centre compiled from grant-awarded laboratories. This approach would produce "cDNA markers" that, combined with the ongoing physical map, would become "a valuable tool for researchers in human genetic disease". The cDNA component was adopted as a "strategy" aimed at yielding sequence information "in a relatively short time span", thus being "more practicable than mega sequencing of the human genome". It was regarded as a "flagship for the UK" and "essential" for achieving "international credibility" and taking "the lead" among the competing genome efforts.[33]

This mode of operation meant that the HGMP pursued a similar strategic approach to the EC's genome programmes. As the EC was doing for yeast and *H. sapiens* (Chap. 2), the MRC sought to involve existing genetics research laboratories in its genome project and distribute the HGMP grants among them as inclusively as possible.[34] This differed from the more selective funding regime of the US-HGP and the wider distance between the large-scale genome centres and genetics research institutions. More fundamentally, the two genome projects differed in their overall

[33] Anonymous: "Directed Programme Strategy Meeting held on 7th March 1990: discussion and development of a strategy by the Directed Programme Committee", National Archives of the UK at Kew (London), Medical Research Council Series, file number FD 7/2749, quotes from pp. 3 and 4. The main advocate for the cDNA strategy was Edwin Southern, a prominent HGMP grant awardee at the Oxford University Department of Biochemistry and inventor of the Southern Blot, a technique that allowed the probing of DNA fragments and the detection of certain variants among them, including mutations connected to diseases. He was supported by Duncan McGeoch, a genetic virologist based in the University of Glasgow and member of the HGMP Directed Programme Committee.

[34] The networks resulting from this distributed funding often overlapped, as in the HGMP Resource Centre participation in an international cDNA consortium sponsored by the EC's Human Genome Analysis Programme: T. Vickers (1991) "The UK Human Genome Mapping Project: project manager's report", p. 24 (courtesy of Tony Vickers). See also Chap. 2 on Horst Domdey and Brigitte Obermaier's Munich-based contribution to both international cDNA consortia and the EC's Yeast Genome Sequencing Project.

goals: whereas the US-HGP aimed for a reference sequence of the whole human genome—something that its much larger budget and timespan allowed—the HGMP restricted its remit to the genome regions on which its *user communities* were working. These human and medical geneticist users would develop catalogues of variation from the resulting mapped and sequenced regions.

3.3   REFERENCE SEQUENCE VS CATALOGUES OF VARIATION

Historically, the production of a reference sequence of the whole human genome was not an objective of the human and medical genetics communities. These communities had indeed engaged in the mapping of the human genome and had done so at an increasing scale since the start of the chromosome workshops, in 1973. However, they had always limited the scope of their efforts to the regions of interest to the genome mappers: geneticists studying specific diseases or biological traits who pooled their results on the chromosomal locations of genes or genetic markers with other community members. The HGMP and the EC's Human Genome Analysis Programme (HGAP) had built on this collective endeavour and sought to foster it with ring-fenced funding, international networking and resource centres that provided technical assistance and shared mapping technologies, as well as cDNA sequence data. Yet, as the support of these programmes was tailored to human and medical geneticists, the mapping and sequencing results were constrained to the genes and markers they were pursuing, rather than covering the entire human genome.

Human and medical geneticists would deem these genes and markers to be mapped at sufficient resolution when they could be assigned to a precise DNA fragment. Once this happened, the fragment would often be sequenced and compared with equivalent genome regions. These comparisons were made between humans and closely-related non-human species, or between healthy individuals and patients suffering the condition with which the gene or marker was associated. The mapping and sequencing processes combined collaboration—at chromosome workshops and more specific groupings, often deploying cDNA techniques—with competition for being the first to determine the chromosome locus or sequence of a gene or marker. A source of inter-species comparison was the growing number of databases with map and sequence information from simpler organisms, such as *S. cerevisiae* or *C. elegans,* that were being compiled through either their own specific programmes or as a result of funding from human genome efforts. In this regard, both the HGMP and HGAP supported the consolidation of mouse data repositories, an organism evolutionarily much closer to *H. sapiens* than yeast or a worm, and from which both medical and developmental inferences could be made.[35] To access data from patients, medical geneticists created consortia—some of them also sponsored by the HGAP (Table 3.1)—that enabled them to uncover genes involved in diseases and compile catalogues of genetic variants associated with the conditions.[36]

These catalogues of variation were often curated at hospitals with strong genetics departments. They formed repositories to which the rest of the community could contribute data, and from which they could access it. The HGMP Resource Centre and other similar central facilities that the HGAP developed shared this philosophy through the community-built and collectively-accessible probe banks, YAC libraries and map and sequence databases they offered to their users.[37] These shared resources were themselves the product of collaborative projects that the resource centres and genetics research laboratories jointly undertook with funding from the HGMP or HGAP (Table 3.1).

In the mid-1960s, before the arrival of DNA sequencing techniques, McKusick had pioneered these types of collections in *Mendelian*

---

[35] See the section "Mouse genetics", in T. Vickers (1992) "MRC Review of the UK Human Genome Mapping Project: Project Manager's Report", pp. 7-8 (courtesy of Tony Vickers). The mouse *Mus musculus* is furnished with a longstanding history of use as a model by both human and animal geneticists, due to its tractability in the laboratory and close evolutionary relatedness to larger mammals such as humans (García-Sancho & Myelnikov, 2019; Rader, 2004).

[36] While map and sequence data tended to be published, the methods to detect variants were often patented, so pharmaceutical companies could obtain licenses and produce diagnostic tests. Due to this, some human and medical geneticists approached the open access agenda for sequence data—that the US-HGP and other genome centres worldwide implemented during the mid-to-late 1990s—with reservations (Chap. 4).

[37] As Table 3.1 shows, the HGAP supported the establishment of three resource centres, two of them providing different libraries of human DNA and one acting as a centralised data platform. The European Data Resource was based in the Centre for Cancer Research at Heidelberg and the repositories of DNA libraries were housed at ICRF and Centre for the Study of Human Polymorphism, the institution from which Généthon emerged as a large-scale mapping centre in France (see note 26). HGAP funding also enabled the HGMP Resource Centre in Britain to host a database with international contributions of cDNA sequences.

Table 3.1 An example of a consortium of institutions pursuing medical genetics goals: the European Gene Mapping Project (EUROGEM), supported by the European Commission's Human Genome Analysis Programme (HGAP). The consortium included institutions involved in genome mapping activities and resource centres. None of these institutions participated in the determination of the human reference sequence nor in the whole-genome physical mapping that aided the sequencing (compare with Chap. 4, Table 4.1). Elaborated by Miguel García-Sancho and Jarmo de Vries, from data collected by Hallen and Klepsch (1995, esp. p. 20)

Institution	Role in European Gene Mapping Project (EUROGEM)
University of Marburg (Germany)	Mapping institution
University of Kiel (Germany)	Mapping institution
Cancer Research Centre at Heidelberg (Germany)	Resource centre (centralised data facility in Europe)
University of Aarhus (Denmark)	Mapping institution
Hospital Ramon y Cajal (Spain)	Mapping institution
Hospital de la Santa Creu i Sant Pau (Spain)	Mapping institution
Laboratoire de Genetique Moleculaire at Vert le Petit (France)	Mapping institution
Pasteur Institute (France)	Mapping institution
Centre for the Study of Human Polymorphism (France)	Mapping institution and resource centre (shared DNA libraries)
Institute of Molecular Biology and Biotechnology at Heraklion (Greece)	Mapping institution
University of Cagliari (Italy)	Mapping institution
University of Rome (Italy)	Mapping institution
University of Dublin (Ireland)	Mapping institution
University College Cork (Ireland)	Mapping institution
University of Leiden (Netherlands)	Mapping institution
University of Groningen (Netherlands)	Mapping institution
Universidade Nova de Lisboa (Portugal)	Mapping institution
MRC Human Genetics Unit at Edinburgh (UK)	Mapping institution
University of Cambridge (UK)	Mapping institution
University College London (UK)	Mapping institution
University of Wales (UK)	Mapping institution
St Mary's Hospital Medical School (UK)	Mapping institution
Imperial Cancer Research Fund (UK)	Resource centre (shared DNA libraries)
HGMP Resource Centre (UK)	Resource centre (cDNA sequence data bank)

Inheritance in Man, a catalogue of annotated chromosome maps that was first published as a series of printed volumes and later as an electronic database (*Online Mendelian Inheritance in Man*). Both the volume series and database incorporated updates with new data stemming from the chromosome mapping workshops and other disease-specific consortia, as well as clinical information about the underlying genetic conditions (Lindee, 2005, Ch. 3; Hogan, 2016, Ch. 3).

With the growth and development of physical mapping and sequencing techniques across the genetics community from the late-1980s onwards, both the workshops and variation catalogues became more specific: the former devoted to single chromosomes and the latter to individual diseases. An early example of this followed from the mapping of the cystic fibrosis gene in 1989, the first condition to be assigned to a physical location, in this case in human chromosome 7. One of the mapping scientists, Lap-Chee Tsui, was subsequently appointed as co-convenor of the chromosome 7 mapping workshops. Tsui also established the Cystic Fibrosis Genetic Analysis Consortium and coordinated the compilation of sequence variants connected to different forms of the disease that were determined by researchers all around the world. The results were gathered in a database that is still active at the University of Toronto Hospital for Sick Children—Tsui's home institution until 2004—and used to diagnose the condition.[38]

During the mid-to-late 1990s, Tsui's endeavour developed into a map encompassing the whole of chromosome 7. A younger member of the Toronto team, Stephen Scherer, built on the networks around the cystic fibrosis consortium and chromosome workshops to create a growing map with assignments associated with other conditions and *loci*. Scherer's collaborators included both medical geneticists and institutions working on the comprehensive mapping and sequencing of chromosome 7, among them the genome centre at Washington University. Yet the objective of Scherer's map was not to serve as a platform for the sequencing of the entire chromosome. Rather than pursuing a single reference sequence—as Washington University and the other genome centres did—Scherer and his fellow medical geneticists sought a way of detecting, mapping and cataloguing variation. Their map was a means of obtaining a set of ordered DNA fragments, some of which could be compared to data derived from patients. That way, differences in both fragment size and pattern, or

---

[38] See http://www.genet.sickkids.on.ca/ (last accessed 14th December 2022).

underlying DNA sequence, could be connected to particular conditions and assigned to specific chromosomal locations.[39]

The pursuit of variation by medical geneticists contrasted with other communities working on non-human organisms. Compared to the HGAP, the EC used a different strategy for yeast and sought a full reference sequence of its genome (Chap. 2). Apart from the extreme discrepancies in genome size, this divergent strategy was due to the aims and necessities of yeast geneticists, biochemists and cell biologists being distinct from those of the communities working on human DNA. While human and medical geneticists were interested in sequence differences underlying disease or other traits, the consortium of laboratories that undertook the EC's Yeast Genome Sequencing Project aimed to use this organism to model the functioning of the eukaryotic cell. Each community, therefore, approached its target genome in a different fashion. In the case of the human genome, the focus was on comparing specific regions—those where genes were located—across either different species or hospital patients versus controls. In the case of yeast, the laboratories in charge of the sequencing project used this organism as a "wild type" (Holmes, 2017) and pursued a standardised description of its genome, in order to relate the sequence data to functional aspects of cell genetics and metabolism. For this reason, they targeted a specific strain—S288C of *S. cerevisiae*—as representative of the yeast species as a whole and did not address variants until the full reference sequence was completed (Szymanski et al., 2019).

Similarly, within the history of molecular biology, substantial efforts had been devoted to achieve comprehensive descriptions of "exemplary" model organisms: viruses and bacteria first and further unicellular and multicellular organisms from the 1970s onwards (quote from Strasser & de Chadarevian, 2011; see also: Creager, 2002; Kay, 1993; Ankeny & Leonelli, 2020). The hope was that, as with the S288C strain of *S. cerevisiae*, those organisms would enable researchers to connect genes to different biological mechanisms and processes, and their effects. This, therefore, paralleled the goal of Brenner's *C. elegans* project, and Sulston's mapping and sequencing of the full genome of the worm. Like the yeast communities, molecular biologists would use the exemplary descriptions and descriptive models (Ankeny, 2000) as the basis of comparative practices.

---

[39] Elsewhere, we have referred to these two different approaches as horizontal sequencing—determining a single sequence representative of the entire human genome—and vertical sequencing: finding multiple variants in a specific genome region (García-Sancho, Leng et al., 2022).

Unlike *S. cerevisiae*, however, the reference sequence of *C. elegans* could not be traced to a specific population.[40]

Brenner considered the human genome to be too large and complex for an equivalent description to that being pursued for *C. elegans*, and so aligned with the human and medical genetics communities through the proposal of the HGMP. Yet, on the other side of the Atlantic, Watson found in the US-HGP the timeframe and resources needed to export the exemplary descriptive approach to the human genome. His genome centre model sought to fully describe the human genome as a standard or wild type, by producing a reference sequence rather than selectively tackling and comparing regions, as human and medical geneticists had traditionally done. This is what has led Hilgartner to identify Watson with a "vanguard" that consolidated genomics as an independent field, one that could be distinguished from other life sciences disciplines (Hilgartner, 2017).[41] In this differentiation, however, the large-scale centres that produced the reference sequence became both separated and distant from the genetics laboratories that would use the data and that were often involved in other forms of conducting genomics, more aligned with the approaches of the HGMP and the EC's programmes.[42]

The US-HGP dominates the historiography of genomics. As we have argued, however, its model of organisation was the exception rather than the rule during the formative years of genomics research. In the previous chapter, we conveyed the heterogeneous array of institutions, genomicists and organisational models involved in yeast genome sequencing. In this chapter, we have documented the diversity that also characterised human genomics. Taken together, both chapters show that the model of the US-HGP—with its large-scale centres and comprehensive sequencing regime—falls short in representing not only the history of genomics but also of the more specific subfield of human genomics (Fig. 3.5).

---

[40] Brenner specifically bred the *C. elegans* variant that was used in the descriptions of the cell lineages and neuron connectivity through complex genetic crossing experiments (Brenner, 1974). Yet this variant was never labelled or attributed to a specific strain, as in the case of yeast. In *H. sapiens*, the next chapter details the protocol by which the DNA to be included in the reference sequence was chosen, and later in the book we compare this process with the production of the yeast and pig reference genomes (Chaps. 4 and 5).

[41] In a similar vein, Michael Fortun has argued that genomics is nothing else than genetics research imbued with high-throughput technologies at accelerated speed (Fortun, 1999, esp. pp. 26-27).

[42] Elsewhere, we have identified different degrees of sequence production—from more proximate to more distal—and argued that there is considerable diversity and gradation across institutions that are outside the confines of the large-scale centre model (García-Sancho, Lowe et al., 2022).

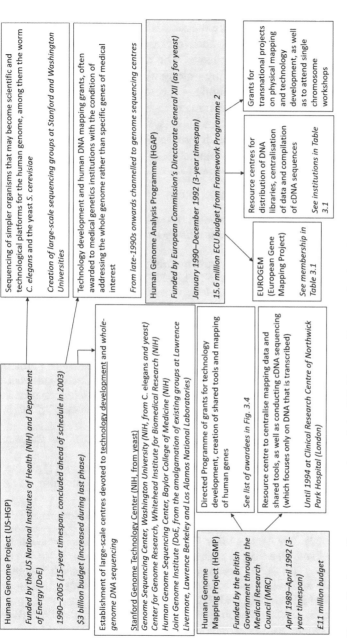

Fig. 3.5 An outline representation of the US Human Genome Project, UK Human Genome Mapping Project and European Commission's Human Genome Analysis Programme. Only aspects that have been discussed in the chapter are included and there are some notable absences, such as the programmes on ethical, legal and social aspects of genomics research that the three initiatives supported. Elaborated by both authors. For a larger version of this figure that can be zoomed in and out, see https://www.pure.ed.ac.uk/ws/portalfiles/portal/314798928/Fig_3_5_increased_final.pdf

In the next chapter, we identify the factors that led to a growing concentration of institutions and productive capacity during the determination of the human reference sequence. The transition of the *C. elegans* project from mapping to sequencing—along with the rise of the Wellcome Trust as an influential, proactive funder—spread the genome centre model beyond the USA and made it dominant in human reference genomics towards the mid-to-late 1990s. This process, we argue, not only affected scientific practice and organisation: it also occluded other historical trajectories in favour of the canonical winners' story based on the US-HGP.

REFERENCES

Aicardi, C. (2016). Francis Crick, cross-worlds influencer: A narrative model to historicize big bioscience. *Studies in History and Philosophy of Biological and Biomedical Sciences, 55*, 83–95.

Ankeny, R. (2000). Fashioning descriptive models in biology: Of worms and wiring diagrams. *Philosophy of Science, 67*, S260–S272.

Ankeny, R., & Leonelli, S. (2020). *Model organisms.* Cambridge University Press.

Ayala, F. J. (1987). Two frontiers of human biology: What the sequence won't tell us. *Issues in Science and Technology, 3*(3), 51–56.

Balmer, B. (1996). The political cartography of the Human Genome Project. *Perspectives on Science, 4*(3), 249–282.

Balmer, B. (1998). Transitional science and the Human Genome Mapping Project Resource Centre. In P. Glasner & H. Rothman (Eds.), *Genetic imaginations: Ethical, social and legal issues in human genome research* (pp. 7–19). Ashgate.

Baltimore, D. (1987). A small-science approach. *Issues in Science and Technology, 3*(3), 48–50.

Brenner, S. (1973). The genetics of behaviour. *British Medical Bulletin, 29*(3), 269–271.

Brenner, S. (1974). The genetics of *Caenorhabditis elegans. Genetics, 77*(1), 71–94.

Brenner, S. (1990). The human genome: The nature of the enterprise. *CIBA Foundation Symposia, 149*, 6–17.

Cook-Deegan, R. M. (1994). *The gene wars: Science, politics, and the human genome.* WW Norton & Company.

Creager, A. (2002). *The life of a virus: Tobacco mosaic virus as an experimental model, 1930–1965.* University of Chicago Press.

de Chadarevian, S. (1998). Of worms and programmes: *Caenorhabditis elegans* and the study of development. *Studies in History and Philosophy of Biological and Biomedical Sciences, 29*(1), 81–105.

de Chadarevian, S. (2002). *Designs for Life: Molecular biology after World War II.* Cambridge University Press.

de Chadarevian, S. (2020). *Heredity under the microscope: Chromosomes and the study of the human genome.* University of Chicago Press.

Dujon, B. (2019). My route to the intimacy of genomes. *FEMS Yeast Research, 19*(3), 1.

Dusyk, N. (2007). The political moral economies of science: A case study of genomics in Canada and the United Kingdom. *Health Law Review, 15*(3), 3–5.

Fortun, M. (1999). Projecting speed genomics. In M. Fortun & E. Mendelsohn (Eds.), *The practices of human genetics* (pp. 25–48). Kluwer.

Fujimura, J. (2000). Transnational genomics: Transgressing the boundary between the modern West and the premodern East. In D. Reid & S. Traweek (Eds.), *Doing Science + Culture* (pp. 71–92). Routledge.

García-Sancho, M. (2012). From the genetic to the computer program: The historicity of 'data' and 'computation' in the investigations on the nematode worm *C. elegans* (1963–1998). *Studies in History and Philosophy of Biological and Biomedical Sciences, 43*(1), 16–28.

García-Sancho, M., Leng, R., Viry, G., Wong, M., Vermeulen, N., & Lowe, J. (2022). The Human Genome Project as a singular episode in the history of genomics. *Historical Studies in the Natural Sciences, 52*(3), 320–360.

García-Sancho, M., Lowe, J., Viry, G., Leng, R., Wong, M., & Vermeulen, N. (2022). Yeast sequencing: 'Network' genomics and institutional bridges. *Historical Studies in the Natural Sciences, 52*(3), 361–400.

García-Sancho, M., & Myelnikov, D. (2019). Between mice and sheep: Biotechnology, agricultural science and animal models in late-twentieth century Edinburgh. *Studies in History and Philosophy of Biological and Biomedical Sciences, 75*, 24–33.

Glasner, P. (1996). From community to 'collaboratory'? The Human Genome Mapping Project and the changing culture of science. *Science and Public Policy, 23*(2), 109–116.

Glasner, P., & Rothman, H. (Eds.). (1998). *Genetic imaginations: Ethical, legal and social issues in human genome research.* Routledge.

Glasner, P., Rothman, H., & Yee, W. C. (1998). The UK Human Genome Mapping Project Resource Centre: A user analysis. In P. Wheale, R. von Schomberg, & P. Glasner (Eds.), *The social management of genetic engineering* (pp. 63–75). Routledge.

Guralnick, R. (2002). A recapitulation of the rise and fall of the cell lineage research program: The evolutionary-developmental relationship of cleavage to homology, body plans and life history. *Journal of the History of Biology, 35*(3), 537–567.

Hallen, M., & Klepsch, A. (Eds.). (1995). *Human Genome Analysis Programme.* IOS Press.

Heeney, C. (2021). Problems and promises: How to tell the story of a Genome-Wide Association Study. *Studies in History and Philosophy of Science, 89*, 1–10.

Hilgartner, S. (2017). *Reordering life: Knowledge and control in the genomics revolution.* MIT Press.

Hogan, A. J. (2016). *Life histories of genetic disease: Patterns and prevention in postwar medical genetics.* Johns Hopkins University Press.

Holmes, T. (2017). The wild type as concept and in experimental practice: A history of its role in classical genetics and evolutionary theory. *Studies in History and Philosophy of Biological and Biomedical Sciences, 63,* 15–27.

International Human Genome Sequencing Consortium. (2001). Initial sequencing and analysis of the human genome. *Nature, 409*(6822), 860–921.

International Human Genome Sequencing Consortium. (2004). Finishing the euchromatic sequence of the human genome. *Nature, 431*(7011), 931–945.

Jiang, L. (2013). *Degeneration in miniature: History of cell death and aging research in the twentieth century.* Arizona State University.

Jones, E. M., & Tansey, E. M. (Eds.). (2015). *Human Gene Mapping Workshops c.1973–c.1991: The transcript of a Witness Seminar held by the History of Modern Biomedicine Research Group.* Queen Mary University and Wellcome Trust.

Jordan, B. (1993). *Travelling around the human genome: An in situ investigation.* INSERM.

Kaufmann, A. (2004). Mapping the human genome at Généthon laboratory: The French Muscular Dystrophy Association and the politics of the gene. In J. P. Gaudilliere & H. J. Rheinberger (Eds.), *From molecular genetics to genomics: The mapping cultures of twentieth century genetics* (pp. 147–175). Routledge.

Kay, L. E. (1993). *The molecular vision of life: Caltech, the Rockefeller Foundation, and the rise of the new biology.* Oxford University Press.

Kay, L. E. (2000). *Who wrote the Book of Life: A history of the genetic code.* Stanford University Press.

Kevles, D., & Hood, L. (Eds.). (1992). *The code of codes: Scientific and social issues in the Human Genome Project.* Harvard University Press.

Kuska, B. (1998). Beer, Bethesda, and biology: How "genomics" came into being. *Journal of the National Cancer Institute, 90*(2), 93.

Lenoir, T., & Hays, M. (2000). The Manhattan Project for biomedicine. In P. R. Sloan (Ed.), *Controlling our destinies. historical, philosophical, ethical, and theological perspectives on the Human Genome Project* (pp. 29–62). University of Notre Dame Press.

Leonelli, S. (2016). *Data-centric biology: A philosophical study.* University of Chicago Press.

Lewin, R. (1984). Why is development so illogical? *Science, 224*(4655), 1327–1329.

Lindee, M. S. (1994). *Suffering made real: American science and the survivors at Hiroshima.* University of Chicago Press.

Lindee, M. S. (2005). *Moments of truth in genetic medicine.* Johns Hopkins University Press.

Lowe, J. W. E. (2016). Normal development and experimental embryology: Edmund Beecher Wilson and *Amphioxus. Studies in History and Philosophy of Biological and Biomedical Sciences, 57,* 44–59.

Löwy, I. (1987). The impact of medical practice on biomedical research: The case of human leucocyte antigens studies. *Minerva, 25*(1-2), 171–200.

Maienschein, J. (1978). Cell lineage, ancestral reminiscence, and the biogenetic law. *Journal of the History of Biology, 11*(1), 129–158.

Maienschein, J. (1990). Cell theory and development. In R. C. Olby, G. N. Cantor, J. R. R. Christie, & M. J. S. House (Eds.), *Companion to the history of modern science* (pp. 357–373). Routledge.

McLaren, D. (1991). *Human genome research: A review of European and international contributions.* Medical Research Council.

Müller-Wille, S., & Charmantier, I. (2012). Natural history and information overload: The case of Linnaeus. *Studies in History and Philosophy of Biological and Biomedical Sciences, 43*(1), 4–15.

New Haven Conference (1974). First International Workshop on Human Gene Mapping. 1974. *Cytogenetics and Cell Genetics, 13*(1–2), 1–216.

Palca, J., & Roberts, L. (1992). The Genome Project: Life after Watson. *Science, 256*(5059), 956–958.

Parolini, G. (2018). Building human and industrial capacity in European biotechnology: The Yeast Genome Sequencing Project (1989–1996). Technical University of Berlin preprint series. Retrieved December 14, 2022, from https://depositonce.tu-berlin.de//handle/11303/7470 (last accessed December 14 2022).

Powell, A., O'Malley, M. A., Muller-Wille, S., Calvert, J., & Dupré, J. (2007). Disciplinary baptisms: A comparison of the naming stories of genetics, molecular biology, genomics, and systems biology. *History and Philosophy of the Life Sciences, 29*(1), 5–32.

Rader, K. A. (2004). *Making mice: Standardizing animals for American biomedical research, 1900-1955.* Princeton University Press.

Ramillon, V. (2007). Le deux génomiques. Mobiliser, organiser, produire: du séquençage à la mesure de l'expression des gènes. PhD dissertation, École des Hautes Études en Sciences Sociales.

Rheinberger, H.-J. (2006). The notions of regulation, information, and language in the writings of François Jacob. *Biological Theory, 1*(3), 261–267.

Rosenberg, D. (2003). Early modern information overload. *Journal of the History of Ideas, 64*(1), 1–9.

Silverman, G. A., Overhauser, J., Gerken, S., Aburomia, R., O'Connell, P., Krauter, K. S., et al. (1996). Report of the fourth international workshop on human chromosome 18 mapping 1996. *Cytogenetics and Cell Genetics, 75*, 111–131.

Sinsheimer, R. L. (1989). The Santa Cruz Workshop—May 1985. *Genomics, 5*(4), 954–956.

Sloan, P. R. (Ed.). (2000). *Controlling our destinies: Historical, philosophical, ethical, and theological perspectives on the Human Genome Project.* University of Notre Dame Press.

Stevens, H. (2013). *Life out of sequence*. University of Chicago Press.

Strasser, B. J. (2019). *Collecting experiments: Making big data biology*. University of Chicago Press.

Strasser, B. J., & de Chadarevian, S. (2011). The comparative and the exemplary: Revisiting the early history of molecular biology. *History of Science, XLIX*, 317–336.

Szymanski, E., Vermeulen, N., & Wong, M. (2019). Yeast: One cell, one reference sequence, many genomes? *New Genetics and Society, 38*(4), 430–450.

Venkatesh, B. (2019). Sydney Brenner—A personal perspective. *Genome Research, 29*(6), vii–ix.

Vermeulen, N. (2016). Big Biology: Supersizing science during the emergence of the 21st century. *NTM Zeitschrift für Geschichte der Wissenschaften, Technik und Medizin, 24*(2), 195–223.

Wang, K., Shen, X., & Williams, R. (2021). Sequencing BGI: The evolution of expertise and research organisation in the world's leading gene sequencing facility. *New Genetics and Society, 40*(3), 305–330.

Watson, J. D. (1990). The Human Genome Project: Past, present, and future. *Science, 248*(4951), 44–49.

Weston, K. (2014). *Blue skies and bench space: Adventures in cancer research*. Cold Spring Harbor Laboratory Press.

Yoshikawa, A. (1990). Japanese biotechnology: Government, corporations, and technology transfer. *The Journal of Technology Transfer, 15*(1–2), 53–60.

PART II

Communities and Reference Genomes

CHAPTER 4

The Funnelling Effect of the Sanger Institute

In 1993, the Sanger Institute—initially named the Sanger Centre—was founded in Britain as an institution to carry out large-scale genome mapping and sequencing.[1] It represented a significant departure from the strategy of the UK Human Genome Mapping Project (HGMP) and from the prior contribution of British laboratories to the European Commission's Yeast Genome Sequencing Project. The Sanger Institute, instead, aligned more with the objectives of the US Human Genome Project (US-HGP) and the genome centres that both the US Department of Energy and National Institutes of Health were establishing to fulfil them. Rather than mapping and sequencing modest amounts of DNA—chiefly at the request of other laboratories—the Sanger Institute undertook to sequence substantial parts of the whole genomes of yeast, *Homo sapiens* and the worm *Caenorhabditis elegans* at its own initiative.

This chapter shows how the emergence of the Sanger Institute changed the landscape of genomic science internationally. We start by situating its origins in the rise of the Wellcome Trust from a modest British charity to

---

[1] Although the original proposal referred to the new institution as the Sanger Institute, this denomination was not adopted until 2001; between 1993 and 2001, Sanger Centre was the name that was used. For ease of reading and given that in the different chapters we narrate events that occurred before and after the change of name, we only refer to the Sanger Institute throughout the book.

© The Author(s) 2023
M. García-Sancho, J. Lowe, *A History of Genomics across Species, Communities and Projects*, Medicine and Biomedical Sciences in Modern History, https://doi.org/10.1007/978-3-031-06130-1_4

119

a major biomedical funder, one that crucially impacted on the contestation between state-supported and commercial institutions over the ownership of the human genome map and sequence. The Wellcome Trust allied with the UK Medical Research Council (MRC) in supporting the establishment of a large-scale centre in the UK where John Sulston could finish the sequencing of *C. elegans* and contribute to the completion of the yeast and human reference genomes.

We stress the pivotal role of Sulston, along with Wellcome Trust manager Michael Morgan, in the conception and ethos of the Sanger Institute. This institution consolidated the factory-style operation that Sulston had envisaged for the sequencing of *C. elegans*, as a departure from the way he and his collaborators had constructed the physical map of this organism (Chap. 2). The Wellcome Trust, as the main funder of the Sanger Institute, provided Sulston with the necessary financial and organisational flexibility to overhaul the *à la carte*, community service approach embodied in the mapping of *C. elegans* and later in the HGMP. As a result, the Sanger Institute avoided establishing itself as an academic institution connected to a university or research council, and instead became a scientific centre that was managed by a charitable company, which was called Genome Research Limited. Due to this, an emphasis on efficiency and accountability in map and sequence data production became central to the identity of the Sanger Institute: these objectives were prioritised over contributing to answering research questions.[2]

The chapter finishes by showing how, three years after the Sanger Institute was launched, Sulston and the Wellcome Trust decisively pushed for the unrestricted release of human sequence data at a conference—the First International Strategy Meeting on Human Genome Sequencing—held in Bermuda. The conference was attended by representatives from established genome centres in the USA and UK, as well as other institutions that were increasingly aligning with this form of conducting and organising human genomics. It led to the formation of what became known as the International Human Genome Sequencing Consortium (IHGSC), the institutional alliance that between 2001 and 2004

---

[2] On the professional values and identity of the Sanger Institute, see Andrew Bartlett's ethnographic work, conducted after the conclusion of the human reference sequence (Bartlett, 2008). This chapter builds on Bartlett's research to show the impact that the institutional configuration and work ethos of the Sanger Institute—and other genome centres—has had on the history of genomics.

published the reference sequence of the whole human genome in the scientific literature and made it available in freely-accessible databases.[3]

We stress how the concentration of mapping and sequencing operations in these institutions—via the channelling of grants to the IHGSC members—was key to how they coordinated the production of a single reference sequence. To this end, the IHGSC led the construction of a bespoke physical map that was designed to aid in the determination and assembly of the sequence. As a result, both the reference sequence and the map were shared products of consensus across the IHGSC, albeit in a way that involved restricted participation compared to the wider institutional involvement in the HGMP, the European Commission's Human Genome Analysis Programme (HGAP) and earlier stages of the US-HGP. The IHGSC reference genome was also narrower than other preceding maps and sequences in terms of the variability across different individuals and human populations that it captured.

The publication of the IHGSC reference sequence, which was announced worldwide at a ceremony in the White House in 2000, led to the widespread belief that the Human Genome Project was a single, international initiative that had always sought to make the entire human genome sequence available in the public domain. The IHGSC members coordinated a number of factory-style genome centres to sequence the whole genome, and presented the reference sequence as successfully and rapidly completed in draft form in the ensuing *Nature* article.[4] Yet as we document in this book, the determination of the human reference sequence was but one of a plethora of initiatives and models that co-existed in the early-to-mid 1990s, some of them converging in the IHGSC effort and some being sidelined. Some sidelined models and initiatives, we argue, continued as distinct lines of genome research during and after the IHGSC endeavour. We conclude this chapter by showing how documenting

---

[3] On the Bermuda conference and formation of the IHGSC, see Maxson Jones et al. (2018) and Hilgartner (2017, Ch. 6). This literature shows the transformative power of the genome centre model on human genome programmes internationally.

[4] The first article in which the IHGSC described its results, published in 2001, quoted 1990—the year in which the US-HGP was launched—as the start date of the production of the reference sequence. This led the authors to refer to the "Human Genome Project" as their collective and international whole-sequencing endeavour, rather than as a US-born initiative (International Human Genome Sequencing Consortium, 2001, p. 862). Since the timeframe of the US-HGP was fifteen years, this and other subsequent publications were considered to have arrived ahead of schedule.

these sidelined lineages allows us to move beyond the canonical history of genomics that portrays the human reference sequence as its totemic outcome. Our history also illustrates the funnelling effect that the Sanger Institute—and, more generally, the IHGSC—has had on the organisation and practice of genomics.

4.1   *C. ELEGANS* SEQUENCING
AND THE PATENTING CONTROVERSY

Sulston's *C. elegans* sequencing project continued a longstanding line of research on this organism at the Laboratory of Molecular Biology (LMB) in Cambridge, UK. As we showed in the previous chapter, *C. elegans* had been proposed as a model for investigating the genetics of development and behaviour by Sydney Brenner, and this became an early line of research at the LMB, which was founded in 1962. The worm *C. elegans* has since consolidated as a widely-adopted model organism. The requirements of the growing community of researchers investigating *C. elegans* genetics provided the basis for an international project to construct a physical map of its genome and later to determine its reference sequence, which was led by Sulston and Robert Waterston of Washington University in St Louis (WU). Yet the sequencing project, which started in 1989, substantially differed from the mapping exercise that the same scientific team had initiated five years earlier.

Firstly, Brenner had left the LMB in 1986 to establish a Molecular Genetics Unit at Addenbrooke's Hospital in Cambridge, where he first proposed the HGMP. This had led Sulston and his LMB associate, Alan Coulson, to gain control of the *C. elegans* mapping project. Under Brenner's leadership and before the start of the mapping, Sulston had compiled the lineages describing every cell division during the worm's embryonic and post-embryonic development. Coulson had joined the *C. elegans* team from another LMB division headed by the inventor of the first DNA sequencing techniques, Frederick Sanger. Just before the start of the mapping effort, Coulson had applied Sanger's sequencing methods to the determination of the full genome of various microorganisms, among them the PhiX174 and lambda bacteriophages (García-Sancho, 2010, p. 306ff). For Sulston and Coulson, these prior experiences cultivated a different vision of how to organise the description of the worm's genome. Rather than following the requests of other laboratories seeking to locate genes—as had been the case with the *C. elegans* mapping project and would be so with the HGMP—they sought a comprehensive characterisation of the worm's genome in which sequencing would be conducted at their own initiative.

Secondly, Sulston and Coulson's vision aligned with the strategy that James Watson was starting to formulate at the US National Institutes of Health (NIH) Office for Human Genome Research. The start of the *C. elegans* sequencing project was preceded by a meeting between these three scientists and Waterston, who had commenced his involvement in the worm mapping project from his position at WU. By the end of the 1980s, Watson was starting to deploy the model of large-scale genome centres through which the NIH would contribute to the US-HGP, alongside the national laboratories of the US Department of Energy. One way in which he sought to increase the scale of genomic work was by supporting groups willing to tackle the full genomes of other organisms, in order to transfer the know-how developed in these efforts to the sequencing of *H. sapiens*. Boosting the groups at the LMB and WU was thus a priority for Watson.

Sulston, Coulson and Waterston's presence at the 1989 biennial symposium of the *C. elegans* community was seen as an opportunity by Watson. The meeting was held at Cold Spring Harbor Laboratory, an institution that Watson directed alongside his newly-acquired NIH role. Watson approached the three worm researchers and invited them to apply for funding through the NIH Office, which that same year became the National Center for Human Genome Research; it was later redesignated as the National Human Genome Research Institute (NHGRI).[5] The funding would jump-start a comprehensive sequencing operation of the genomes of *C. elegans* and yeast: the former at WU and the LMB, and the latter at WU and Stanford University (Chap. 2). The NIH would support the sequencing of several yeast chromosomes and 3 million of the 100 million nucleotides of the worm's genome, as long as the UK Government committed to provide two-thirds of the funding on the LMB side.

Watson's plan was that this initial funding could subsequently be extended to broader areas of these and other genomes. As discussed earlier, Sulston describes the Cold Spring Harbor meeting in his memoirs as a "prison door" moment that shaped his scientific life forever: from the mapping and sequencing of the worm to the sequencing of the human genome (Chap. 2, see also Sulston & Ferry, 2002, p. 13). In what follows,

---

[5] On the naming history of the NIH Office, the National Center for Human Genome Research and the NHGRI, see Chap. 3, especially Sect. 3.1 and note 9.

we argue that the alignment of Sulston and Watson's visions enabled the genome centre model to expand and gradually acquire international influence. More importantly, this alignment started narrowing the array of practices and institutional configurations that were considered to be genomic science.[6]

Sulston and Coulson accepted Watson's proposal and approached the MRC for the British tranche of funding. In their application, they presented a three-phase operation of which the MRC grant—if awarded—would only support the first two. These first two phases comprised "testing technical and managerial procedures" and the sequencing of about one million nucleotides of the *C. elegans* genome over three years. The work would develop at "extensions" of Sulston and Coulson's existing laboratories and require the purchase of automatic instrumentation. The team would comprise three technicians and four scientists—including Sulston and Coulson—that would combine large-scale sequencing with the continued refinement of the physical map. The third phase would extend the sequencing endeavour to the whole *C. elegans* genome and be conducted at a "factory setting" to be established at "industrial estates" or other areas outside of traditional academic centres. By then, the team members of phases one and two would "move forward as team leaders" and additional "relatively unskilled" junior staff would need to be hired.[7]

Sulston and Coulson explicitly stated in the application that they had been "encouraged to expect extensive support" from the US-HGP, this

---

[6] The consequences of this funnelling effect were not restricted to the way of conducting and organising genomics: Watson and the worm researchers also created a powerful, winner's narrative that has shaped the perception of the history of this field (Chap. 1, especially Sect. 1.2).

[7] J. Sulston and A. Coulson (1989) "Mapping and sequencing the genome of *Caenorhabditis elegans*", application to the UK Medical Research Council's Human Genome Mapping Project, quotes from Appendix 1, p. 1. Obtained through the Medical Research Council and available at Wellcome Library, London, Papers and Correspondence of Sir John Sulston, file number PP/SUL/A/2/1/3. This was at a time when 'science parks' were being established, instantiating a shift from Fordist factory-style production that was declining as part of the accelerated programme of deindustrialisation promoted by the policies of the Thatcher government. The intention was to replace disappearing older heavy industries with new industries, for example biotechnology and university spin-offs. Indeed, as Hallam Stevens has argued, although there was interest on the part of the promoters of genomics in establishing factory-style settings, implying a kind of Fordist production, the actual organisational model of sequencing that resulted was anything but (Stevens, 2013, Ch. 3; see also Bartlett, 2008; Ramillon, 2007).

being the reason for a whole phase of their proposed operation starting after the MRC grant had concluded. In this regard, they argued that the infrastructure and "scale" required for the third phase could not be funded "in its entirety through the normal granting procedures".[8]

The MRC funded the proposal through the first cycle of grants awarded by the HGMP (1989–1992). Yet Sulston and Coulson anticipated that there would be no funding mechanisms—not from the HGMP or any other usual biomedical grant-giving body—to further develop the sequencing of C. elegans once phase two had finished. The worm sequencing project had already been an outlier compared to the other HGMP grants (see examples in Chap. 3). This divergence increased in 1992, when the already expanded facilities of Sulston and Coulson needed to be transformed into an industrial sequencing facility. The duration and level of support that Sulston and Coulson required, as well as the factory-style institution they envisaged, was incompatible with the HGMP grants and their remit of funding targeted mapping and sequencing work at existing academic laboratories. Furthermore, by the early-to-mid 1990s, the HGMP was becoming increasingly involved in controversy.

One layer of controversy was between the interests of human and medical geneticists on the one hand, and the goals of the HGMP on the other. Following the first round of grants, the MRC and the advisory committees of the HGMP recommended tightening the funding criteria, in order to ensure that work supported by the Directed Programme of awards would address areas of the human genome that had not already been targeted by gene mapping projects. Given that the diseases and other traits on which British laboratories worked implicated a limited number of human genes, their research priorities started to diverge from the requirements of the Directed Programme. As a result of this, unless the applicants artificially expanded the areas they worked on, the HGMP Resource Centre would either face duplicate mapping requests from the grant holders or receive data on gene loci that did not fill the gaps in the human genome database (Balmer, 1996).

---

[8] J. Sulston and A. Coulson (1989) "Mapping and sequencing the genome of *Caenorhabditis elegans*", application to the UK Medical Research Council's Human Genome Mapping Project, quotes from Appendix 1, p. 2. Obtained through the Medical Research Council and available at Wellcome Library, London, Papers and Correspondence of Sir John Sulston, file number PP/SUL/A/2/1/3.

The second controversial issue was less internal to the HGMP and affected the ownership and patentability of DNA sequences. Following the launch of the US-HGP in 1990, the NIH allowed patent applications to be filed for DNA sequences comprising or connected to human genes. This move aligned with the scientific policies that had led to the emergence of the biotechnology industry the decade before, spurred during Ronald Reagan's Administration and continued in George H. W. Bush's (Rasmussen, 2014; Yi, 2015). The patenting of sequences triggered a heated debate, with some scientists and administrators vehemently opposing the creation of proprietary rights on such fundamental data. Among the fiercest critics of these practices was Watson, who in 1992 resigned from his US-HGP leadership position in protest. Although the NIH subsequently changed its policy and increasingly discouraged the patenting of the results of sequencing that it funded, other scientists and institutions welcomed the exercise of property rights on DNA sequences.

This was the case for Craig Venter, a biochemist initially based at the NIH Institute of Neurological Disorders and Stroke. Like many yeast genomicists, in the 1980s he realised that emergent DNA techniques would enable him to turn his attention from examining functionally-relevant proteins to identifying and analysing the genes involved in their synthesis. In 1992, Venter left the NIH due to his growing frustration with what he perceived as a conservative attitude: neurogeneticists and administrators in his home institute continued to focus on a set of pre-defined brain conditions rather than using recombinant and DNA sequencing techniques to find genes on a larger scale (Venter, 2007). He founded a charitable organisation called The Institute for Genomic Research (TIGR), which would go on to generate a large number of human DNA sequences that were potentially linked to diseases with a genetic basis.

The sequences were patented by TIGR and licensed exclusively to its partner biotechnology company, Human Genome Sciences (Jackson, 2015). The method that Venter used to locate and determine them involved producing Expressed Sequence Tags (ESTs). It yielded similar results to the complementary DNA (cDNA) sequencing approach that Ross Sibson was pursuing at the HGMP Resource Centre (Chap. 3). This activity led the MRC to also patent its cDNA sequences, in spite of the growing scientific and public controversy. Both MRC officers and HGMP staff justified the patents as a defensive move that would "protect the sequences" from proprietary commercial exploitation by Venter or any other entrepreneur.[9]

The growing commercial interest in DNA sequences also affected the *C. elegans* project. In 1992, when the first two phases of the sequencing operation were close to completion, Sulston and Waterston were approached by Frederic Bourke, a US entrepreneur who wanted to enter the biotechnology market after a successful career in the retail industry. Bourke proposed the creation of a company that would complete the sequence of *C. elegans* and tackle the human genome. The firm would be based in Seattle, where yeast genome mapper Maynard Olson was moving after the University of Washington had established a Molecular Biotechnology Department with funding from IT tycoon Bill Gates (Chap. 2). Waterston and Sulston were never fully convinced of the feasibility of Bourke's proposal. They both preferred to continue to be state-supported scientists, but the end of the MRC and NIH grants was causing uncertainty about their next move.[10]

This led Sulston to discuss his future prospects with Aaron Klug, a structural biologist who had succeeded Brenner as director of the LMB. They both believed that in order to undertake the third stage of the worm project, Sulston would need a funding scheme committed to large-scale and comprehensive genome sequencing. Given the much more specific remit of the HGMP, Klug offered to mediate between the MRC and the Wellcome Trust, a charity that by the early-1990s was substantially reconfiguring its strategy and involvement in genomic science. These conversations led to the realisation of Sulston's envisaged factory-style operation, in the form of a genome centre that undertook significant chunks of the whole-genome sequencing of *C. elegans*, yeast and *H. sapiens*. The

---

[9] T. Vickers (1992) "MRC Review of the UK Human Genome Mapping Project: Project Manager's Report", p. 7 (courtesy of Tony Vickers). By that time—and following the aggressive biotechnology policies in the USA and Japan—the MRC and other funding bodies were being subjected to increasing pressure from the UK Government to commercialise their results (de Chadarevian, 2011; Owen & Hopkins, 2016). Within the Resource Centre, some staff accepted the patents reluctantly, while others considered them a logical and adequate move. Among the latter group of staff was Sibson, who had previously worked in the radio-pharmaceutical company Amersham (interview with Miguel García-Sancho, Royal Liverpool University Hospital, March 2014).

[10] Correspondence between John Sulston and Robert Waterston on Frederic Bourke's proposal. Papers and Correspondence of Sir John Sulston, Wellcome Library, London, file PP/SUL/B/5/14.

new British centre aligned with the large-scale genome centre model of the US-HGP, and distanced itself from the distributed approach of the HGMP and the European Commission.

4.2   THE WELLCOME TRUST AND ITS ADVOCACY FOR A 'NEW GENETICS'

The Wellcome Trust's status as a biomedical funder predates genomics research, the establishment of the LMB and the emergence of molecular biology. This charity was created in 1936, following the death of Henry Wellcome, owner of the British-based pharmaceutical company Burroughs Wellcome, which was later renamed as the Wellcome Foundation. The Wellcome Trust took ownership of the pharmaceutical company with the charitable mission of using its revenues to advance medicine through support for research (Hall & Bembridge, 1986). In 1986, it began a new strategy for its charitable work that consisted of gradually selling the shares of the pharmaceutical company and reinvesting the income. This strategy, engineered by the Trust's new director of finance Ian Macgregor, meant that if the investments were successful, the revenue would generate potentially endless capital. This capital could then be used by the Trust to fund research and operate independently from the Wellcome Foundation. As the sale of shares increased throughout the late-1980s and early-1990s, so did the independence of the Trust, its resources to invest and, ultimately, its ability to function as a funding body.

This period of considerable growth coincided with important developments in genetics, a substantial part of which derived from the application of recombinant DNA techniques to medical research. In 1991, the Wellcome Trust appointed an expert group to advise on how best to support and seed the new genetic medicine. One of its first interventions was funding, along with the European Commission's HGAP, the holding of chromosome mapping workshops in Europe. Yet as their investment revenues rose, the Trust became keen to distinguish itself from other charities focused on specific diseases and conditions such as the Imperial Cancer Research Fund (ICRF). During the early-to-mid 1990s, at the same time that the ICRF became a main driver and participant in the HGMP, the Wellcome Trust developed a strategy with its advisory group to fund the establishment of research centres bridging genetics and medicine across the UK.[11]

This was the context in which Klug's mediation between the Wellcome Trust and the MRC took place. During the first half of 1992, he brokered a series of meetings between teams headed by Dai Rees and Bridget Ogilvie, chief executive of the MRC and director of the Wellcome Trust, respectively. They agreed that the Trust's strategy of supporting new institutional settings for genetic medicine squared with Sulston's aspiration for a factory-style genome centre. Furthermore, they concurred that in the light of their remits and available resources, the MRC was prepared to fund work on model organisms with smaller genomes—such as *C. elegans*—while the Wellcome Trust would financially support the mapping and sequencing of the human genome.

The next step was to visit Sulston's group at the LMB and ask him for a detailed proposal that would be presented to the Wellcome Trust's genetics advisory group, and then be externally reviewed. Michael Morgan, the Trust's director of research partnerships and ventures, acted as the liaison between Sulston and the advisory group.[12] In July 1992, as the proposal was being reviewed, the financial capacities of the Wellcome Trust increased significantly when its chairman, Roger Gibbs, sold another tranche of Wellcome Foundation shares, leaving their holding now below 50%.

The proposal, submitted in the summer of 1992, argued for the establishment of a "new centre" that would be named after Frederick Sanger, the inventor of the first sequencing techniques and Coulson's first line manager at the LMB. This new institution would grow "out of the *C. elegans* sequencing project" and become "a facility" that would "sequence

---

[11] Martin Bobrow and Nick Hastie (both members of the Wellcome advisory group), interviews with Miguel García-Sancho in Cambridge and Edinburgh, June 2015 and July 2015 respectively. In 1989, twelve British medical charities formed the Genetic Interest Group (now called Genetic Alliance UK) to join forces in support of genetics research (Kent, 1999; Mikami, 2020, pp. 153ff). The Wellcome advisory group was an independent body; at times, it also used the name Genetic Interest Group (see Papers and Correspondence of Sir John Sulston, Wellcome Library, London, file number PP/SUL/B/1/1/4/1). After the establishment of the Sanger Institute, the Wellcome group recommended the creation of a new Centre for Human Genetics in Oxford, which was founded in 1994.

[12] For a general outline, see: Sulston & Ferry, 2002, pp. 108ff. A more detailed paper trail can be found at the Wellcome Trust Corporate Archive, Wellcome Library, London, Sanger Institute files, reference WT/C/2/3/8.

and interpret a substantial part of the human genome". As an "interim" goal, the Sanger Institute would "contribute heavily to the sequencing of the yeast genome" to help finish that project "within two to three years"; ahead of the European consortium's schedule. Another key difference to both the European consortium's approach to yeast sequencing, and the HGMP and Venter's approach to human sequencing, was that rather than setting an embargo period for the release of the data or patenting the results, the Sanger Institute would aim for rapid dissemination of the maps and sequences it produced "to the public domain".[13]

The structure of the institute would comprise a "technology core" conducting DNA mapping and sequencing on a "large-scale" and "quasi-industrial basis". This would be headed by a senior scientist and run by technical staff, many of which would be "unskilled". The core would be at the centre of operations, serving distinct sections devoted to *C. elegans*, yeast and human genome work, as well as informatics and cDNA sequencing (Fig. 4.1). The informatics section would assemble sequences from various DNA fragments, annotate genes within the strings of nucleotides (Chap. 6) and organise and store the information in databases. The cDNA section would "test" the value of the genome projects by using some of the mapping results to generate sequence data and address "biological research topics", especially in the field of neuroscience. Biological research, however, was planned to represent only a small fraction of the overall activity of the Sanger Institute: about 10%. The remaining 90% would focus on large-scale mapping and sequencing across the whole genome rather than targeting smaller areas using the cDNA approach.[14]

Once the Sanger Institute was established, Coulson was appointed head of the *C. elegans* section, while Bart Barrell, a researcher who had also worked with Frederick Sanger on the development of sequencing techniques at the LMB, led the yeast genome effort. Sulston became responsible for the technology core and, from this position, was able to coordinate the whole institute (Fig. 4.1).[15]

---

[13] Sanger Institute proposal (undated and untitled), Papers and Correspondence of Sir John Sulston, Wellcome Library, London, file number PP/SUL/B/1/1/1/2, quotes from pp. 1 and 2.

[14] Sanger Institute proposal (undated and untitled), Papers and Correspondence of Sir John Sulston, Wellcome Library, London, file number PP/SUL/B/1/1/1/2, quotes from pp. 2 and 3.

[15] On Sulston's role, and Coulson and Barrell's appointments, see: Sulston and Ferry (2002, pp. 117ff). Barrell and Coulson had joined the LMB as technical assistants to help Frederick Sanger with the development of sequencing methods (García-Sancho, 2010, pp. 296ff). Although both of them later pursued an academic career, Coulson did not complete his PhD until 1994, well into the *C. elegans* sequencing effort.

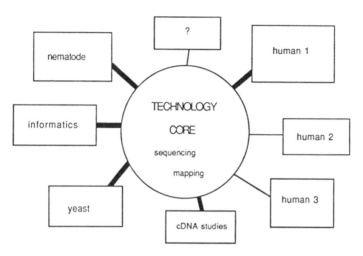

Fig. 4.1 Above, a diagram included in the 1992 application to establish the Sanger Institute, showing how its structure and the distribution of work between the different mapping, sequencing and bioinformatics sections was envisaged. Below, a picture of the Board of Management of the Sanger Institute, including

(*continued overleaf*)

Another early recruit who decisively contributed to the Sanger Institute's human genome work was David Bentley. Bentley had also started his career in Frederick Sanger's LMB division, where he spent the first year of his PhD. At the same time Bentley was conducting research there, Coulson and other co-workers were sequencing bacteriophage viruses using the newly-developed DNA sequencing techniques. Following the institutional migration of his supervisor, Bentley completed his PhD at Oxford and obtained his first academic positions in London. By the time of the Sanger Institute proposal, he was based in the Division of Medical and Molecular Genetics of Guy's Hospital, London. The head of Bentley's department was Martin Bobrow, a reputed medical geneticist and member of the Wellcome Trust advisory group. Bobrow's research focused on the mapping of genes connected to different conditions, among them haemophilia and muscular dystrophy.[16]

Bentley used a promising technique called positional cloning to identify new gene–disease associations, most notably for immune disorders arising from mutations in the X chromosome (X-linked disorders). This technique had emerged in the late-1980s and enabled researchers to find disease-associated genes based on progressively homing in on the genetic location of the putative gene. The identification and mapping of DNA markers surrounding the gene on the chromosome enabled researchers to narrow down the search to specific genomic regions. This was

----

Fig. 4.1 John Sulston (front row, second-from-left), Alan Coulson (back row, second-from-right) and Bart Barrell (back row, far-left). The roles of David Bentley (front row, second-from-right), Jane Rogers (front row, far-left), Murray Cairns (back row, second-from-left), Mike Stratton (back row, far-right) and Richard Durbin (front row, far-right) are discussed later in this chapter, and also in Chaps. 5 and 6. Above image: retrieved from Papers and Correspondence of Sir John Sulston, Wellcome Library, London, file number PP/SUL/B/1/1/1/2, p. 3; reproduced with permission from the Wellcome Library. Below image: retrieved from Waterston and Ferry (2019, p. 437, Figure 7); reprinted with permission from Wellcome Sanger Institute

[16] Martin Bobrow and David Bentley, interviews with Miguel García-Sancho, Cambridge (UK) and Cold Spring Harbor Laboratory, June and July 2015 respectively. Bentley's PhD supervisor was George Brownlee, Sanger's biographer and right-hand person at the LMB during the development of RNA sequencing techniques (Brownlee, 2014).

followed by sifting through the handful of genes in the region to find mutations that presented in patients suffering the disorder. Positional cloning avoided the need for gathering the information that geneticists had traditionally needed—such as details about the protein involved in a genetic condition—in order to deduce the function of a gene and then identify it. At Guy's Hospital, with funding from the HGMP, Bentley also initiated studies to develop shared resources for genome mapping, concentrating first on chromosomes 22 and X. These projects later became central to the production of the human reference sequence at the Sanger Institute.[17]

Positional cloning, as with Venter's EST method, enabled geneticists to scroll human chromosomes in search of different genes. A driving force behind this technique was Francis Collins, who started developing it at Yale University, the home institution of Frank Ruddle, co-founder of the journal *Genomics* and promoter of the chromosome workshops (Chap. 3). Collins later moved to the University of Michigan, where he used positional cloning in the mapping of various disease genes, including the one responsible for cystic fibrosis to chromosome 7 (Sferra & Collins, 1993).

Positional cloning and the EST method thus enabled Collins, Bentley and Venter to shift from one gene or chromosomal region to another. Rather than being constrained by a focus on a specific disease and having to determine the presence of particular proteins or other biomolecules, these three researchers could now move across conditions and research problems more easily. Yet the traditional funding regime and institutional organisation of medical genetics limited this multi-locus chromosome scrolling approach. In 1993, just months after Venter left the NIH to found TIGR, Bentley moved to the Sanger Institute with his Guy's Hospital colleague, Ian Dunham. That same year, the NIH appointed Collins as director of the National Center for Human Genome Research (later NHGRI) following Watson's resignation. In these three institutions, the researchers would address the whole human genome rather than looking for specific genes or regions connected with particular conditions.

---

[17] David Bentley, interview with Miguel García-Sancho, Cold Spring Harbor Laboratory, July 2015. Bentley regards the reference genome produced by the Sanger Institute and the IHGSC as a resource that "led to solving of many genetic disease diagnoses and has since underpinned progress in genomic medicine."

Dunham and Bentley brought some of their Guy's Hospital collaborations to the Sanger Institute, as well as their interest in chromosomes 22 and X. Yet, their remit at the Sanger Institute was to map and sequence the whole chromosomes rather than helping to locate specific disease-associated genes. This created tensions with some of the community of medical geneticists, especially those working on conditions or biological problems connected to chromosomes 22 and X (Sulston & Ferry, 2002, p. 131). The early-to-mid 1990s had witnessed a continuation of the policies with which Reagan in the USA and Margaret Thatcher in the UK had attempted to nurture the biotechnology industry the decade before: the encouragement of private sector investment in developing applications of publicly and charitably-funded biomedical research (Myelnikov, 2017). Due to this, some geneticists regarded the comprehensive efforts at the Sanger Institute as a threat to their means of acquiring scientific credit and funding prospects. By releasing their results in the public domain, Bentley, Dunham and their whole-chromosome teams could devalue the publication of sequence data in medical journals or the patenting of gene detection techniques that could be licensed to companies manufacturing diagnostic kits.

The key difference underlying these tensions was the Sanger Institute's ambition of sequencing the whole genome rather than restricting their efforts to the traditional target of medical geneticists: protein-coding regions of the chromosomes corresponding with genes implicated in diseases. This difference was manifested by the emphasis that Sulston had placed on his commitment to determine the yeast, *C. elegans*, and human sequences from one end to the other and restrict the more targeted, cDNA approach to just one unit of his envisaged new institute. Although avoiding the abundant non-coding regions may have seemed to be an "advantage" of the cDNA approach, Sulston's proposal argued that sequencing the chromosomes in full was "self-evidently the only route to a complete understanding" of the genomes. This was, to a large extent, due to "transcription control elements" regulating the activation of genes and the synthesis of proteins being found "largely in non-coding regions".[18]

---

[18] Untitled and undated application document, Papers and Correspondence of Sir John Sulston, Wellcome Library, London, file number PP/SUL/B/1/1/1/2, quotes from p. 6.

Sulston and Coulson had become aware of the importance of these regulatory regions during the prior physical mapping of *C. elegans*. At that early stage, during the mid-to-late 1980s, the majority of mapping requests that they received came from laboratories working on the developmental biology of the worm. Investigating the switching of genes on and off during development had been a main objective of Brenner's initiation of *C. elegans* as a model organism and Sulston's descriptions of the worm's embryonic and post-embryonic cell lineages (de Chadarevian, 1998). These mechanisms of gene regulation and their salience for worm researchers encouraged and justified Sulston and Coulson's 1989 proposal to sequence the whole *C. elegans* genome, a proposal that later informed their vision for the Sanger Institute.

The Sanger Institute was thus the result of a confluence of interests and strategic visions. In the first instance, the genome centre model that Watson and other US-HGP champions were deploying converged with Sulston and Coulson's will to address the whole *C. elegans* sequence in the 1989 *prison door* meeting that led to the formulation of the factory-style sequencing operation. Second, the commitment to tackling full genomes made by both the US-HGP and the *C. elegans* sequencing project, aligned with the imaginaries and aspirations of three different communities: an established base of molecular biologists seeking comprehensive descriptions of the processes underlying life; a fledging group of developmental biologists interested in regulatory as well as protein-coding regions of DNA; and a new breed of medical geneticists who, like Bentley and Dunham, sought to move beyond the traditional focus on individual disease genes. Thirdly, this novel and comprehensive ambition, one that promised a much broader set of beneficiaries and stakeholders, persuaded the Wellcome Trust: an emergent funder that aimed to support distinctive new models of genetics research. As the US-HGP did with Watson's genome centre model, the Wellcome funding operationalised Sulston and Coulson's factory-style vision. This vision would acquire a life of its own during the mid-to-late 1990s and would reposition—and eclipse—existing genome programmes.

4.3   MANAGERIAL OPTIMISATION
AND THE WHOLE-GENOME COALITION

Following favourable reports from its advisory group and external reviewers, the Wellcome Trust, along with the MRC, agreed to establish the Sanger Institute with a start-up grant of 40 to 50 million pounds.[19] The renewal of this initial funding would be subject to financial and scientific review following the first five years of operation. The main, and almost only, criterion for this review would be progress with the map and sequence data: a strong indication of this was that the cDNA unit requested in the initial proposal was not implemented.[20]

While the rationale of the MRC in starting the conversations leading to the funding of the Sanger Institute had been to avoid losing its flagship *C. elegans* project to the USA—given the impossibility of continuing Sulston's sequencing initiative at the LMB—the top priority of the Wellcome Trust was making a substantial contribution to the elucidation of the human genome. This difference largely stemmed from the disparate positions of each agency in terms of funding policy. The MRC needed to support a variety of biomedical disciplines using more rigid grant schemes, while the absence of other funding commitments placed the Trust in the enviable position of being able to make a larger award that would support the mapping and sequencing of human chromosomes. This was by far the most onerous expenditure item of the Sanger Institute and led the worm and yeast work to be subordinated to the human genome, as was the case in the US-HGP.

Morgan would help Sulston with the logistic and administrative details of setting up the new institute. During the second half of 1992, they toured a number of potential locations with the clear aim of avoiding traditional academic settings. After visiting various industrial parks on the outskirts of Cambridge, London and Edinburgh, they chose the country

---

[19] According to the UK Retail Price Index measure of inflation, the equivalent sum as of November 2022 would be just over double this amount, about 101 million pounds for the top-line figure of 50 million pounds in 1992: https://www.bankofengland.co.uk/monetary-policy/inflation/inflation-calculator (last accessed 16th December 2022).

[20] Jane Rogers, interview with Miguel García-Sancho, Cold Spring Harbor Laboratory, July 2015. Martin Bobrow, one of the members of the Wellcome Trust advisory group, highly approved of the Sanger Institute building on the previous *C. elegans* work at the LMB (Martin Bobrow, interview with Miguel García-Sancho, Cambridge, June 2015). Both the structure and the organisation of work at the intended new institute developed the *factory setting* that Sulston and Coulson had envisaged for the last phase of the worm sequencing project, which would focus on the rapid and large-scale production of data.

estate of Hinxton Hall, which included several large buildings and surrounding lands. The site, ten miles south of Cambridge, had served multiple purposes since the eighteenth century, the last of them being the
hosting of a suite of laboratories for a metallurgical company (Fletcher &
Porter, 1997, Ch. 3). Once the Wellcome Trust purchased the site, refur-
bishment works ensued to develop provisional facilities where operations
could be quickly started and then expanded in the longer term (Fig. 4.2).
Sulston and Morgan considered that the Hinxton location would benefit
from its proximity to Cambridge, London and Oxford, while keeping
it independent from academic environments.[21]

This institutional independence was perceived as crucial for the smooth
running of the Sanger Institute. From the proposal stage, Sulston had
envisaged a radically different structure from that of any academic research
institution. When Morgan sought to implement this vision, his belief was
that merely avoiding university campuses would not suffice, and innova-
tive day-to-day forms of operation would need to be added to the equation. After considering different options with MRC officials and Wellcome
trustees, it was decided that the legal nature of the Sanger Institute would
be that of a research institution funded and managed by a non-profit company called Genome Research Limited (GRL).[22] This led to a dual governance structure with a scientific manager and a separate head of corporate

---

[21] For a general outline, see: Sulston and Ferry (2002, pp. 112ff). A more detailed paper
trail can be found at the Wellcome Trust Corporate Archive, Wellcome Library, London,
Sanger Institute files, reference WT/C/2/3. Stephen Hilgartner has argued that, while
independence and distinctiveness from other life science disciplines was crucial in shaping the
identity of genomic science, this field always positioned itself as a continuation of progress in
molecular biology: the completion of the human genome sequence in 2003 was intentionally
scheduled to coincide with the fiftieth anniversary of the elucidation of the double helix of
DNA by James Watson and Francis Crick (Hilgartner, 2017, pp. 215ff). The Sanger Institute
acknowledged its proximity—both geographical and scientific—to the double helix achievement, which had taken place at the Cavendish Laboratory of Cambridge, see: Fletcher and
Porter (1997, p. 3).

[22] For a general outline, see: Sulston and Ferry (2002, pp. 112ff). A more detailed paper
trail can be found at the Wellcome Trust Corporate Archive, Wellcome Library, London,
Sanger Institute files, reference WT/C/2/3. The model of a company limited by guarantee
was one of the many options considered for transforming publicly-funded research establishments in the UK in the early- and mid-1990s, encouraged by New Public Management
ideas. As there were no shareholders in this model, surpluses would have to be reinvested,
but the institution had to be governed in the manner of a private company (Boden et al.,
2003). Some existing research establishments such as the Roslin Institute adopted this model
(in 1995 in this case; see Lowe, 2021).

Fig. 4.2 Above, the first building in which the Sanger Institute operated, located in the grounds of Hinxton Hall, an eighteenth century estate ten miles south of Cambridge. The building had previously housed metallurgical laboratories. Below, the Wellcome Trust Genome Campus that has developed at the Hinxton site from 1993 onwards, with early buildings on the left-hand side of the image. Above image: reproduced from Fletcher and Porter (1997, p. 9) with permission from the Wellcome Library. Below image: reproduced with permission from Wellcome Sanger Institute

services. Jane Rogers, a senior LMB administrator, was recruited for the scientific manager position, with the remit of coordinating the mapping and sequencing projects. Murray Cairns, formerly a manager in the brewing industry, became head of corporate services and liaison between the Sanger Institute and GRL (Fig. 4.1). Both the MRC and the Trust were represented on GRL's board of governors and oversaw the progress of the Sanger Institute (Sulston & Ferry, 2002, Ch. 3).

Sulston's view was that the sequencing of the human genome could at least be started without the need for further technological developments. This was at odds with the initial goals of the US-HGP, which advocated a focus on mapping until the performance of automatic sequencing instruments had improved.[23] By the time the Sanger Institute opened, in 1993, its collaboration with Waterston's WU group around the sequencing of *C. elegans* had entered a new phase, working towards characterising the whole genome of that worm. Yet the only comprehensive sequencing work that WU and other US genome centres were conducting, was on organisms with substantially smaller genomes than that of *H. sapiens*.

At the same time, Venter's patents on human sequences were growing in number. Several institutions—among them competing pharmaceutical companies—joined forces with US genome centres to determine cDNA sequences that were then released into the public domain (Hilgartner, 2017, pp. 149ff). For Sulston and Waterston, however, the best way of countering proprietary ambitions was launching a concerted effort that would sequence and make the entire genome freely-available, rather than targeting specific fragments or waiting for "some magic new" sequencing technology (Sulston & Ferry, 2002, p. 140).

Waterston was the first to articulate this urgency in a 1994 email to Sulston entitled "an indecent proposal". Written shortly after he visited Britain, it outlined a strategy by which the genome centre at WU and the Sanger Institute could both tackle and complete the human sequence with a small number of collaborators. The plan required a commitment by funding agencies to concentrate support in a handful of carefully-chosen institutions with large-scale sequencing capacities. This implicitly challenged the more inclusive, distributed approach of the European Commission and UK HGMP, which allotted a greater degree of independence in conducting sequencing to the institutions that they funded. As

---

[23] Jane Peterson, interview with Miguel García-Sancho, National Human Genome Research Institute (NHGRI), November 2018.

Sulston put it in a meeting with Wellcome trustees in which he strongly defended Waterston's approach, the underlying message was to "stop fiddling around" and realise that a concerted whole-genome project would be "cheaper" than "pour[ing] the budget into half efforts".[24]

Waterston's proposal led to what began to be called the "megalomaniac" genome project (Sulston & Ferry, 2002, Ch. 4). In the months following the email, he and Sulston started pressing their funders to increase their grants either immediately, or during the next cycle of support of their centres, so they could meet their unprecedentedly ambitious sequencing targets. They also entered into correspondence with potential partners that could join them in the sequencing enterprise. These included another fledgling genome centre, the NIH-funded Whitehead Institute in Boston, as well as Généthon, based on the outskirts of Paris. Généthon was devoted to large-scale human genome mapping, and was the only European institution on a par with the Sanger Institute (Chaps. 2 and 3).[25] In Britain, the Hinxton site where the Sanger Institute was based attracted two other institutions in 1994, with the transfer of the HGMP Resource Centre from Northwick Park Hospital, and a successful bid to house the European Bioinformatics Institute (EBI).

The EBI was the result of the expansion of the first centralised database to store DNA sequences, which had been based for 14 years in the European Molecular Biology Laboratory (EMBL) at Heidelberg and was due to move to a building of its own (García-Sancho, 2011). Sulston and Michael Ashburner, a Cambridge-based computational biologist, submitted a proposal to incorporate it at the Hinxton site with support from the Wellcome Trust. The growing importance of the Sanger Institute in DNA sequencing and the advantages of having the EBI next door to this major sequence producer led the proposal to unexpectedly beat rival and less

[24] Papers and Correspondence of Sir John Sulston, Wellcome Library, London, file number PP/SUL/B/2/1/1, quotes from email from Robert Waterston to John Sulston, 10th September 1994 and Sulston's manuscript notes ahead of a meeting with Wellcome trustees in late 1994.

[25] Papers and Correspondence of Sir John Sulston, file numbers PP/SUL/B/2/1/3 and PP/SUL/B/2/1/4. The 1992 Sanger Institute proposal already mentioned Généthon, and argued that existing genome centres, as well as those in the planning stage, "shall not compete", in the sense of avoiding the "duplication" of efforts. See: Papers and Correspondence of Sir John Sulston, Wellcome Library, London, file PP/SUL/B/1/1/1/2.

logistically-demanding bids from, among others, Heidelberg University.[26] As we show later in the book, input from the EBI was crucial in the assembly, annotation and curation of the reference sequences that the Sanger Institute and other genome centres produced (Chap. 6).

The other institution moving to Hinxton—the HGMP Resource Centre—was transformed into an interface providing training, computer access and other support for users of genomic data. This was the vision of the MRC, which developed the databases that the Resource Centre housed, and appointed new personnel to take on the fresh mission.[27] The coordination between the Resource Centre, the EBI and the Sanger Institute was, however, challenging at times, due to the differences in approach and scale between the three institutions.[28]

In 1996, the Wellcome Trust convened an international meeting in Bermuda where it invited scientists and administrators from institutions active in DNA mapping and sequencing, including many of Waterston and

---

[26] For a general outline, see: Sulston and Ferry (2002, pp. 112ff). A more detailed paper trail can be found in: Papers and Correspondence of Sir John Sulston and Michael Ashburner, and Wellcome Trust Corporate Archive, Wellcome Library, London, references PP/SUL/B/4/2, PP/MIA/C and WT, accession 2320, box A2159, file 005307/A. Ashburner became joint-director of the EBI and a main figure in the sequencing of the genome of the fruit fly *Drosophila melanogaster,* an effort that was partially funded by the European Commission. His cooperation with Sulston and preparation of the EBI bid started in December 1992, when the Sanger Institute had been funded but not yet physically established in Hinxton.

[27] When the Resource Centre moved, Ross Sibson left the HGMP and became director of molecular genetics at the Clatterbridge Cancer Research Trust, an offshoot of the Merseyside Oncology Treatment Centre. He used his prior experience in cDNA sequencing—and the data available at the British and European resource centres—to set up a new initiative, detecting and gathering sequence variants implicated in cancers of the treatment centre patients. Tony Vickers, the first HGMP manager, had left the project in 1992: Ross Sibson, interview with Miguel García-Sancho, Royal Liverpool University Hospital, March 2014; Tony Vickers, two-part email interview with Miguel García-Sancho, September and December 2013. Sibson's post-HGMP career reflects the considerable overlaps between the interests of personnel at genome project resource centres and those of medical geneticists (Chap. 3).

[28] Ross Sibson, interview with first author, Royal Liverpool University Hospital, March 2014. Tony Vickers, two-part email interview with Miguel García-Sancho, September and December 2013. Apart from these user support activities, the Resource Centre became involved in the sequencing of pufferfish *Fugu,* a whole-genome project that was started by Brenner after he left the LMB and established a Molecular Genetics Unit at Addenbrooke's Hospital in Cambridge (Chap. 3). This project, however, was not incorporated into the operations of the Sanger Institute.

Sulston's correspondents, as well as their funders.[29] One year before, and again at chairman Roger Gibbs' initiative, the Wellcome Trust had considerably increased its financial capacity by selling its remaining Wellcome Foundation shares to Glaxo Laboratories, which had been a rival pharmaceutical company. The Bermuda meeting, as well as a number of preparatory and follow-up gatherings, have been carefully reconstructed by Kathryn Maxson Jones, Rachel Ankeny and Robert Cook-Deegan, who have documented the complex negotiations leading to the establishment of a set of principles for the free release of sequence data. These principles and further refinements have since shaped the practice of genomic science (Maxson Jones et al., 2018). Here, what we emphasise is how the meeting and its concluding principles enabled the Wellcome Trust and the NIH to operationalise Waterston and Sulston's ambitions.

A critical mass of the Bermuda attendees agreed with the principles of making the sequence data that they determined rapidly available in open-access databases (Guyer, 1998). These databases were housed in three international repositories, located at the EBI, NIH National Center for Biotechnology Information and the National Institute of Genetics of Japan.[30] The Bermuda agreement cemented a longer-term commitment by its signatories to start a comprehensive and coordinated attack on the whole human genome sequence. The geography of data repositories, along with Sulston and Waterston's pivotal role in promoting the rapid and unrestricted release of sequence data, placed the Wellcome Trust and the NIH in a strong position to lead this concerted effort. Venter, who attended the Bermuda meeting with TIGR colleagues, became increasingly isolated as he was one of the few participants defending proprietary rights on DNA sequence data.

Some Bermuda attendees suggested the Human Genome Organisation (HUGO) as a potential coordinator of an international whole-genome sequencing initiative. This organisation had been launched in

---

[29] A full list of participants can be retrieved from pp. 9–15 of the record "1996 Bermuda Roster & Agenda", available at DukeSpace, research data from the project led by Kathryn Maxson Jones, Rachel Ankeny and Robert Cook-Deegan. Permanent link of the record: https://dukespace.lib.duke.edu/dspace/handle/10161/7716 (last accessed 16th December 2022). Information on the project researching the Bermuda meetings: https://dukespace.lib.duke.edu/dspace/handle/10161/7407 (last accessed 16th December 2022).

[30] From the late-1980s onwards, the entries of these three repositories were synchronised, so users could access the same sequence results regardless of which database they queried (Stevens, 2018).

1988, after an encounter between Brenner and medical geneticists Victor McKusick and Walter Bodmer at Cold Spring Harbor Laboratory. Its objective was to coordinate scientists involved in human DNA mapping and sequencing, so they would collaborate and avoid duplicating efforts. HUGO, however, did not itself provide funding for mapping and sequencing enterprises. More importantly, it had been explicitly created as independent from any government-funded or transnational human genome programme. Its main activity throughout the 1990s had been organising the human chromosome workshops, which most HUGO member scientists attended with support from the Wellcome Trust and the March of Dimes, a US charity committed to pre-natal genetic screening (Bodmer, 1991). HUGO was thus too close to the distributed approach to mapping and sequencing that Sulston and Waterston had sought to overcome with the concerted, whole-genome effort stemming from the Bermuda agreement.[31]

Due to this, the whole-genome coalition and their programme of work developed "more organically". According to Mark Guyer—who became director of the NHGRI's extramural (grant-funding) programme—the coordination was brokered by the funding agencies of the sequencing grants rather than being left to HUGO or any other entity that lacked the legal authority to manage grant funds. The necessary distribution of work and quality assessment was thus achieved through frequent meetings and even more frequent telephone calls, and involved effective collaboration among the participating scientists and funding agency staff.[32]

This form of collaboration also allowed the convergence of the different human genome programmes, therefore encouraging the coalescing of scientists and funders into a single sequencing effort. As a result, the International Human Genome Sequencing Consortium (IHGSC) emerged with an initial membership of 20 large-scale sequencing centres from the USA, UK, France, Germany, China, and Japan, plus

[31] HUGO had been a target of Sulston's criticism following Waterston's 'indecent proposal'. In 1994, the same year of Waterston's email, HUGO convened a meeting to discuss how to deal with the patenting of protein-coding sequences by Venter and other researchers. Sulston considered that he and Waterston "should hijack" the meeting by proposing their whole-genome effort as an alternative to partial sequencing. Papers and Correspondence of Sir John Sulston, Wellcome Library, London, file PP/SUL/B/2/1/1, quote from Sulston's manuscript notes prior to the meeting.
[32] Mark Guyer, personal communication with Miguel García-Sancho, National Human Genome Research Institute (NHGRI), November 2018.

bioinformatics institutions and administrative agencies.[33] The remit of this alliance was to produce a reference sequence encompassing the full human genome, something that they did in draft form in 2000 and in more final form in 2003. The results were published in the journal *Nature*—the former as an initial draft in 2001 and the latter as a more finished version in 2004—and the data were released to the three open-access international repositories: the EMBL-EBI Nucleotide Sequence Database, NIH GenBank and the DNA Data Bank of Japan.

The IHGSC embodied the funnelling effect that genomics was experiencing by the turn of the millennium. The composition, size and remit of this coalition was much narrower than the diversity of institutions, programmes and modes of organisation that had proliferated in both human and non-human genomics from the late-1980s onwards. The distributed, piecemeal and more inclusive approaches of the HGMP and European Commission contrasted with this selective club which pursued the sequencing of the whole human genome to prevent TIGR and Venter patenting it. Watson, Sulston and Waterston's visions converged and materialised in the IHGSC, which had the Wellcome Trust and NHGRI as its largest funders. This was reflected in the geography of the resulting coalition: twelve of the twenty large-scale centres of the IHGSC and four of the five top sequence contributors were based in the USA. The other top sequence contributor—second overall—was the UK's Sanger Institute.[34]

---

[33] This grouping did not have a fixed name until 2001, when the IHGSC designation was coined to sign the 2001 *Nature* paper announcing the draft human reference sequence. Cold Spring Harbor Laboratory, the site of the *prison door* encounter between Watson, Waterston, Sulston and Coulson, became a major location of IHGSC deliberations, along with two follow-up meetings in Bermuda in 1997 and 1998. This laboratory, and Watson as its long-serving leader, has played a key role in the construction of master narratives of molecular biology, biotechnology and genomics.

[34] The Sanger Institute, along with the DoE Joint Genome Institute and NIH genome centres at Washington University, Whitehead Institute, and Baylor College of Medicine, contributed more than 80% to the draft reference sequence that the IHGSC described in *Nature* in 2001, according to press releases following publication: http://www.sanger.ac.uk/news/view/first-draft-book-humankind-has-been-read and https://www.genome.gov/10001457/2000-release-working-draft-of-human-genome-sequence (last accessed 16th December 2022).

4.4   FROM BESPOKE MAP TO REFERENCE SEQUENCE

Venter's response to the emergent coalition was to formulate a strategy to determine the whole human sequence ahead of the IHGSC. In 1998, he became CEO of Celera Genomics, a company that launched a parallel effort to produce a whole human genome sequence. Unlike the IHGSC endeavour, this sequence would be temporarily stored in a private database, so patents could be sought. Venter used a sequencing approach that he had devised before the Bermuda meetings and successfully applied to the genome of the bacterium *Haemophilus influenzae* in 1995. This technique, called whole-genome shotgun sequencing, sought to enable the determination and assembly of full genomes without constructing a prior physical map, contrary to the hierarchical map-based sequencing that the IHGSC intended to execute. Along with the powerful automatic sequencing instrument that Venter had at his disposal—the ABI 3700 sequencer, produced by a company belonging to the same corporate group as Celera—the speed of this shotgun approach was a new threat to the open release agenda (Hilgartner, 2017, Ch. 7).

The NHGRI and Wellcome Trust reacted by increasing their financial support, as Waterston and Sulston had been requesting over the preceding four years. While the Trust awarded a substantially higher grant to the Sanger Institute for the period 1998 to 2003, the NHGRI channelled the US-HGP funding to the Whitehead Institute, WU, and Baylor College of Medicine, the latter hosting a new genome centre established in 1996. In 1997, the three large-scale mapping and sequencing centres that the US Department of Energy (DoE) sponsored merged into the Joint Genome Institute. These five institutions, which started to be called the Genomic 5 or G5, took the lead of the IHGSC operation (see note 34).

A problem presented by this funding boost and the advent of the IHGSC more generally, was how to combine this rapid, concentrated and comprehensive sequencing endeavour with the genetic linkage and physical mapping that the US-HGP and other funders of the coalition had been supporting. Despite having grown in resolution throughout the 1990s, most of the resulting genetic linkage and physical maps had not been produced for the specific goal of aiding whole-genome sequencing. This was due to the majority of human genome programmes combining the objective of improving maps with that of supporting medical genetics communities (Chap. 3). The maps were thus focused on a limited range of chromosomal regions that contained the loci of genes connected to diseases.

Additionally, rather than creating a consensus representation that could be used to build a reference sequence, these maps had been produced with the goal of uncovering variation: the differences in the mapped regions that presented across healthy and diseased individuals. This had been the case for the maps produced by the US-HGP in its early years, as well as those funded by the French and German national human genome programmes, which shared the community support ethos of the HGMP and HGAP. Another difficulty was that those maps had been generated by communities of medically and evolutionary-inclined geneticists that were only marginally represented in the G5 institutions and in the IHGSC as a whole.

In the face of this—and the pressing competition of Celera's approach—the IHGSC decided to produce their own bespoke maps for whole-genome sequencing. This decision enabled the development of tools that were specifically designed to support the determination of the reference genome. The maps were intended to encompass the full set of human chromosomes at sufficient resolution, to enable the identification of the ordered DNA fragments that were needed to sequence all chromosomal regions and then assemble the results into a complete reference genome. These comprehensive bespoke maps could, however, also work as platforms to which prior maps—and the information contained in them about clinically relevant variation—could be linked. The whole-genome maps were mainly produced by the same institutions that undertook the sequencing. Making connections to the more detailed maps incorporating variation, however, required collaboration with the medical and human genetic groups that had been previously funded by the national and European genome programmes.

A first step towards the construction of these bespoke maps was obtaining a library or collection of DNA fragments encompassing the whole human genome. These fragments needed to be cloned: multiplied after their insertion into a reproducing organism, so they would be available in sufficient quantity for the sequencing operation. The Yeast Artificial Chromosomes (YACs) that had been used in the mapping of *C. elegans* (Chap. 2) were discarded due to their tendency to contaminate the foreign DNA inserted into them. Bacterial Artificial Chromosomes (BACs) or those derived from bacteriophage virus P1 (PACs) were preferred for their greater stability, despite only allowing smaller inserts.

The size of the human genome—considerably larger than any other species sequenced thus far—made the production of the library a complex endeavour requiring expert knowledge and technical dexterity. This led the IHGSC to rely on an external collaborator: Pieter de Jong's laboratory at Roswell Park Cancer Institute (RPCI). Prior to his appointment at RPCI, de Jong had trained as a biochemical engineer in Europe and worked at the Lawrence Livermore National Laboratory—one of the DoE genome centres—during the early years of the US-HGP. This had furnished him with expertise in large-scale, whole-genome mapping technologies, which he applied to the detection of mutations involved in genetic diseases (Buxton et al., 1992).

Like Venter, Collins and Bentley, de Jong belonged to a community of younger and technologically-savvy biomedical researchers who were pushing the boundaries of medical genetics from specific single-locus diseases to broader areas of the human genome. At RPCI, where he moved in 1993, de Jong and his team distributed libraries to both genetic research institutions and large-scale genome centres.

Both the IHGSC reference sequence and its bespoke physical map were largely based on the RPCI-11 library, produced by de Jong's team. This library was obtained from the blood of an anonymous male donor chosen from a set of ten men and ten women who came forward in answer to an advert placed in *The Buffalo News* on 23rd March 1997—RPCI is located in Buffalo, a city in upstate New York, close to the border with Canada marked by the Niagara Falls. Although the initial IHGSC policy was to use a wider range of DNA sources,[35] in the end almost three-quarters of the total number of nucleotides comprising the draft sequence published in 2001—over 74%—came from the RPCI-11 library, with a further 17% derived from seven other libraries, four of which were produced by de Jong's group. Overall, more than 90% of the human reference sequence was therefore derived from these eight libraries, all of them produced using DNA sourced from male donors (International Human Genome Sequencing Consortium, 2001, p. 866).

One reason for this relatively small pool of DNA was that, as Adam Bostanci has shown, the IHGSC relied on data that suggested that the sequence similarity between any two humans was 99.9%, which to

[35] According to de Jong's group, the original protocol was "to sequence the human genome from a composite of ~10 BAC clone resources each contributing ~10% of the donor's DNA to the final genome sequence" (Osoegawa et al., 2001, p. 484).

them made the choice of donor irrelevant, and the use of samples from people of different ethnicities and sexes scientifically meaningless (Bostanci, 2006). Given this, it was believed that reducing the number of libraries and mainly using one would substantially simplify the task of assembling the DNA fragments, while not affecting the representativeness of the results.

The bespoke mapping effort yielded physical maps of individual chromosomes and one comprising the whole human genome. It mainly used the RCPI-11 library to ease cross-referencing between the different mapping operations, and also across the mapping and sequencing projects.[36] Yet clones from other libraries and data from other mapping endeavours were also incorporated to enhance the content of the maps for particular chromosomes. This was the case for the X-chromosome map, which used fragments from the RPCI-13 library produced by de Jong's team from a female donor who answered the *Buffalo News* advert (Bentley et al., 2001). The IHGSC members also collaborated with other institutions that had experience of mapping specific "known regions" of the chromosomes they were assigned. This provided the mappers with detailed knowledge and data that complemented and, at times, corrected the results of the maps they were devising with more generic protocols (The International Human Genome Mapping Consortium, 2001, p. 935). The compilation of these maps of individual chromosomes was, however, subordinated to the task of producing an overall one, and was therefore directed towards the drive to represent the whole genome rather than constituting tailored resources for medical or human genetics.

This hierarchy was reflected in the February 2001 issue of *Nature* in which the first full draft of the IHGSC reference sequence was published. Along with the sequence, the journal issue included a physical map of the whole human genome and ten maps of individual chromosomes. In all of them, the authors emphasised that the purpose of the maps had been easing the IHGSC operation via the creation of a "tiling path" of ordered DNA fragments that could then be sequenced and assembled into the reference genome. One of the teams stressed that the "only prerequisite" for devising those maps was having a "centralised repository" of data about the BAC clones that were used in the sequencing. Other resources, despite providing "useful information" for the selection of clones and

---

[36] On the iterative, back-and-forth relationship between DNA mapping and sequencing, see Lowe (2018). In this book, we assess different ways of operationalising these mapping-sequencing relationships and their underlying power dynamics.

"validation" of the results, "were auxiliary" to the centralised library (Brüls et al., 2001, p. 948; see also Tilford et al., 2001; Montgomery et al., 2001; Bentley et al., 2001). The whole-genome map article emphasised that, although human genome mapping had been an ongoing exercise for over a decade, its scale had "increased approximately tenfold" since 1998 to "keep pace with the ramping up of the sequencing effort" (The International Human Genome Mapping Consortium, 2001, p. 934).

The whole-genome map article was signed by an International Human Genome Mapping Consortium (IHGMC) that incorporated fourteen institutions from the IHGSC, including four of the large-scale sequencing centres from the G5. The rest of the mapping consortium membership comprised de Jong's RCPI group and five consolidated teams of cancer geneticists (Table 4.1, below). The inclusion of these geneticists was partly driven by the Cancer Chromosome Aberration Project, an initiative funded by the NIH National Cancer Institute that sought to integrate markers of the disease across different human genome maps. This led to the markers and other results of the project being nested in the IHGMC maps.[37] One of the cancer genetics teams, based in the Albert Einstein College of Medicine, coordinated the physical mapping of chromosome 12, described in the 2001 *Nature* issue. The mapping of the rest of the chromosomes published that year was led by the Sanger Institute, the Whitehead Institute and Genoscope, all prominent members of the IHGSC.[38]

The IHGMC strategy differed from previous mapping initiatives that the human and medical genetics communities had pursued during the 1980s and 1990s. Compared to the chromosome 7 mapping led by the University

---

[37] Thomas Ried, one of the promoters of the Cancer Chromosome Aberration Project, had started his career at Yale University under the mentorship of the *Genomics* journal co-founder Frank Ruddle. The 2001 *Nature* issue in which the draft reference sequence was published included, after the bespoke physical maps, contributions relating this information to pre-existing genetic and cytogenetic maps, as well as telomeric region at the end of the chromosomes (Cheung et al., 2001; Riethman et al., 2001; Yu et al., 2001).

[38] The Sanger Institute was the largest individual mapper and led chromosomes 1, 6, 9, 10, 13, 20 and X, the latter being one of the initial targets of Bentley and Dunham's teams. Their other early objective, chromosome 22, had become the first one to be fully sequenced at the Sanger Institute in 1999, following its bespoke physical mapping (Dunham et al., 1999). The Whitehead Institute and Genoscope led chromosomes Y and 14, respectively. Généthon provided the linkage map to aid the assignment of sequenced DNA fragments to chromosomes (Deloukas et al., 1998). The connection of the genetic and physical maps used another technique—radiation hybrid mapping—that was also employed in pig genomics (Lowe, 2022).

Table 4.1 Table reflecting the overlaps between the institutions represented in the First International Strategy Meeting on Human Genome Sequencing (Bermuda, 1996; left-hand column), those forming the International Human Genome Mapping Consortium (IHGMC; middle column) at the time of the 2001 publication of the reference sequence in *Nature* and those listed as genome centres of the International Human Genome Sequencing Consortium (IHGSC; right-hand column) in the same publication

Bermuda, 1996	IHGMC	IHGSC
The Institute for Genomic Research [later gives rise to Celera Genomics]	**Washington University School of Medicine, Genome Sequencing Center**	**Whitehead Institute for Biomedical Research, Center for Genome Research**
Lawrence Livermore National Laboratory [later merged into Joint Genome Institute]	**Wellcome Trust Genome Campus** [including Sanger Institute]	**Sanger Institute**
European Molecular Biology Laboratory	**National Center for Biotechnology Information (*)**	**Washington University Genome Sequencing Center**
Wellcome Trust	**National Human Genome Research Institute**	US DoE Joint Genome Institute
University of Cambridge	Albert Einstein College of Medicine (*)	**Baylor College of Medicine Human Genome Sequencing Center**
Merck Research Laboratories	**Baylor College of Medicine,** Human Genome Sequencing Center	RIKEN Genomic Sciences Center
Applied Biosystems	**Roswell Park Cancer Institute**	**Genoscope** and CNRS UMR-8030
Sanger Institute	**Multimegabase Sequencing Center**	GTC Sequencing Center
National Center for Human Genome Research [later renamed National Human Genome Research Institute]	Fred Hutchinson Cancer Research Institute	**Department of Genome Analysis, Institute of Molecular Biotechnology**
University of Texas Southwestern Medical Centre	The Children's Hospital of Philadelphia	Beijing Genomics Institute/Human Genome Center
National Center for Genome Resources, Santa Fe	**Genoscope**	**Multimegabase Sequencing Center,** The Institute for Systems Biology
Baylor College of Medicine, Department of Molecular and Human Genetics [later gives rise to the Human Genome Sequencing Center]	**US DOE Joint Genome Institute**	Stanford Genome Technology Center

(*continued*)

Table 4.1 (continued)

Institut de Génétique et de Biologie Moléculaire et Cellulaire	**Stanford Human Genome Center and Department of Genetics**	**Stanford Human Genome Center**
Washington University School of Medicine, Genome Sequencing Center	University of California, Santa Cruz (*)	University of Washington Genome Center
European Commission	British Columbia Cancer Research Centre	Department of Molecular Biology, Keio University School of Medicine
Department of Molecular Biotechnology, University of Washington (later gives rise to Multimegabase Sequencing Center)	**Department of Genome Analysis, Institute of Molecular Biotechnology**	**University of Texas Southwestern Medical Center at Dallas**
Rosewell Park Cancer Institute	Departments of Human Genetics and Pediatrics, University of California	**University of Oklahoma's** Advanced Center for Genome Technology
Human Genome Centre, University of Tokyo	RIKEN Genomic Sciences Center	**Max Planck Institute for Molecular Genetics**
Whitehead Institute/MIT Centre for Genome Research	Department of Molecular Biology, Keio University School of Medicine	**Cold Spring Harbor Laboratory**, Lita Annenberg Hazen Genome Center
German Federal Ministry for Research and Technology	**Max-Planck-Institute for Molecular Genetics**	GBF, German Research Centre for Biotechnology
Max Planck Institute for Molecular Genetics		
Los Alamos National Laboratory [later merged into Joint Genome Institute]		
National Center for Biotechnology Information, National Library of Medicine		
Stanford University, Department of Genetics and Human Genome Center		
Cold Spring Harbor Laboratory		
Nara Institute of Science and Technology		
Department of Genome Analysis, Institute of Molecular Biotechnology, Jena		

(*continued*)

Table 4.1   (continued)

California Institute of
Technology

Medical Research Council

University of Oklahoma

Health Effects and Life
Sciences Research Division,
US Department of Energy

University of Oxford

HUGO Americas

Centre for Medical Genetics,
Marshfield Research
Foundation

Généthon [later gives rise to
Genoscope]

Consortia institutions represented at the Bermuda meeting are in bold in the middle and right-hand columns. Overlapping consortia institutions are shaded in grey. Institutions from the IHGMC marked with an asterisk (*) were listed as author affiliations in the 2001 reference sequence publication, but not as genome centres (International Human Genome Sequencing Consortium, 2001; The International Human Genome Mapping Consortium, 2001; list of Bermuda attendees from https://dukespace.lib. duke.edu/dspace/handle/10161/7716, last accessed 16th December 2022). The National Human Genome Research Institute of the NIH, the Wellcome Trust and the US Department of Energy—all represented in Bermuda—were listed in the reference sequence publication as institutions that had a leading managerial role. Table elaborated by both authors.

of Toronto Hospital for Sick Children that we discussed earlier in the book (Chap. 3), the maps described in the 2001 *Nature* issue were constructed in a less inclusive and collective way, and they were also less attentive to variation. While the chromosome 7 effort involved the collation of contributions made by a wide range of medical genetics laboratories, the IHGMC was a more selective club formed of a smaller number of institutions—mainly genome centres—that mapped larger chromosomal areas. The reasons for undertaking the mapping were also different: positioning genes or markers implicated in diseases in the case of the Toronto-led chromosome 7 initiative, and preparing the genome for sequencing in the case of the IHGMC. This meant that once genes or markers were positioned—and regardless of the rest of the chromosome being mapped—the institutions coordinated by the Toronto hospital would turn to identifying variation in these target regions, and then investigating differences in the sequences of the pertinent segments of DNA across healthy and diseased individuals. The IHGMC, by contrast, prioritised the mapping of entire chromosomes and only used variation as a second layer of information to help verify and add detail to the whole map and, ultimately, to the reference genome sequence.

Celera's sequencing strategy was actually more sensitive to variation in the genome, despite their whole-genome shotgun technique not requiring initial physical mapping. It was based on five blood donors selected from a pool of twenty-one, three of whom were female and two male—one of them was Venter. The company stated its commitment to a sequence that should be "a composite derived from multiple donors of diverse ethnic backgrounds"—one of the five selected volunteers was African-American, one Asian-Chinese, one Hispanic-Mexican and there were two Caucasians (Venter et al., 2001, p. 306). Celera's commercial orientation and its plan of devising a restricted-access database required that variation be easily related to the sequence. The company's potential customers, mainly in the biotechnology and pharmaceutical industries, needed to find the sequence data useful for biomedical research.

This became especially vital when, in 2000 and after continued pressure and mediation, Celera agreed to publish its draft sequence in the journal *Science* and make some of the data publicly available (Hilgartner, 2017, Ch. 7). This they did against the background of the full and open release of data from the IHGSC. Consequently, Celera decided to refocus its business strategy, increasingly emphasising the development of diagnostic and therapeutic tools using the sequence data, over charging for access to its databases (Rabinow & Dan-Cohen, 2005). This resulted in an alliance with the Toronto-led chromosome 7 effort and an alignment of their collectively produced physical map and associated medical annotations with Celera's sequence. In 2003, Celera, the Toronto team and more than 40 other institutional co-authors—mainly from medical schools and hospitals—described the sequence of chromosome 7 in detail, including examinations of regions containing clinically-relevant variation (Garcia-Sancho, Leng et al., 2022, see also Chap. 6). At the same time as this collaboration, the IHGSC and IHGMC published fully mapped and polished sequences of each human chromosome—sometimes in collaboration with other co-authors when specific knowledge was required—ahead of the more 'complete' version of their reference genome, which appeared in *Nature* in 2004.

All of this shows that both Celera and the consortia regarded their sequences as platforms to which further information could be linked: about chromosomal positions, inter-individual and inter-species variation, and the biological implications of these. Yet their large-scale mapping and sequencing endeavours and the publicity around them, especially after the 2001 draft publications, led to the consolidation of a success narrative that has emphasised the production of the reference sequence, and overlooked the ways in which this sequence was related and contextualised to other forms of biological data. In stressing and praising the abstracted reference sequence, this master narrative has also abstracted away the prior diversity of genomics to a few participants, modes of organisation and forms of representing variation within the resulting maps and sequences. This funnelling effect has narrowed the public perception of what genomics was, and what it has produced. It has led to the marginalisation of those programmes and institutions that did not converge in the IHGMC and IHGSC endeavours. The Resource Centre became the only surviving component of the HGMP and was increasingly overshadowed by the Sanger Institute and EBI, while the distributed model of the European Commission dissipated around the turn of the millennium (Chap. 2).

This lost—or forgotten—diversity of genomics can be retrieved by examining the processes by which reference sequences were produced and what data were incorporated in—and linked to—them. Historicising the model of genomics instantiated by the IHGMC and IHGSC endeavours has enabled us to uncover the journeys made by different forms of genomic data towards either their incorporation in—or linkage to—the human reference sequence. By reconstructing the historicity of these journeys—as fellow scholars have done with other scientific fields (Leonelli & Tempini, 2020)—this chapter has resurfaced the contribution of human and medical geneticists to the human reference genome, despite these communities being only peripherally represented in Celera, the IHGMC and the IHGSC. The next chapter examines a different historical instantiation of large-scale, concentrated sequencing: the determination of the reference genome of the pig, *Sus scrofa*, which largely took place at the Sanger Institute. Here, the journeys of the data underlying that reference genome present greater continuities between the production of earlier maps and the reference sequence, than was exhibited by the IHGMC and IHGSC.

References

Balmer, B. (1996). Managing mapping in the Human Genome Project. *Social Studies of Science, 26*(3), 531–573.
Bartlett, A. (2008). *Accomplishing sequencing the human genome*. PhD dissertation, Cardiff University.
Bentley, D., Deloukas, P., Dunham, A., French, L., Gregory, S., Humphray, S., et al. (2001). The physical maps for sequencing human chromosomes 1, 6, 9, 10, 13, 20 and X. *Nature, 409*(6822), 942–943.
Boden, R., Cox, D., Nedeva, M., & Barker, K. (2003). *Scrutinising science: The changing UK government of science*. Palgrave Macmillan.
Bodmer, W. F. (1991). HUGO: The Human Genome Organization. *The FASEB Journal, 5*(1), 73–74.
Bostanci, A. (2006). Two drafts, one genome? Human diversity and human genome research. *Science as Culture, 15*(3), 183–198.

Brownlee, G. G. (2014). *Fred Sanger, double Nobel Laureate: A biography.* Cambridge University Press.

Brüls, T., Gyapay, G., Petit, J.-L., Artiguenave, F., Vico, V., Qin, S., et al. (2001). A physical map of human chromosome 14. *Nature, 409*(6822), 947–948.

Buxton, J., Shelbourne, P., Davies, J., Jones, C., Van Tongeren, T., Aslanidis, C., et al. (1992). Detection of an unstable fragment of DNA specific to individuals with myotonic dystrophy. *Nature, 355*(6360), 547–548.

Cheung, V. G., Nowak, N., Jang, W., Kirsch, I. R., Zhao, S., Chen, X. N., et al. (2001). Integration of cytogenetic landmarks into the draft sequence of the human genome. *Nature, 409*(6822), 953–958.

de Chadarevian, S. (1998). Of worms and programmes: *Caenorhabditis Elegans* and the study of development. *Studies in History and Philosophy of Biological and Biomedical Sciences, 29*(1), 81–105.

de Chadarevian, S. (2011). The making of an entrepreneurial science: Biotechnology in Britain, 1975–1995. *Isis, 102*(4), 601–633.

Deloukas, P., Schuler, G., Gyapay, G., Beasley, E., Soderlund, C., Rodriguez-Tome, P., et al. (1998). A physical map of 30,000 human genes. *Science, 282*(5389), 744–746.

Dunham, I., Hunt, A., Collins, J., Bruskiewich, R., Beare, D., Clamp, M., et al. (1999). The DNA sequence of human chromosome 22. *Nature, 402*(6761), 489–495.

Fletcher, L., & Porter, R. (1997). *A quest for the code of life: Genome analysis at the Wellcome Trust genome campus.* Wellcome Trust.

García-Sancho, M. (2010). A new insight into Sanger's development of sequencing: From proteins to DNA, 1943–1977. *Journal of the History of Biology, 43*(2), 265–323.

García-Sancho, M. (2011). From metaphor to practices: The introduction of "Information Engineers" into the first DNA sequence database. *History and Philosophy of the Life Sciences, 33*(1), 71–104.

García-Sancho, M., Leng, R., Viry, G., Wong, M., Vermeulen, N., & Lowe, J. (2022). The Human Genome Project as a singular episode in the history of genomics. *Historical Studies in the Natural Sciences, 52*(3), 320–360.

Guyer, M. (1998). Statement on the rapid release of genomic DNA sequence. *Genome Research, 8*(5), 413.

Hall, A. R., & Bembridge, B. A. (1986). *Physic and philanthropy: A history of the Wellcome Trust 1936–1986.* Cambridge University Press.

Hilgartner, S. (2017). *Reordering life: Knowledge and control in the genomics revolution.* The MIT Press.

International Human Genome Sequencing Consortium. (2001). Initial sequencing and analysis of the human genome. *Nature, 409*(6822), 860–921.

Jackson, M. W. (2015). *The genealogy of a gene: Patents, HIV/AIDS, and race.* The MIT Press.

Kent, A. (1999). The role of voluntary consumer organisations in genetic services in the United Kingdom. *Public Health Genomics, 2*(4), 156–161.

Leonelli, S., & Tempini, N. (Eds.). (2020). *Data Journeys in the sciences.* Springer Nature.

Lowe, J. W. E. (2018). Sequencing through thick and thin: Historiographical and philosophical implications. *Studies in History and Philosophy of Biological and Biomedical Sciences, 72,* 10–27.

Lowe, J. W. E. (2021). Adjusting to precarity: How and why the Roslin Institute forged a leading role for itself in international networks of pig genomics research. *The British Journal for the History of Science, 54*(4), 507–530.

Lowe, J. W. E. (2022). Humanising and dehumanising pigs in genomic and transplantation research. *History and Philosophy of the Life Sciences, 44,* 66.

Maxson Jones, K., Ankeny, R., & Cook-Deegan, R. (2018). The Bermuda triangle: The pragmatics, policies, and principles for data sharing in the History of the Human Genome Project. *Journal of the History of Biology, 51,* 693–805.

Mikami, K. (2020). Citizens under the umbrella: Citizenship projects and the development of genetic umbrella organizations in the USA and the UK. *New Genetics and Society, 39*(2), 148–172.

Montgomery, K. T., Lee, E., Miller, A., Lau, S., Shim, C., Decker, J., et al. (2001). A high-resolution map of human chromosome 12. *Nature, 409*(6822), 945–946.

Myelnikov, D. (2017). Cuts and the cutting edge: British science funding and the making of animal biotechnology in 1980s Edinburgh. *The British Journal for the History of Science, 50*(4), 701–728.

Osoegawa, K., Mammoser, A. G., Wu, C., Frengen, E., Zeng, C., Catanese, J. J., et al. (2001). A bacterial artificial chromosome library for sequencing the complete human genome. *Genome Research, 11*(3), 483–496.

Owen, G., & Hopkins, M. M. (2016). *Science, the state and the city: Britain's struggle to succeed in biotechnology.* Oxford University Press.

Rabinow, P., & Dan-Cohen, T. (2005). *A machine to make a future: Biotech chronicles.* Princeton University Press.

Ramillon, V. (2007). *Le deux génomiques. Mobiliser, organiser, produire: du séquençage à la mesure de l'expression des gènes.* PhD dissertation, École des Hautes Études en Sciences Sociales, Paris.

Rasmussen, N. (2014). *Gene Jockeys: Life science and the rise of biotech enterprise.* Johns Hopkins University Press.

Riethman, H. C., Xiang, Z., Paul, S., Morse, E., Hu, X. L., Flint, J., et al. (2001). Integration of telomere sequences with the draft human genome sequence. *Nature, 409*(6822), 948–951.

Sferra, T. J., & Collins, F. S. (1993). The molecular biology of cystic fibrosis: The work of Francis Collins. *Annual Review of Medicine, 44*(1), 133–144.

Stevens, H. (2013). *Life out of sequence: A data-driven history of bioinformatics.* The University of Chicago Press.

Stevens, H. (2018). Globalizing genomics: The origins of the International Nucleotide Sequence Database Collaboration. *Journal of the History of Biology, 51*(4), 657–691.

Sulston, J., & Ferry, G. (2002). *The common thread: A story of science, politics, ethics and the human genome.* Bantam Publishers.

The International Human Genome Mapping Consortium. (2001). A physical map of the human genome. *Nature, 409*(6822), 934–941.

Tilford, C. A., Kuroda-Kawaguchi, T., Skaletsky, H., Rozen, S., Brown, L. G., Rosenberg, M., et al. (2001). A physical map of the human Y chromosome. *Nature, 409*(6822), 943–945.

Venter, C. (2007). *A life decoded: My genome, my life.* Penguin.

Venter, C., Adams, M. D., Myers, E. W., Li, P. W., Mural, R. J., Sutton, G. G., et al. (2001). The sequence of the human genome. *Science, 291*(5507), 1304–1351.

Waterston, R. H., & Ferry, G. (2019). Sir John Edward Sulston CH. 27 March 1942–6 March 2018. *Biographical Memoirs of Fellows of the Royal Society, 67*, 421–447.

Yi, D. (2015). *The recombinant university: Genetic engineering and the emergence of Stanford Biotechnology.* The University of Chicago Press.

Yu, A., Zhao, C., Fan, Y., Jang, W., Mungall, A. J., Deloukas, P., et al. (2001). Comparison of human genetic and sequence-based physical maps. *Nature, 409*(6822), 951–953.

CHAPTER 5

The Pig Community and Their Reference Genome

In 2006, two years after the human reference genome was deemed 'completed', one of its key contributors, the Sanger Institute, became involved in another initiative to produce a full reference sequence: the Swine Genome Sequencing Project (SGSP). This project drew upon a variety of different funders and contributing laboratories, and was led by the Swine Genome Sequencing Consortium (SGSC), which involved many institutions that had conducted pig genome mapping in the 1990s. The SGSC designated the Sanger Institute as the large-scale centre that would conduct most of the sequencing effort: determining the 2.7 billion nucleotides of the reference genome of the pig *Sus scrofa*, slightly smaller than that of *Homo sapiens*. By the time of the start of the SGSP, the Sanger Institute had moved from its original, provisional accommodation to a purpose-built facility in what had then become the Wellcome Genome Campus in Hinxton, Cambridgeshire. At this new location, it had dramatically increased its sequencing capacity and become one of the most productive genome centres worldwide. By the mid-2000s, the continuous decline of sequencing costs due to improved instrumentation, more experienced personnel, and ever-refined pipelines and modes of organisation, enabled a single genome centre—in this case the Sanger Institute—to complete a large reference sequence without needing to ally with other genome centres.

© The Author(s) 2023   159
M. García-Sancho, J. Lowe, *A History of Genomics across Species,
Communities and Projects*, Medicine and Biomedical Sciences in
Modern History, https://doi.org/10.1007/978-3-031-06130-1_5

The SGSP evolved from prior genome mapping programmes on *S. scrofa*. Different institutions seeking to locate genes and markers on pig chromosomes for a variety of purposes—from agricultural breeding to immunology and transplantation biology—converged in coordinated swine mapping efforts between the late-1980s and early-1990s. Some of them were conducted within a single country, including ones supported by the US Department of Agriculture (USDA): for instance, the in-house (intramural) mapping operation at the USDA Agricultural Research Service Meat Animal Research Center (USDA MARC). Based in Clay Center (Nebraska), USDA MARC operated with a factory-based model of mapping, analogous to the large-scale genomics facilities that the National Institutes of Health (NIH) and the US Department of Energy (DoE) were instituting (Chap. 4).

Another major effort sponsored by the USDA was the Pig Genome Coordination Program (PGCP), launched in 1993 under the leadership of Iowa State University animal geneticist Max Rothschild as part of the National Animal Genome Research Program. The PGCP was—and is—an extramural programme, and as such was conducted under the auspices of the USDA's Cooperative State Research, Education and Extension Service (CSREES) from 1994 to 2009.[1] The PGCP has performed a coordinating and community-building function, funding and distributing shared resources such as mapping tools, contributing to the development of other community resources such as mapping databases, and helping to forge collaborations both within the USA and beyond. As in the contemporary UK Human Genome Mapping Project (Chap. 3), the USDA also disbursed grants to individual researchers and laboratories seeking to map areas of the pig genome.

Other swine genome programmes were funded by transnational institutions, such as the European Commission (EC). This was the case for the Pig Gene Mapping Project (PiGMaP), which by the close of its second iteration had established a network of 29 laboratories coordinated by the Roslin Institute in Scotland.[2] Between 1991 and 1996, these labora-

---

[1] And, since 2009, the USDA's National Institute of Food and Agriculture.

[2] Not all of these formal collaborators received EC funds for their participation; some were members of a wider 'European Laboratory Without Walls' that facilitated the sharing of materials and mapping data. This included institutions beyond Europe, such as Iowa State University and the National Institute of Animal Industry in Japan. See Lowe (2021) on the institutional participation in PiGMaP and Roslin's role in it. There were other smaller collaborations, such as one involving Scandinavian countries that overlapped the outset of PiGMaP, but many of these institutions effectively became part of the wider network of PiGMaP itself, while retaining some efforts focused on particular breeds of local interest.

tories pooled and exchanged data and materials—such as DNA samples from carefully-bred reference families of pigs—to generate genetic markers, assign them to specific chromosomes and map their positions (Archibald et al., 1995; Yerle et al., 1995). The purpose of mapping such markers was to provide signposts to researchers so that they could narrow down the location of genes or other functionally-relevant regions in chromosomes.

As with the Human Genome Analysis Programme and unlike the Yeast Genome Sequencing Project—both of them also sponsored by the EC's early Framework Programmes (Chap. 2)—PiGMaP was mainly focused on *mapping* the chromosomes of *S. scrofa* and did not seek to determine its full genome sequence. Indeed, the participants were quite adamant that at that point this was neither a feasible nor necessary task. The community was well aware that the number of mapped markers and genes at the outset of the 1990s was tiny, and that further populating those maps with additional markers was the immediate priority. This would enable more refined mapping based on the landmarks provided by these initial maps.

Throughout the 1990s, the community of pig genomicists that had formed around the mapping efforts continued to produce ever-more refined maps, including those using new kinds of markers and mapping methods. Developing the means by which these maps and mapping data could be exploited for a range of possible applications was a major focus. Completely integrated genetic linkage or physical maps were never produced in this period, in part because the primary interest of the community was in generating *useful* data rather than *complete* maps. But some integrated maps were developed. Significantly, one brought together the USDA MARC efforts with the growing alliance of PiGMaP and the network of institutions in its orbit, including some American institutions (e.g. for chromosome 6, Paszek et al., 1995).

At the turn of the millennium, these communities had not pursued significant sequencing of large stretches of the pig genome, with most sequencing efforts instead directed towards the focused characterisation of particular genes and their neighbourhoods. The funding that the pig mappers had access to was not sufficient for a whole-genome sequencing effort like the one that was being undertaken by the International Human Genome Sequencing Consortium (IHGSC). The immediate research needs of the pig genomicists did not require a reference sequence. This, as we show below, changed in the space of a few years. So did the wider situation in genomics, which made the prospects of sequencing the genome of a livestock species like *S. scrofa* more realistic and worthwhile for the community.

Individual researchers and laboratories, as well as the community as a whole, pursued a variety of different avenues of potential support and funding. This drew upon strategies of diversification and enabled different pots of funding to be accessed for particular tasks that could contribute towards the wider effort of sequencing the genome (Lowe, 2018). Like the IHGSC, the SGSC was supported by national public funding agencies—among them the USDA, the UK's Biotechnology and Biological Sciences Research Council (BBSRC) and the Danish Government—but also sub-national administrations, such as funders from specific US states, as well as industry bodies. The Sanger Institute operated as a contractor for the community, drawing largely on funds acquired from the USDA. The relationship between the Sanger Institute and the community was far more integrated than such an arrangement might suggest, though, with both parties working together on defining the sequencing effort and shaping its products. The SGSC made the *S. scrofa* sequence data available in the global, open-access databases in 2009 and described the sequence in *Nature* in 2012 (Groenen et al., 2012).

The prominent role of the Sanger Institute in the SGSP and the sequencing of the human genome suggests that the production of the swine reference sequence was configured in a similar manner to the IHGSC-led project.[3] At first glance, both initiatives seem to have emerged from the formation of selective groupings and the channelling of several lines of funding into the concentrated and comprehensive production of a whole-genome sequence. The SGSP would appear to be even more concentrated and narrow than the human genome project. The development of technologies and fall of associated costs made the funnelling effect of the large-scale sequencing model more pronounced: one genome centre undertook the sequencing of the whole pig genome, while for the human one the Sanger Institute needed to pool its efforts with nineteen other institutions. The pig genome endeavour was also deeply informed by the experiences of the human genome sequencing that had preceded it. Yet a

---

[3] This sense of continuity is reinforced by Carol Churcher, who in 2008—two years after the start of the SGSP—succeeded Jane Rogers as Head of Sequencing Operations at the Sanger Institute. Churcher had been a member of staff at the institute since its foundation, when in 1993 she joined the yeast sequencing effort led by Bart Barrell (Chaps. 2 and 4).

more detailed examination of the historicity of both reference genomes shows that the communities involved in the prior swine mapping programmes were much more represented in the SGSC than human and medical geneticists were in the IHGSC, and more heavily involved in shaping the reference genome that was produced.[4] In other words, when the trajectories of the communities involved in pig mapping are taken into account and the emphasis is not placed exclusively on large-scale sequencing, the funnelling effect caused by the advent of reference genome sequencing is less pronounced in pig genomics than in human. Indeed, some of the diversity of actors, practices and modes of organisation of the mapping phase survived during the production of the reference sequence of *S. scrofa*.

This chapter explores the means by which the pig mappers remained involved in the production of the reference sequence. In line with earlier parts of the book and prior scholarship (Szymanski et al., 2019), we show this by portraying the genome as a rhetorical and practical space in which pre-existing communities involved in DNA mapping and sequencing could converge or fragment. The next section of this chapter documents how the *S. scrofa* genome, as an object to be mapped, fostered an alliance dominated by animal geneticists oriented towards the problems of the animal breeding industry with whom they had regular contact. This community also included immunogeneticists pursuing research on the potential use of the pig as a source of organs for human transplantation. Many of the immunogenetics researchers were themselves institutionally associated with the agriculturally oriented animal geneticists. A substantial fraction of these animal geneticists were also interested in what we describe as *systematic* research, meaning an appreciation of diversity and evolutionary relationships. This line of research ran parallel to mapping and sequencing from the mid-1990s onwards, and as we address in Chap. 7, led to new collaborators participating in the pig genome community following the release of the reference genome.

---

[4] It is also worth noting that the SGSC was formed prior to the sequencing project, as a body intended to unite the community to corral the resources to conduct it, and then lead it. By contrast, the IHGSC was largely a post-facto creation, intended to give some unity to an effort that had coalesced internationally from the mid-1990s onwards, but had not assumed a definite unitary organisational form.

The alliance of animal geneticists and immunogeneticists drove the production of successive genomic resources, methods and tools from the 1990s onwards. A key example of this was the creation of comprehensive libraries of DNA fragments, which would be used in the concerted physical mapping of the pig genome and its subsequent sequencing. As the IHGSC had done a decade earlier, the SGSC commissioned a specialist laboratory to construct some libraries that were used for the reference genome effort. These were produced by the same team led by Pieter de Jong that had assisted the human sequencers (Chap. 4). Yet in the case of *S. scrofa*, other libraries created by the pig mappers acted as additional DNA sources for the reference genome and were thus repurposed from their original agricultural and immunogenetic goals.[5]

We conclude by observing that the previous trajectories of the pig genomicists, and their redeployment of tools and resources, made them acutely aware of the affordances and limitations of their reference sequence. Similar to the case of yeast (Chap. 7), they were cognisant of what variation was included in—and excluded from—their reference sequence. This allowed them to appropriately interpret what the reference sequence represented, and consequently to generate new genomic resources linked to it to compensate for the variation known—or reasonably suspected—to be absent. The pig reference sequence, however, differed from the *Saccharomyces cerevisiae* one, in being a conglomeration of DNA from different breeds and populations as opposed to being sourced from a single yeast strain (Chap. 2). Consequently, it was conceived more as a provisional resource than something definitive, reflecting satisficing disposition of the pig community and the kinds of research purposes that they conceived that their data could contribute towards (on 'satisficing', see Wimsat, 2007).

5.1   MAPPING MARKERS AND THE USES OF PIG GENOMICS

In the 1990s, mapping the pig genome and finding ways to use the data they produced became a key task of the community of institutions and researchers that investigated the genetics of the pig. A substantial part of this community was oriented towards the problems of livestock breeding.

---

[5] Elsewhere, we have referred to this repurposing process as "bricolage" and stressed the importance of such processes for historians: they allow the reconstruction of the distinct and evolving trajectories that triggered the emergence of the genome as both a research object and a resource for various communities (Lowe, Leng, et al., 2022).

As this had, prior to the 1990s, been dominated by quantitative genetics, pig genome mapping represented an intersection between the newer molecular genetics and the long-established quantitative genetic tradition. The latter involved formulating statistical approaches to enable breeders to make use of the plethora of data on a multitude of traits of interest to farmers—such as litter size or lean meat content—to inform selective breeding decisions for populations of farm animals. These breeding pro-grammes were and are conducted by private sector breeding companies (such as the Pig Improvement Company, or PIC, which we encounter many times in the rest of the book), farmers' cooperatives or state organ-isations. From the 1980s onwards, there has been a shift away from publicly-funded research institutions conducting many aspects of the breeding process, and towards these bodies concentrating on providing the scientific basis and data to inform private sector breeders (Agar, 2019, Ch. 3; Myelnikov, 2017).

The advent of genomics in the late-1980s provided an opportunity for agriculturally-oriented research institutions to recalibrate their work in this way. These institutions could now produce genomic data and statisti-cal and computational tools for their potential application in breeding pro-grammes, with the breeders themselves taking on the further development and incorporation of these data and tools in their own operations.

In the case of *S. scrofa*, as with other farm animal species, the 1990s represented a period in which maps became ever more populated with increasing numbers—and new kinds—of genetic markers: Restriction Fragment Length Polymorphisms, Amplified Fragment Length Polymorphisms, minisatellites and microsatellites, to name a few of the most significant (Table 5.1).

New mapping assignments were made, databases for storing mapping and related data were developed, and statistical and computational tools were constructed for the detection of chromosomal loci associated with variation in traits of interest: Quantitative Trait Loci, or QTL. These loci were normally markers laying nearby genes. They could also be genes themselves, or parts thereof. Initial mapping relied on the extraction of DNA from cross-bred reference families of pigs, with DNA samples dis-tributed across many laboratories in collaborative projects. PiGMaP, and the other national and international mapping collaborations, enabled a coordination—and in some respects, a division—of labour that made use of the capabilities and resources of particular laboratories to contribute to common resources such as maps and mapping databases. This was vital in

Table 5.1   Descriptions of four main types of genetic markers used in pig genome mapping. Adapted by both authors from Lowe and Bruce (2019)

Genetic marker	Description
Restriction Fragment Length Polymorphisms	Produced by selective cutting of DNA by specific restriction enzymes. Different sequences will result in differing lengths of the fragments produced by this digestion. Not as variable as minisatellites and microsatellites.
Minisatellites	'Motifs' or patterns of over 10 base pairs that are repeated up to 10 times. Highly variable. Unevenly distributed across the genome and much less prevalent than microsatellites.
Microsatellites	'Motifs' or patterns of under 10 base pairs that are repeated dozens of times. Highly variable. Spaced out across the genome, but mainly in noncoding regions.
Amplified Fragment Length Polymorphisms	Produced by selective cutting of DNA by specific restriction enzymes. Different sequences will result in differing lengths of the fragments produced by this digestion. The fragments are amplified by Polymerase Chain Reaction (PCR) to make the process of obtaining them easier—requiring only small amounts of DNA. Not as variable as minisatellites and microsatellites.

a community where, USDA MARC apart, no one institution possessed the capacity to take on the tasks of genome mapping alone. Alan Archibald at the Roslin Institute was an instrumental figure in brokering these collaborative projects on the European side and in linking up the European efforts with US groups.[6]

At this stage, there was no conception of producing a reference sequence—or even of mapping the whole pig genome—on the part of most pig geneticists. One reason for this was the increasingly difficult funding environment that this community had endured since the 1980s. The decreasing economic and social importance of agriculture had led most Western governments to expect that industry would become the main funder of food-related research. State support was channelled towards projects and tools that held promise for achieving more efficiency—rather than more quantity—in food production, such as genetic engineering (García-Sancho & Myelnikov, 2019). As a result, the pig genomics community developed a suite of approaches to use and adapt data, knowledge, methods and tools produced for the genomes of species

---

[6] Due to tensions between some leading university-based groups, Archibald also usefully linked different elements of the US pig genome community.

such as human and mouse, which had a longer and more-established history of mapping (Hogan, 2016; Lyon, 2002; Paigen, 2003a; Paigen, 2003b; Rader, 2004) and more resources than were available to pig geneticists. The development of infrastructures and data for the genomes of other species, such as the human, therefore became a key resource for pig genomics, and a comparative inferential apparatus was articulated to make full use of it (Lowe, 2022).

Yet achieving an equivalent level of resolution to the human or mice maps was not an objective of pig geneticists *per se* nor an inevitable outcome of their activities. This was due to their predominantly agricultural orientation as opposed to the biomedical goals of most human and mouse geneticists. For the majority of researchers mapping *S. scrofa*, as well as their associates in the breeding industry, the identification of particular genes with known biological mechanisms and phenotypic effects was desirable, but not essential. In the early-to-mid 1990s, it was presumed that knowledge of the presence or absence of particular markers known to be *linked* to a locus associated with variation in traits would be sufficient for improving the effectiveness of selective breeding. The goals of the research did not, therefore, require that the molecular genetic basis of observed phenotypic variation be discerned, contrary to the needs of medical genetics research where this is imperative (Lowe & Bruce, 2019). Because of this, comprehensive sequence data was not seen as a necessity for informing breeding decisions, in the same way that it was felt to be a key resource that would radically advance the understanding of the genetic basis of disease—to cite the justification of the likes of James Watson for completely sequencing the human genome—or the identification and characterisation of genes responsible for key cellular processes: André Goffeau's motivation for sequencing the yeast genome (Chaps. 2 and 3).

Several developments around the turn of the millennium changed this perspective. First of all, the maps were becoming extremely well-stocked with markers of different kinds, and were arrayed across the chromosomes at increasingly higher resolutions. The payoffs from incremental improvements to these maps were therefore diminishing. It also became increasingly apparent that using a panel of even dozens of markers linked to variation in traits that breeders wanted to select for was not yielding results that matched the high expectations some had for this approach.[7] Soon, statistical models were articulated by quantitative geneticists that required the use of many magnitudes more markers across the genome, an approach known as genomic selection (Haley & Visscher, 1998; Meuwissen et al., 2001). A particular kind of marker, abundant and available across the genome, the Single Nucleotide Polymorphism or SNP, was particularly valuable for this approach. Whole-genome sequencing efforts were a good source of the data that was needed for the identification of these.

Another significant area where it was becoming increasingly apparent that a fully sequenced genome would be valuable was in research on the immunogenetics of the pig. This was tied to the decades-long history of using the pig as a model for transplantation research and surgery, and more recently as a potential source of organs for humans—xenotransplantation. Researchers working in this field had been mapping the Major Histocompatibility Complex (MHC), a region in chromosome 7 of pigs (and chromosome 6 in humans) that is densely populated with genes involved in immune response. Incompatibilities between the products of different genes—and versions of genes—in this genome region are the cause of adverse reactions leading to the immune rejection of a transplanted organ. Identifying these genes and their different variants is therefore a crucial task for effecting transplantations, both within species and across them.

The mapping of the swine and human MHCs—since the 1970s and 1960s, respectively—was an extremely tricky task given how densely packed and highly variable the genes are in this region (on the history of human MHC research, see: Thorsby, 2009). In the 1990s, the task of using pig organs for transplantation was complicated by the discovery of retroviruses embedded in the pig's DNA—Porcine Endogenous RetroViruses, or PERVs. It was feared that viruses could become activated if pig organs were transplanted into immuno-compromised humans who had not co-evolved with the viruses like the pigs had. For these reasons, it became imperative to sequence the pig genome: to further characterise the

---

[7] An example of this were the hopes that PIC had for Marker-Assisted Selection. For more on this and other innovations at the intersection of publicly-funded research and the breeding sector, see Bruce and Lowe (2022).

swine MHC and its differences to the human MHC and to assess the presence or absence of PERVs (Rohrer et al., 2002).[8]

Immunogenetics was thus one area of research that motivated the creation of a pig genomic library, a set of *S. scrofa* sequence fragments stored in the DNA of microorganisms such as viruses, yeast and bacteria. The natural proliferation processes of these vectors were used to clone and multiply the pig DNA fragments. First a Yeast, and then a Bacterial Artificial Chromosome library (a YAC and a BAC) of *S. scrofa* were constructed by a team at Laboratoire Mixte CEA-INRA de Radiobiologie Appliquée (hereafter, CEA-INRA).[9] This institution was based on the campus of the Institut National de la Recherche Agronomique (INRA) in Jouy-en-Josas, south-west of Paris.

CEA-INRA was originally set up in 1964 with funding from two state agencies: INRA, the multi-branch French agricultural research body, and the French atomic energy agency (Commissariat à l'Énergie Atomique; CEA).[10] It was led by Marcel Vaiman from its inception and pursued research on the genetics of immune response in the pig, in order to improve the efficacy of transplantations of organs. Another early member was Christine Renard, who joined at the outset of the 1970s and developed serological methods for immunological analysis (see below). Patrick Chardon joined the team in the 1970s and Claudine Geffrotin in the 1980s. They both implemented new molecular biology-based approaches in the group. A key addition in the 1990s was Claire Rogel-Gaillard, who was vital in developing and deploying the new genome libraries (Fig. 5.1).

The team's early research led to the successful development of the pig as a surgical model for transplantation procedures. Researchers at CEA-INRA co-discovered the pig's MHC (the Swine Leucocyte Antigen complex, SLA) in 1970, and then went on to pioneer the mapping—and later sequencing—of this region. Initially, this was achieved by serological methods, a core immunology technique that uses immune reactions between antibodies in blood serum and white blood cells from a different

---

[8] More recently, the reference sequence has been used to validate the deletion of PERVs from the genome effected by genome editing (Niu et al., 2017; Yang et al., 2015).

[9] It became known as INRA-CEA Laboratoire de Radiobiologie et d'Etude du Génome from 1999.

[10] On the history and the transformations of INRA, see Bonneuil and Thomas (2009).

Fig. 5.1 Picture of four key members of the CEA-INRA team over the course of its history from 1964. From left to right: Marcel Vaiman, Claire Rogel-Gaillard, Christine Renard and Patrick Chardon. Photograph taken by James Lowe, Paris, November 2017

individual as a mapping indicator.[11] CEA-INRA was a participant in PiGMaP from the start of the project in the early-1990s. In it,

---

[11] This lymphocytotoxicity test, primarily performed by Renard, consisted in creating antisera, blood serum samples containing antibodies produced in pigs subject to skin grafts. They used the ones found to be specific in identifying variants underpinning immune rejection as test substances to see if they created an immune reaction when exposed to the molecules (antigens) on the surface of white blood cells (lymphocytes) from the animal to be tested. This allowed them to discern different sets of variants possessed by the test animals, to detect and map haplotypes (sets of particular combinations of genetic variants) and identify potentially compatible donors as a result. However, the range and availability of sufficiently-specific reagents was limited, and they were better suited to defining haplotypes rather than individual variants (Lunney et al., 2009).

they performed flow cytometry, a technique that sorts chromosomes and therefore aided the mapping of markers to specific pig chromosomes. They also serologically analysed pigs from reference families across Europe that were used in the mapping, as well as developing tools for the further characterisation of chromosome 7.

Through this, CEA-INRA used the funding and networking opportunities of PiGMaP to advance their ongoing survey of the SLA complex by employing physical mapping techniques.[12] This mapping endeavour involved the creation of DNA libraries and the use of probes to identify coding sequences. This work was conducted in the first year of the second round of PiGMaP, which ran from December 1994 to November 1996. The CEA-INRA team created a library using Yeast Artificial Chromosomes (YACs) as vectors. Here we focus on the source of DNA for this, the ways in which the creators of the libraries evaluated them, and the uses to which they were put. We then describe how and why they created DNA libraries in Bacterial Artificial Chromosomes (BACs), showing how they became community resources that aided the mapping of increasingly larger areas of the pig genome, as well as other forms of genome analysis.

For their library construction, the CEA-INRA workers drew on techniques used by a group led by Daniel Cohen in Paris, who constructed YAC libraries to contain clones of the MHC in *H. sapiens*: the Human Leucocyte Antigen complex (HLA). They had already collaborated with Cohen, who was a former student of Jean Dausset. Dausset had discovered the HLA, for which he won the Nobel Prize, and his team had been working with the CEA-INRA group since 1968. With Dausset, Cohen was a co-founder of the Centre for the Study of Human Polymorphism (Centre d'Etude du Polymorphisme Humain; CEPH), the institution from which Généthon had arisen in 1990 with funding from the French Muscular Dystrophy Association (Association Française contre les Myopathies). Généthon was founded to systematise their attempts at mapping the loci of different genetic diseases. This new institutional base had enabled Cohen to scale up from the HLA to the whole human genome: as we discussed in previous chapters, Généthon was a leading institution during the early stages of human genomics and produced the first comprehensive linkage and physical maps using high-throughput automated approaches (Chaps. 3 and 4; Kaufmann, 2004). In assisting in the

---

[12] PiGMaP progress report March 1993, in "PiGMaP Reports 91-92 92-93" partition, obtained from Alan Archibald's personal archive, 24th March 2017.

construction of pig libraries to aid in the mapping of the SLA region, Cohen also contributed towards the scaling up to the eventual tackling of the whole pig genome.

To produce their pig library, the CEA-INRA group extracted DNA from peripheral blood lymphocytes, a kind of white blood immune cell, from two boars (males) of the Large White breed. The laboratory had long used Large White pigs in their immunogenetic research, dating back to the 1960s. A hardy and adaptable pink-skinned pig that is amenable to crossbreeding in livestock breeding programmes, it was also an internationally prevalent breed for commercial food production. The very thing that had made it useful for agriculture therefore also made it useful for conducting and applying pig genetics research. For instance, it was used in the crossing experiments of PiGMaP as well as in the production of the CEA-INRA YAC library.

The boars used for this library each had a distinct homozygous SLA haplotype, meaning that the genes making up the haplotype (see note 11) were the same on both strands of DNA. The construction of this library rested on decades of prior mapping of the SLA complex to determine these haplotypes: sets of specific combinations of genetic variants. This mapping first used serological methods combined with cytogenetic techniques, and then from the 1980s involved genomic approaches. An early example of the latter was an experiment, published in 1985, in which the CEA-INRA team applied restriction enzymes to pig DNA samples and hybridised the resulting fragments to human cDNA probes acquired from CEPH. They showed that this technique had greater specificity than serological analysis, revealing different haplotypes within ones that serological methods had identified as the same. The new SLA variants were considered to be sub-types of those detected with the preceding mapping techniques. The nesting of the newly determined haplotypes in the older serologically identified ones adduced credibility by conforming to previous classifications while partitioning them still further. As a result, they concluded that these genomic methods offered the prospect of "increasing knowledge concerning SLA genetic organization and complexity" (Chardon et al., 1985, p. 170).

Once CEA-INRA had constructed the YAC library, they needed to test—or validate—the new resource. They performed tests to discern the average size (and range of sizes) of the clones, how many YACs were chimeric (pig DNA contaminated with yeast DNA) and the presence or absence of particular sequences. For the latter, they used primers—which trigger the amplification of specific stretches of DNA—of particular known genes. These primers were either produced locally or acquired from ten other laboratories in the wider PiGMaP network. As well as inspecting the accuracy of their library, the CEA-INRA team examined whether there were enough overlapping sequences present in the clones to build them into larger sets of ordered fragments or contigs, and therefore be able to encompass broader areas of the pig genome. In these ways, they were assessing the utility of the YACs themselves (through evaluation of size and proportion of chimeras), whether the library provided sufficient coverage (through examining the presence of known genes) and the extent to which it could be applied to larger-scale physical mapping (Rogel-Gaillard et al., 1997). By the time of this evaluation, in 1997, the team at Jouy-en-Josas managed a library of some 18,000 clones that had been tested using underlying sequence information and DNA fragments from other pig breeds.[13]

During the evaluation, they screened the library to identify clones containing parts of the SLA, using primers for four genes and finding three of these represented. They also screened the library for repeat sequences, as a starting point for being able to characterise the organisation of centromeres—regions that link the two halves of chromosomes and feature abundant repetitive sequence patterns. This task was crucial to their studies of the SLA, which spans the centromere of chromosome 7. Some of the YAC clones flagged in this screening were then sequenced and compared to previously identified centromeric repeat sequences.

While 85% of the sequences screened for were found, this percentage was lower than expected given prior knowledge of the prevalence of these repetitive sequences in the yeast genome (Chap. 2). These divergences, which could point to biases in the coverage of the library, were explained with reference to "the influence of the cloning system on the selection of specific regions", the screening procedures they adopted and the quality

[13] Partition—"EC PiGMaPII—Final Report"—folder 1, p. 31. From Alan Archibald's personal archive, obtained 15th May 2017.

and range of the primers used.[14] Additionally, they swapped samples from their own library with clones contained in two other pig genome libraries created in Göttingen and Berlin. By using different libraries and cloning systems in conjunction with their own, as well as refining methods and tools, they hoped to advance the coverage and utility of their YAC library (Rogel-Gaillard et al., 1997).

YAC libraries were favoured at this stage because of the large insert size they allowed, of clones up to 1Mb, a million bases or nucleotides. Once libraries were needed for fine-grained physical mapping, however, the disadvantages of YACs—such as the risk of chimerism due to contamination by yeast DNA—outweighed the storage capacity advantage. As with human genome mapping (Chap. 4), the CEA-INRA team therefore decided to produce a library stored in BACs, as their lack of chimerism made up for their smaller storage capacity of up to 300Kb: 300,000 bases, or nucleotides. A BAC library of *S. scrofa* was created at Jouy-en-Josas in 1999 with DNA from one of the Large White boars that had been employed before. This time, the DNA was extracted from skin fibroblasts, connective tissue cells that synthesise collagen and other fibres.

Once again, the library was primarily constructed to address the immunogenetic interests of the group, in particular assessing the presence of PERVs in the DNA of pigs. As with the prior YAC library, it was also validated by assessment of its coverage, levels of chimerism and insert sizes. It was screened using known markers to test the extent to which it replicated known genomic features, and contigs were built using overlapping sequences that were identified. So validated, the library could now be screened using primers for known PERV sequences, and the clones thus identified were isolated and analysed. This enabled the researchers at Jouy-en-Josas to both satisfy their more immediate goals—probing the clones with known PERV sequences and identifying their chromosomal position—and to build larger contigs using overlaps between the library's

---

[14] Partition—"EC PiGMaPII—Final Report"—folder 1, p. 31. From Alan Archibald's personal archive, obtained 15th May 2017. Coverage is a metric that is calculated by multiplying the total number of reads (in the case of libraries, the number of fragments or clones) by the quotient of the average read length (the size of the clones in number of nucleotides) divided by the total genome length (expressed in nucleotides). The higher the coverage, the higher the number of genomes that are theoretically represented either in a library or a sequencing operation, e.g., 4X, or 10X. Greater coverage means that it is supposedly more likely that errors or absences are ironed out, and consensus sequences based on multiple reads covering the same nucleotide should be more reliable.

DNA fragments. In other words, the BAC library, as its YAC predecessor, overflowed its original SLA focus and could be used to map increasingly larger areas of the pig genome (Rogel-Gaillard et al., 1999).

The team at CEA-INRA screened the library on request from researchers across the world, distributing clones for free. They saw this as a key service to their fellow pig genomicists and other researchers. It also helped them to forge new connections in a network of laboratories that they perceived as becoming ever denser and more international.[15] Screening the library was a laborious process, involving manual rather than automated picking and analysis of clones. In the long-term, it would have been far too strenuous and costly for it to continue to be conducted by the same researchers mapping the SLA complex. Consequently, the BAC-YAC Resource Center was formed with technicians and engineers placed in charge of managing and screening the library. The mapping team, therefore, transferred their libraries to the Resource Center, a technical laboratory also belonging to INRA that distributed clones on request to the wider research community.[16] Its rationale and operation resembled the Resource Centre that the UK Medical Research Council had created in the early 1990s within its Human Genome Mapping Project (HGMP, Chap. 2).

Other DNA libraries were established for concrete research purposes: a YAC library at USDA MARC was constructed and characterised in collaboration with a researcher at the University of Otagu in New Zealand (Alexander et al., 1997); a PAC library was created by a German collaboration using an artificial chromosome derived from P1 bacteriophage,[17] and a BAC library (PigEBAC) was developed at the Roslin Institute.

PigEBAC was created over 1997 and 1998 with funding from the EC and the UK's BBSRC. It was then further processed and housed at the HGMP Resource Centre, which by the mid-to-late 1990s had been

[15] Interview conducted over Skype with Claire Rogel-Gaillard by James Lowe, May 2017.

[16] Interview conducted by James Lowe with Patrick Chardon, Christine Renard and Marcel Vaiman, in the presence of Claire Rogel-Gaillard, in Paris, November 2017. This oral history should be taken as the primary support for the foregoing historical account—if not, necessarily, for our interpretation of it—in places where another citation has not been indicated.

[17] This involved a collaboration between workers at the Institute of Veterinary Medicine (Georg August University of Göttingen), Institute of Animal Breeding (Technical University of Munich) and a laboratory in a medical genetics clinic in Bavaria (Al-Bayati et al., 1999).

relocated to the same campus near Cambridge where the Sanger Institute was based (Chap. 4). The clones were distributed to the wider community from the Resource Centre. The DNA used in PigEBAC, as with the French YAC library, was acquired from the peripheral blood lymphocytes of a boar. Yet in this case, the boar was the offspring of a cross between a Large White female and a Meishan breed male. This hybrid origin was considered to be appropriate to the stated motivation of producing the library: it was explicitly intended to aid specific genetic research as well as more general genome mapping efforts (Anderson et al., 2000).

Indeed, many of the reference populations used in PiGMaP had been constructed by crossing Large White and Meishan pigs. These two breeds of pig—though the Meishan is typically classified as a sub-breed of the Taihu pig—were geographically distinct in their origins: Yorkshire in the case of the Large White, and the Chinese province of Jiangsu for the Meishan. The two pigs were also quite dissimilar: the Meishan is darker, with wrinkled rather than smooth skin, and is fatter and more reproductively prolific (Fig. 5.2). The latter characteristic made it of interest to Western breeders and allied researchers, who aimed to boost this quality in their local pig populations by crossbreeding with Meishans. For this reason, efforts were made to import these pigs, which resulted in transplantations of small populations to France in 1979, the UK in 1987 and the USA in 1989.

The presumed genetic distance of the two breeds was deemed an additional advantage to their use in mapping. Polymorphisms or variability at particular loci in the cross-bred offspring could be used to calculate genetic linkage between pairs of genetic markers: the frequency at which they are jointly inherited. A BAC library based on the same kind of genetic material as used in the prior mapping of markers could help refine and evaluate the existing assignments of loci still further. This diversity of uses shows that although the Roslin library was designed in part for genome mapping, it reflected the trajectories, networks, and evolving goals and interests of the communities that had coalesced around the pig genome (Fig. 5.3). Apart from its use in aiding mapping, it was also intended to be used, more immediately, as a resource for the identification of QTL: chiefly those of value to the pig breeding industry that much of this community was oriented towards.

DNA libraries such as the ones produced at CEA-INRA and Roslin are shared reference resources. They constitute validated and progressively characterised collections with known and described provenance that can be consulted by the wider community, and for which the potential uses are

Fig. 5.2 Top: A Meishan sow at the Roslin Institute. Bottom: A Large White pig, also at the Roslin Institute. Photographs taken by the Roslin Institute photographer Norrie Russell, and provided to us courtesy of Alan Archibald

not narrowly prescribed or channelled by the sources and means of their construction. In this way, they are similar to cell lines (Landecker, 2010), mouse strains (Rader, 2004) and seeds held in banks (Curry, 2017; Curry, 2019; Peres 2016). In the case of the *S. scrofa* libraries, their production,

Fig. 5.3 Photograph of an early PiGMaP meeting in Toulouse, December 1991. It features the communities of agriculturally-oriented geneticists and immunogeneticists that coalesced around the mapping of the pig genome. Note Alan Archibald of the Roslin Institute (ninth from right, wearing a Scottish kilt), Max Rothschild of Iowa State University (eleventh from left, with a beard) and Lawrence Schook of the University of Illinois (tall and at the back in the centre of the group). Many other key figures who continued to play a significant role in pig genomics were also in attendance, such as Marcel Vaiman (ninth from left, with light-grey hair and a tie on) and Patrick Chardon (brown hair, just to the left of centre at the back, behind the woman with the brown bag) of CEA-INRA, and Louis Ollivier (thirteenth from left, holding a coat) of the INRA station at Jouy-en-Josas, who we meet in Chap. 7. Photograph courtesy of Lawrence Schook

circulation and validation from the late-1990s onwards helped to intensify the connections that had begun to be forged in projects to improve maps of the pig genome. This convergence reflected, and was further fostered by, the ongoing mapping and by successive projects that aimed to produce other resources and tools of use for the community and the breeding industry.[18] The community dimension of pig mapping and its concomitant

---

[18] In spite of its growing cohesiveness, this community was not homogeneous—it varied across specific initiatives, institutions and geographical settings. USDA MARC maintained its own significant independent activity, for example, while Japanese researchers would not be integrated as much as the European and North American ones. There were also distinct lines of research that involved different subsets of the community. Some concentrated more on breeding-related research, or on genetic diversity, or immunogenetics. A core of institutions focused heavily on producing more general-purpose genomic data and tools like genome libraries. There was, though, considerable cross-over between these subsets, reflecting the diversification and varied collaborations of these researchers and institutions, and the need for collaborative efforts and international divisions of labour to produce the platforms that would be of use for the various research aims and activities of an increasingly close—and still small—community.

concern with variation persisted when, at the turn of the millennium, the opportunity arose to characterise the full *S. scrofa* genome.

5.2   THE GENEALOGIES OF THE MAP AND THE SEQUENCE

One of the ironies of pig genome sequencing was that, although physical mapping preceded and informed the whole-genome sequencing operation in a manner that faithfully replicated the original strategy of the IHGSC (Chap. 4), the creation of this physical map was a separate project designed to contribute to the community's more proximate research goals. This physical mapping constituted a continuation, albeit in a more comprehensive way, of the preceding mapping activities of the pig community. As a result, it naturally used the DNA libraries that two of the more prominent mapping institutions—CEA-INRA and the Roslin Institute—had produced and distributed to their peers. Yet in the USA, the landscape of concerted mapping programmes was different. One of them, centred around USDA MARC, adopted some of the organisational and logistic aspects of the large-scale production model that was becoming increasingly pervasive in human genomics.

USDA MARC approached the same specialist team from which the IHGSC had commissioned the production of libraries to map and determine the reference sequence of the human genome: Pieter de Jong's group at the Roswell Park Cancer Institute. The resulting pig BAC library was named after de Jong's institutional acronym (RCPI) and given serial number 44, since the team had been involved in the construction of preceding libraries for other organisms. RPCI-44's creation was paid for by USDA MARC, with several researchers from there involved in its characterisation and analysis. The library was derived "from four crossbred male pigs (breed composition: 37.5% Yorkshire [Large White], 37.5% Landrace, and 25% Meishan)" (Fahrenkrug et al., 2001, p. 472).[19] It therefore overlapped with the two breeds that provided the DNA for PigEBAC and, like the Roslin library, was intended to be used in both mapping and identifying further genetic markers of agricultural interest. Yet RPCI-44 was at

---

[19] *Landrace* breeds around the world originate in the Danish Landrace breed and are not to be confused with the term *landrace breeds* that designates locally-specific traditionally grown or reared types of plant or animal. The RCPI-44 library used the American Landrace, which like the other Landrace breeds is bred for food production. As with the Meishan pigs, Landrace breeds were originally imported into the USA in small numbers following sensitive political negotiations, albeit earlier, in 1934: http://afs.okstate.edu/breeds/swine/americanlandrace (last accessed 9th December 2022).

this point the only library whose creators made explicit mention of its possible use for the sequencing of large genomic regions or even the whole genome of *S. scrofa* (Fahrenkrug et al., 2001).[20]

RPCI-44 became publicly available in 1999, the same year as PigEBAC. By that time, the IHGSC effort was approaching its zenith and its participants were looking to the post-human reference sequence world. A potential new horizon for some of them, especially those more narrowly specialised in conducting large-scale mapping and sequencing, was undertaking the genomes of other organisms. Debates and planning took place at the NIH on opening funding streams to sequence non-human genomes that could provide both comparative insights for data validation, as well as knowledge of interest for medicine, agriculture and industry (Chaps. 6 and 7). During the early-2000s, the communities that had converged around pig mapping attempted to take advantage of these opportunities to position *S. scrofa* as a candidate to be sequenced next.

Here, the role of a handful of international conferences in allowing the small and increasingly tight-knit pig genomics community to come together and develop new ideas and strategies was key. Three of these were especially significant: the Plant and Animal Genome conference held annually in January in San Diego, California; the biennial International Society of Animal Genetics (ISAG) conference held in a different location every even-year summer; and the quadrennial World Congress of Genetics as Applied to Livestock Production, that also moves around venues worldwide. These meetings concerned (and still concern) multiple species, in contrast to the typical conferences of human researchers, and are used as occasions for convening working groups and consortia, for holding meetings to discuss the progress of existing projects, and debating the prospects for forming new ones.

Many of the researchers in the pig community work on other organisms as well, usually other farm animals. Contact with colleagues pursuing genomics research on other species raised their horizons, as well as informed their own strategising. For example, pig genomicists learned from developments in cattle genomics, as we shall see in this chapter and Chap. 7. Chicken genome researchers, who like cattle genomicists were pursuing a full reference sequence before the pig, also intersected with the pig genome community, though in a more direct way through the

[20] https://www.animalgenome.org/pigs/newsletter/No.038.html (last accessed 9th December 2022).

involvement of people like Wageningen University's Martien Groenen in both efforts. Industry representatives, particularly from the breeding sector, were (and are) regular participants in these conferences, and the informal sharing of new developments in the academic and private settings has been key to shaping research and industry agendas.

The NIH held a workshop in July 2001 on 'Developing Guidelines for Choosing New Genomic Sequencing Targets', involving key figures in the IHGSC. Richard Frahm, the National Program Leader of Animal Genetics at the USDA's CSREES, advocated for sequencing an agriculturally important species, emphasising its potential economic impact in addition to its value for comparative purposes. However, already at this stage, the phylogenetic position (where a species is located in the tree of life) of the candidate species was being emphasised by other participants at the workshop as a key criterion, and this meant that advocates for sequencing some non-agricultural organisms could stake more convincing claims. Parallel to this, the USDA was exploring its own options for genomics. The new Under Secretary for Agriculture responsible for research, Joseph Jen, requested that the US Government's Office of Science and Technology Policy create an Interagency Working Group on Domestic Animal Genomics.[21]

In 2001, a different group—the Alliance for Animal Genome Research—was established in the USA by agricultural industry bodies and research institutions to advocate for the development of genomics research concerning animals used in agriculture. From the beginning, they were led by Kellye Eversole, and the Alliance used her firm Eversole Associates to lobby public officials and politicians. This would prove useful in acquiring funds in the US Government's budgeting process.[22]

An early success of the Alliance for Animal Genome Research was in getting the US National Academy of Sciences to convene a meeting on domestic (i.e. farm) animal genomics. This was funded to the tune of $100,000 by the two main research arms of the USDA (the Agricultural

---

[21] Jen was the Under Secretary for Research, Education, and Economics from 2001 to 2006. An executive position, an Under Secretary of the USDA is hierarchically below the cabinet-rank Secretary and their Deputy Secretary. A third-tier position (Level III in the US Government's 'Executive Schedule'), in 2001 there were seven Under Secretaries in the USDA (at the time of writing, there are eight).

[22] https://www.animalgenome.org/pigs/newsletter/No.049.html (last accessed 9th December 2022); https://www.animalgenome.org/pigs/newsletter/No.050.html (last accessed 9th December 2022).

Research Service—ARS—and CSREES) and took place on 19th February 2002 in Washington DC. In addition to contributions on comparative genomics, many of the discussions centred on which species should be sequenced (Pool & Waddell, 2002).

Spurred by the impending competition for resources for sequencing, the pig genome community marshalled its own efforts. Arising out of the ISAG meeting in August 2002, a permanent animal genome sequencing committee was created and a working group was tasked with writing a 'White Paper' to submit to the NIH to interest them in sequencing the pig genome. In October, the White Paper was submitted, and a 'Scientific Stakeholders meeting' of the Interagency Working Group (coordinated by Jen) was held, with Rothschild, Gary Rohrer from USDA MARC and Fuller Bazer (a Texas A&M University reproductive biologist) advocating for the pig.[23]

The White Paper was co-authored by Rohrer, Rothschild, Lawrence Schook and Jon Beever of the University of Illinois, together with Richard Gibbs and George Weinstock of the Human Genome Sequencing Center at Baylor College of Medicine. Its arguments for sequencing the pig genome heavily emphasised its potential value for human health through developing the pig as a biomedical model, and in terms of what it could contribute to human genomics.[24] The former contention drew on long-established work in shaping the pig as an animal model of disease. Schook in particular had worked in this vein, and much early research at CEA-INRA had addressed the potential of *S. scrofa* for advancing human medicine. This line of work emphasised the biomedical fruits of pig genetics research to uncover genes relevant to human disease and health, some of which had been conducted by the more agriculturally-inclined scientists and institutions.

---

[23] https://www.animalgenome.org/pigs/newsletter/No.056.html (last accessed 9th December 2022); https://www.animalgenome.org/pigs/newsletter/No.057.html (last accessed 9th December 2022).

[24] In Chaps. 6 and 8, we examine ways in which human genomics has shaped genomic research on non-human species, through the creation of infrastructures and standards as well as the norms and organisational forms to which non-human genomicists have to conform and adapt. This is in addition to the more direct impact of subordinating the genomics of non-human species to the pursuance of human genomics, either by conceiving them as pilot programmes (Chap. 2) or as sources of additional forms of data that could inform the annotation of the human reference genome.

The ability to genetically modify and clone pigs, together with the existence of mapping and DNA library resources, were adduced in support of their contention that pig genomics was sufficiently mature and ready for whole-genome sequencing. This case was further supported by the ongoing construction of a BAC fingerprint map by a consortium of INRA, the University of Illinois, USDA MARC, the Roslin Institute, the BBSRC and the Sanger Institute. The White Paper also stressed the comparative genomics expertise and knowledge built up by pig genomicists, which could provide a conduit for the translation of pig mapping and sequencing data to human genomics (Rohrer et al., 2002; García-Sancho et al., 2017, pp. 13-14).

These efforts culminated in the formation of the SGSC in 2003, with its inaugural meeting held at INRA Jouy-en-Josas in September, hosted by Schook and Patrick Chardon. In addition to researchers from many of the same mapping institutions that had come together in the 1990s on both sides of the Atlantic, representatives from China, South Korea and Japan were also present and played a significant role in the sequencing effort to come. Reflecting their growing importance in this area, agents of the USDA and the Alliance for Animal Genome Research were also in attendance. The basic principles for the operation of the Consortium, estimates for resource requirements and commitments for contributions towards the eventual project, were laid out at this meeting (Schook et al., 2005).

Although at first glance its structure and operation seemed to replicate the IHGSC, the SGSC differed in a number of important respects. While the leading institutions of the IHGSC were large-scale sequencing centres that had been either created *de novo* or considerably enhanced for the determination of the human reference genome, the SGSC's membership included many participants in the prior swine genome mapping programmes that existed long before concerted sequencing appeared on the horizon (Table 5.2).

In terms of funding, the organisations that came to support the SGSC were more agriculturally-inclined and less biomedically-oriented than the ones that underwrote the human genome coalition. The contributors to the SGSC included a lower proportion of charities, but there was a stronger presence of public and private funders connected to local economic interests, such as livestock production and breeding. Finally, the SGSC was a unified body dedicated to garnering the funds needed to sequence the whole genome of the pig, to map out the strategy and means to do so, and also to guide and involve itself in that sequencing. It was a concrete

Table 5.2 List of members of the Swine Genome Sequencing Consortium elaborated by James Lowe with data from: https://www.igb.illinois.edu/labs/schook/sgsc/index.php (last accessed 9th December 2022). Key participants in PiGMaP and USDA mapping initiatives are indicated in bold. The selection criterion for mapping participants included in this table is authorship on at least one of the following papers: Archibald et al. (1995), Yerle et al. (1995) and Rohrer et al. (1996). It should be noted that the selection criterion excludes many scientists who were involved in some way in mapping and/or sequencing. For instance, Timothy Smith was key to resequencing the pig genome (Chap. 7) but was not a member of the Consortium, and Patrick Chardon and Tosso Leeb (to pick only two examples of many possible ones) were involved in mapping and genome library creation in the 1990s, but were not authors on the three papers used for identifying mapping participants to include in this table. Compare the continuity exhibited in this table to the discontinuity in human genomics, as illustrated by the differences between the institutions listed in Table 3.1 (Chap. 3) and those listed in Table 4.1 (Chap. 4)

Members of the Swine Genome Sequencing Consortium

Gerard Albers (Nutreco)
Alan L. Archibald (Roslin Institute)
Craig W. Beattie (University of Illinois at Chicago)
Jonathan E. Beever (University of Illinois)
Mark Boggess (National Pork Board, USA)
Joseph P. Cassady (North Carolina State University)
Patrick Chardon (INRA Jouy-en-Josas)
Bhanu Chowdhary (Texas A & M University)
Kellye Eversole (Alliance for Animal Genome Research)
Merete Fredholm (The Royal Veterinary and Agricultural University, Denmark)
Greg Gibson (North Carolina State University)
Elisabetta Giuffra (Tecnoparco, Lodi, Italy)
Jan Gorodkin (The Royal Veterinary and Agricultural University, Denmark)
Ronnie Green (USDA-ARS)
Martien Groenen (Wageningen University)
Barbara Harlizius (The Institute for Pig Genetics, The Netherlands)
Debora Hamernik (USDA CSREES)
Sean Humphray (The Wellcome Trust Sanger Institute)
Steve Kappes (USDA-ARS)

Bin Liu (Beijing Genome Center)
Pramod Mathur (Canadian Centre for Swine Improvement)
Denis Milan (INRA Toulouse)
Alan Mileham (Sygen)
Sung-Jong Oh (National Livestock Research Institute, Korea)
Anna Palmisano (USDA CSREES)
F.A. Ponce de Leon (University of Minnesota)
Muquarrab Qureshi (USDA CSREES)
Jane Rogers (The Wellcome Trust Sanger Institute)
Gary Rohrer (USDA MARC)
Max F. Rothschild (Iowa State University)
Lawrence Schook (University of Illinois)
Paul Sundberg (National Pork Board, USA)
Tosso Leeb (University of Bern)
Hirohide Uenishi (National Institute of Agrobiological Sciences, Japan)
John Webb (Maple Leaf Foods)
Alan Wildeman (University of Guelph)
Ming-Che Wu (Taiwan Livestock Research Institute)
Hiroshi Yasue (National Institute of Agrobiological Sciences, Japan)

---

entity from the very beginning, something that had been missing from the human genome effort, with the IHGSC largely being a retrospectively established name for a coalition that had emerged during the second half of the 1990s and into the 2000s (Chap. 4).

At its launch, the SGSC believed that a sum in the range of 50 million dollars would be required.[25] Fortunately for the pig genomicists, this proved to not be the case, as the body that could provide funds on such a scale—the NIH National Human Genome Research Institute—did not prioritise *S. scrofa* as a sequencing target, focusing instead on cattle as its chosen agriculturally-important species, in part because of backing from the cattle industry and the rapid progress that was being made as a result (Chaps. 6 and 7).[26]

As they were unsuccessful in attracting NIH funds for whole-genome sequencing, they turned their attention to generating other key resources—such as the BAC fingerprint map and Expressed Sequence Tags. They also focused on further exploiting what they already had, using smaller pots of money and collaborating in a similar way to how they had been doing previously. Alongside this, efforts to secure funds from the USDA to sequence the whole pig genome continued. The existing genomic efforts and the support of companies and pork industrial boards helped, as did the ongoing connections with USDA officials, the assistance of the local congressman for the University of Illinois at Urbana-Champaign, Representative Timothy Johnson, and the lobbying of the Alliance for Animal Genome Research.

This bore fruit in 2006, when Jen, just before leaving his post as Under Secretary at the USDA, approved $10 million for sequencing the pig genome (formally awarded to the University of Illinois), which was signed off by the then Secretary of Agriculture, Mike Johanns. Increasing automation and refinement of sequencing processes had reduced costs and therefore lowered the barriers for the full genome sequencing of less well-funded species, but the required investment was still substantial. The funds from the USDA were complemented by additional resources provided by Iowa State University and North Carolina State University, as well as industry bodies: the National Pork Board, the Iowa Pork Producers Association and the North Carolina Pork Council. The other institutions

---

[25] https://www.animalgenome.org/pigs/newsletter/No.063.html (last accessed 9th December 2022).

[26] Interview conducted with Lawrence Schook over Skype by James Lowe, January 2017.

involved in the SGSC brought their own resources to bear on the overall programme, once again drawing on grants to perform particular pieces of research and create new resources (Lowe, 2018).[27] Key to the USDA's support was the demonstration that the community of pig genomicists was united behind one project and that the initiative had international buy-in. The existing international basis of the community helped, as did the agreement of a separate Sino-Danish collaboration to contribute data from what had threatened to be a rival project.[28]

Schook was appointed co-director of the SGSC alongside Mike Stratton, an expert in cancer genetics at the Sanger Institute. Schook approached pig genomics from the direction of establishing *S. scrofa* as an animal model of disease, but like many of the other members of the community, his genetic and genomic research led him to work towards multiple domains of application. Like the USDA MARC researchers, he relied on de Jong for the construction of the library that was mainly used in the concerted projects to physically map and then sequence the whole pig genome. In 2000, de Jong's team had moved from RCPI to the Children's Hospital Oakland Research Institute (CHORI), located on the other coast of the USA within the San Francisco Bay Area. There they led the BACPAC Resources Center, a unit specialised in the mass production and distribution of DNA libraries.[29]

De Jong had approached Schook at the January 2002 Plant and Animal Genome conference, with news that he had received funding to construct a pig DNA library. De Jong suggested that an inbred female pig be used, as this would have two copies of the X chromosome, no Y chromosome (as these are notoriously difficult to deal with), and reduced heterozygosity— the variation between each chromosome in a pair. Schook had access to a reference family that had been constructed at the University of Illinois at

---

[27] USDA News Release, January 13th 2006. USDA AWARDS $10 MILLION TO SEQUENCE THE SWINE GENOME. In "Pig Sequencing" folder of Alan Archibald's personal archive, obtained 17th May 2017.

[28] Interview conducted with Lawrence Schook over Skype by James Lowe, August 2017.

[29] In 2019, the libraries and other resources started to be distributed by a company, BACPAC Genomics, rather than by CHORI. https://bacpacresources.org/home.htm (last accessed 9th December 2022).

Urbana–Champaign for the purposes of mapping QTL.[30] One of the pigs in that family was more inbred than any other and would therefore be more homozygous and amenable to library production: a Duroc (North American domestic breed) sow born in 2001. Once Schook and his colleague Jon Beever had decided to use her, they sent de Jong 250 millilitres of her blood as requested, and from this the DNA was extracted from white blood cells to produce the CHORI-242 BAC library, as well as another fosmid library also used in sequencing.[31]

The sow chosen by Schook and de Jong had a name: TJ Tabasco (Fig. 5.4, top). Oddly, she was named after her offspring, who were clones of her. TJ Tabasco was an acronym of the first letters of the names given to nine of them, deriving from animated characters: Tinker Bell, Jasmine, Tiana, Aurora, Belle, Ariel, Snow White, Cinderella and Olivia (Fig. 5.4, bottom).

These DNA libraries and other reference resources were used in physical mapping at The Keck Center for Comparative and Functional Genomics of the University of Illinois at Urbana–Champaign, the French national sequencing centre Genoscope, and the Sanger Institute. Both Genoscope and the Sanger Institute were prominent IHGSC members. The Sanger Institute had been the second largest contributor to the draft human reference sequence published in 2001, and Genoscope was the seventh largest contributor (and second most productive European centre). The Sanger Institute was also contracted by the SGSC to determine the genome sequence and undertake some of the work that transformed this string of nucleotides into an assembled and fully annotated reference genome, as we see in Chap. 6.[32] The physical map and the BAC libraries were used as the basis for the main part of the sequencing of the reference genome at

---

[30] The construction of such families had a long history at many of the institutions that form part of the community of pig genomicists: agricultural research institutions like INRA and the Roslin Institute, and the land grant universities of the US. They keep herds and flocks of farm animals for research purposes. This was a legacy of their previously more direct role in breeding, but it became useful with the advent of genomics research.

[31] Interview conducted with Lawrence Schook over Skype by James Lowe, August 2017. A fosmid library uses a bacterial F-plasmid as the vector. The one discussed here was not named, nor became a product more widely distributed by the BACPAC Resources Center.

[32] Genoscope had been created in 1996 with government funding with the remit of building on the prior, charity-sponsored work at Généthon by Cohen and other large-scale genome mappers. Baylor College had been considered to perform the sequencing, but due mainly to the pig genomics community's existing relationship with the Sanger Institute, and the more comprehensive range of services offered there, the latter was chosen.

Fig. 5.4 Top: TJ Tabasco, as preserved on the wall of Lawrence Schook's office at the University of Illinois at Urbana–Champaign. Bottom: Cloned offspring of TJ Tabasco. Cultures of foetal fibroblast cells and tissues at different developmental stages were derived from these pig clones and used to construct whole-genome shotgun libraries at the Sanger Institute and cDNA libraries. The cDNA libraries were used for the annotation of the reference sequence that was derived chiefly from DNA obtained from their mother. Photographs courtesy of Lawrence Schook

the Sanger Institute. This reference sequence and the physical map were largely derived from the CHORI-242 library that de Jong had produced for Schook, alongside the fosmid library from TJ Tabasco and the three other BAC libraries mentioned, produced by CEA-INRA, the Roslin Institute, and de Jong's group at USDA MARC's request (Schook et al., 2005; Humphray et al., 2007; Lowe, 2018). Annotation of the reference sequence made use of the cDNA libraries derived from the cultures of TJ Tabasco's clones (Fig. 5.4).

In its overall strategy, the SGSP operated in a way that reflected the original plan for the determination of the human reference genome: a distinct physical mapping stage that preceded and informed subsequent hierarchical shotgun sequencing. The sequencing of the DNA from the libraries—again, chiefly CHORI-242—was almost exclusively undertaken by the Sanger Institute, using the factory-style methods refined in human genome sequencing. Yet beyond the mere determination of sequence data, there was substantial input from the rest of the SGSC members. First, the participating laboratories played a crucial role in augmenting the initial stitching together—assembly—of the sequenced clones into larger and more contiguous stretches of sequence across the whole genome. Secondly, a number of Chinese and Danish institutions provided key additional data for the assembled genome using next-generation sequencing (Wernersson et al. 2005).[33] Thirdly, this and other supplemental sequencing, along with the prior mapping and knowledge of genetic diversity across breeds possessed by the community, informed the analyses comparing the genomes of different breeds of pig that accompanied the 2012 *Nature* paper describing the reference genome.[34]

There was a distinct role for the Sanger Institute in this project, compared with its roles in *S. cerevisiae* sequencing and the IHGSC. Contrary to the case of the yeast genome project—where the Sanger Institute was a participant—and the IHGSC effort that the Sanger Institute had helped to coalesce, the Sanger Institute here opened up to and worked with a *separate* community. The pig genomicists were able to take advantage of the *repertoires* established at the Sanger Institute to process DNA libraries,

---

[33] These institutions were the Beijing Genomics Institute in China, and in Denmark: the Center for Biological Sequence Analysis at the Technical University of Denmark; University of Aarhus; The Royal Veterinary and Agricultural University; and Danish Institute of Agricultural Sciences. https://rth.dk/resources/piggenome/ (last accessed 9th December 2022).

[34] The supplemental sequencing included the whole-genome shotgun approach that Celera pursued. See Lowe (2018) for an account of the sequencing of the pig genome; here we concentrate on some key aspects, especially those with which we can make salient comparisons to the processes evident in human and yeast sequencing.

sequence DNA, assemble sequences and validate the results. Rachel Ankeny and Sabina Leonelli (2016, p. 19) deploy the term repertoire to mean "the material, social, and epistemic conditions under which individuals are able to join together to perform projects and achieve common goals, in ways that are relatively robust over time despite environmental and other types of changes, and [that] can be transferred to and learnt by other groups interested in similar goals". In this case, however, other groups interested in different goals beyond the production of a reference genome participated in and helped to direct the repertoires established at the Sanger Institute. The availability of a comprehensive physical map and previous genetics research allowed members of the community to request that certain areas of the genome be given special attention, for example, with targeted sequencing of those areas at higher levels of coverage conducted by the Sanger Institute.

Alongside this, the fragments of determined sequence were joined together—assembled—using the previously elucidated physical map, which indicated the order and relative positions of clones derived from the DNA libraries. As had been the case in the IHGSC effort, the software package PHRAP was used to assemble the pig sequence data generated at the Sanger Institute into—in this case—279 sets of overlapping contiguous sequences known as 'contigs'. Using methods and pipelines developed for human genome sequencing, workers at the Sanger Institute then applied automated pre-finishing procedures and closed remaining gaps with selective sequencing of BAC clones that were known to span them. The Sanger Institute made checks of coverage, extent and contiguity, and the pig genomics community themselves contributed to checking and correcting the provisional assembly so produced by the Sanger Institute. One way that they did this was to check the orientation and order of *scaffolds* containing contiguous sequence using a previous physical map.[35] Greater conformity with this map in a newer assembly (or *build*) was adduced as evidence that it constituted an improvement (Groenen et al., 2012, Supplementary Information).

Members of the pig genomics community were granted access to the Genome Evaluation Browser (gEVAL) that had been created and was maintained by the Genome Reference Informatics Team (GRIT) at the Sanger Institute. This enabled them to view the assembly, assess its

---

[35] This physical map was produced using a method called radiation hybrid mapping. On this method and its role in pig genomics, see Lowe (2022).

accuracy and communicate their findings to GRIT. This process drew upon the community's more general knowledge of the structure and nature of the pig genome, detailed knowledge of particular regions, and their facility in exploiting human genome data to inform their assessment of the *S. scrofa* sequence. In some cases, assembly errors identified by members of the community were used to amend algorithms that were deployed in the process of constructing genome assemblies (Lowe, 2018). We explore these relationships and interactions between the members of the pig genome community and the specialist genomic labour and pipelines at the Sanger Institute in more detail in Chap. 6.

The immediate output of this was an imperfect frozen abstraction, the representative reference genome of the pig, which could be continually annotated further, and eventually replaced by other frozen abstractions: newer versions of the representative reference genome. In terms of variation, the reference genome looked to be just as limited in scope as the human and yeast genomes. Yet, due to the nature of its construction—in particular, the involvement of the existing community of pig genomicists—a significant constituency of users were acutely aware of the variation included in—and excluded from—the reference sequence. Like the human genome, it was substantially based on DNA from a single individual, but for the pig genomics community, it constituted a resource to be built on and linked to others, just as with their previous efforts and outputs. It was not supposed to represent the species in all its variation and diversity. They were aware of this and compensated for it.

5.3   Reference Genomes and Their Affordances

The sequencing of the pig genome appears to represent a continuation of the tendency towards the concentration of reference sequence producers that we have outlined in the preceding chapters of this book. While in the early-to-mid 1990s, initiatives to complete the yeast genome combined distributed and concentrated approaches to the determination of the reference sequence—represented by the EC and US programmes, respectively (Chap. 2)—towards the end of the twentieth century the IHGSC consolidated an intensive, large-scale production system that was embodied in the genome centres (Chap. 4). Ten years later, in the mid-to-late 2000s, the role of the Sanger Institute in the SGSC implies an even more concentrated production model in which only one genome centre was sufficient to determine the full reference sequence of *S. scrofa,* as opposed to twenty in the human genome effort.

Yet, the involvement of pig geneticists in the production and continuous adaptation of their reference genome—something that we explore further in Chap. 6—compensated for the delegation of part of the production process to the Sanger Institute. As a result, this community held a different perception of the reference genome that they helped to create than the yeast and human genomicists did for their reference genomes (Table 5.3).

The yeast genome was produced as a community resource to be shared by geneticists, biochemists and cell biologists interested in the study of this single-celled organism. Key to its production was agreement about the suitability of using the S288C strain of *S. cerevisiae* as a model to investigate the workings of genes and cellular processes in eukaryotic organisms such as yeast. S288C had a prior history of use in genetic experimentation and shared genetic linkage and physical maps of it existed when the genome sequencing efforts started in 1989. This eased the convergence of the different communities of yeast geneticists and biologists around the objective of genome sequencing (Szymanski et al., 2019; Vermeulen & Bain, 2014).

The S288C strain thus became the glue that aligned the heterogeneous institutions and differential sequencing practices of the distributed approach promoted by the EC's programme. The presumed—or intended—invariance of this strain also allowed the meshing of the data produced by the European consortium with that generated by the American and Asian institutions also involved in yeast genome sequencing. S288C was used as a fundamental common object for investigating the biology and genetics of yeast by its communities of researchers. Not surprisingly, once its reference sequence was produced in 1996, the further functional explorations of the *S. cerevisiae* genome—among them the EUROFAN project—relied on the same strain and were undertaken by largely the same participants as the genome sequencing programmes (Chap. 7).

The case of the human genome is quite different. Here, the strategy propounded by the leaders of the genome centres that were being established during the 1990s, departed from existing chromosome mapping practices of human and medical geneticists. These genome centre leaders belonged to a new generation of researchers that were supported by both rising biomedical funders—the Wellcome Trust—and influential scientific celebrities such as Nobel laureate molecular biologist James Watson. The younger breed of researchers and their supporters formulated a vision of producing a map and reference sequence of the entire human genome. These reference resources would represent the species as a whole and, as

Table 5.3 A comparison of the yeast, human and pig reference genomes (elaborated by both authors)

	Yeast	*Human*	*Pig*
Mapping communities	Concentrated in a few respected genetics and molecular biology groups (Robert Mortimer, Maynard Olson) that freely disseminated data across the community.	Distributed across human and medical geneticists who pooled data in Chromosome Workshops and collaborative human genome programmes (e.g. Human Genome Analysis Programme).	Distributed across animal geneticists and immunogeneticists who shared data and materials in concerted genome mapping programmes (e.g. PiGMaP).
Sequencing communities	Different from prior mappers. Geographical asymmetries: genome centres in the USA-UK; a heterogeneous consortium of yeast geneticists, biochemists, cell biologists and small companies under the EC.	International Human Genome Sequencing Consortium (formed by genome centres, administrative agencies and bioinformatics institutes; did not include a substantial part of the chromosome mapping communities).	Swine Genome Sequencing Consortium (included mapping communities plus a genome centre as an external contractor).
Relationship of maps to sequence	Previous comprehensive genetic linkage and physical mapping. Bespoke physical maps constructed for each chromosome to aid the sequencing operation.	Whole-genome physical mapping conducted alongside sequencing to aid assembly and annotation. Sequencing drew upon both whole-genome and chromosome-specific physical mapping.	Sequential: whole-genome physical mapping used as the basis for the selection of clones to sequence, and then to guide the assembly of the sequence.

(continued)

Table 5.3 (continued)

	Yeast	Human	Pig
Presumed representativeness of the genome	Specific *Saccharomyces cerevisiae* strain (S288C). The sequencing community were aware of the partiality of this, and the unusual nature of this specialist laboratory strain.	Aimed to represent *Homo sapiens* as a whole. DNA used in the sequencing was derived from a limited number of individuals without an attempt made to reflect human diversity. At the time, this was not thought to be a problem for the representativeness of the resulting reference sequence.	Derived from a single pig (TJ Tabasco) with some additional use of libraries derived from the DNA of multiple breeds. Pig genomicists had an instrumental conception of the extent to which the reference sequence represented *Sus scrofa* as a whole.

with the map and sequence of the yeast genome, would enable researchers to address the molecular basis of fundamental life processes, including pathologies and development from embryo to adult. Unlike the yeast efforts, though, the human genome did not correspond to a specific 'strain' of *H. sapiens*—it was a vaguer abstraction than that. Furthermore, the human reference sequence was determined by a less heterogeneous set of institutions and techniques: the IHGSC coalition of genome centres deploying industrial modes of data production and processing, supported by administrative agencies and bioinformatic infrastructures (Chap. 4).

The vision of the genome centre coalition clashed with the approach of human and medical geneticists, for whom only the genome regions that varied between healthy individuals and those suffering from genetic diseases deserved attention.[36] Prominent and long-serving geneticists based

[36] There were intermediate actors and communities between these two extreme positions. One of them was the French Muscular Dystrophy Association, which provided charitable funding for Généthon. Within its human mapping work, Généthon combined a whole-genome operation with a more targeted, *à la carte* role serving the demands of medical geneticists (Kaufmann, 2004, see also Chap. 3). This dual role was also undertaken in the early years of the US genome centres. Their targeted, disease-oriented projects were, however, increasingly discontinued in the mid-to-late 1990s, and outsourced to associated laboratories or spin-off companies in which genome centre scientists were involved (Hilgartner, 2017).

in hospitals and medical schools, such as Victor McKusick and Walter Bodmer, had dominated early discussions of genomics as a nascent discipline during the mid-to-late 1980s. They had also served as coordinators and advisors in the first concerted initiatives to map and sequence the human genome, which largely adopted the distributed, networked approach of the EC's yeast genome programme—with the notable exception of the national human genome effort in the USA (Chap. 3). Throughout the 1990s, however, the influence that these reputed medical geneticists had in the new sequencing centres declined, which was reflected in their peripheral involvement in the IHGSC. The differences between the two groups inhibited the involvement of a large part of the human and medical genetics communities in the production of the reference sequence and hindered the subsequent clinical exploitation of this resource. For a substantial fraction of medically-oriented geneticists, the lineages of the IHGSC map and sequence—in terms of the human populations they represented, or their associations with previous maps of healthy and diseased individuals—were blurry and difficult to reconstruct. This led to them preferring to keep using their more chromosome or region-specific, locally-produced and clinically-targeted resources: maps and sequences extracted from hospital patients that were compared to control data from persons not affected by the conditions being studied.

Pig genomics represents a third, distinct case. The involvement in the SGSC of the communities that had coalesced during projects to systematically map the pig genome meant that the resulting reference sequence would only ever be seen as an arbitrary abstraction from the known or supposed genetic diversity of pigs.

Crucially, unlike in the yeast and human reference sequence efforts, *S. scrofa* mapping and sequencing was never presumed to be comprehensive or complete. Satisficing according to proximate concrete translational goals was the aim of pig genomicists, as it had been for medical geneticists involved in the HGMP or other early human genome programmes. Due to this, the swine reference sequence was perceived as a dynamic resource that would qualitatively change when updated. This resource would also form the basis for the creation of new reference resources incorporating different variation, as the objectives of its user communities evolved. This continuous iterative adjustment was reflected in practices such as annotation of the reference sequence with data concerning immune response genes or other traits of interest for pig geneticists and the livestock breeding industry and the creation of new datasets cataloguing inter-breed variability. These practices were regarded as part of the ongoing production and use of the reference sequence, rather than as an appendix or postscript to be added once the genome was 'finished'. As we see in the next chapter, this led the collaboration between the Sanger Institute and the other SGSC members to be extended in order to develop the annotation of the pig sequence so it could be aligned with numerous and changing research priorities (Chap. 6).

Key aspects of the production and nature of the pig genome remain invisible if its story is just told from the perspective of the sequencing work conducted at the Sanger Institute using a small set of libraries largely drawn from one highly-inbred pig. The other consortium members contributed an array of practices—from mapping to assembly and annotation—that were crucial in augmenting and transforming this partial sequence into a usable genome.[37] They conducted these in conversation with the Sanger Institute, which opened itself up to the input of this community and changed the way key parts of its specialised data production and processing pipelines worked. This contribution by pig geneticists provided them with a perspective on their reference genome that human geneticists lacked. For human and medical geneticists, their relative absence from the IHGSC effort complicated their ability to link the reference sequence to data they routinely produced about clinically-relevant variation.

These differences suggest that the master narrative of genomics—centred on the production of the human reference sequence but with that presumed to stand in for genomics as a whole—constitutes an artifactual historical representation. When the historical lens goes beyond the mere compilation of reference sequences, genomics research emerges as a broader enterprise that diverges from this canonical trajectory. The history of the production of less well-known—and especially non-human—genomes enables us to better discern connections between the creation of reference sequences and pre-existing practices and communities engaged with genome mapping. In the case of *S. scrofa*, these practices and

---

[37] The way that the Sanger Institute operated can be conceptualised as "horizontal sequencing", a form of work concerned with the determination of a one-dimensional string of nucleotides. As well as human and medical geneticists, the SGSC members engaged in "vertical sequencing", a strategy that addresses smaller areas of the genome but takes into account variation beyond the single dimension of a reference sequence. On this horizontal-vertical distinction as applied to human genomics, see García-Sancho and Leng et al. (2022). Elsewhere, we have conceptualised the range of activities that go into creating a usable reference genome in terms of "thick" as opposed to "thin" sequencing (Lowe, 2018).

communities continued shaping the string of nucleotides produced at the Sanger Institute and its representation in data infrastructures. When it came to the determination of the pig reference sequence, the range of actors and activities did not narrow as radically as they did in the IHGSC effort. The participation of an established community of researchers in the whole-genome sequencing effort with the Sanger Institute—which itself had further evolved as a specialist genome centre—accounts for the more direct and contextualised use of the sequence data by the pig geneticists, especially compared to the more peripheral human and medical geneticists. The next part of the book explores how this community involvement—and its differing extents and nature in yeast, human and pig genomics—shapes annotation and other post-reference genomic practices (Chaps. 6 and 7).

The extreme filtering out of much of the variety of pre-reference genomic research in the IHGSC is shown to be exceptional when lineages and connections between earlier and later genomic research are considered, for instance between the early pig mapping programmes and what has commonly been called "post-genomics" (Richardson & Stevens, 2015). We show that, outside the success narrative of the IHGSC, the advent of a reference sequence does not by itself create the post-genomic world. Furthermore, when other species such as *S. scrofa* are considered, continuities—from before the reference sequence was determined to after it—can be discerned for communities, practices, resources, knowledge and objectives. Taking these points on board transforms the history of genomics into a dynamic and recursive field rather than a dichotomous, linear and teleological space punctuated by the *completion* of reference sequences.

References

Agar, J. (2019). *Science Policy Under Thatcher*. UCL Press.

Al-Bayati, H. K., Duscher, S., Kollers, S., Rettenberger, G., Fries, R., & Brenig, B. (1999). Construction and characterization of a porcine P1-derived artificial chromosome (PAC) library covering 3.2 genome equivalents and cytogenetical assignment of six type I and type II loci. *Mammalian Genome, 10*, 569–572.

Alexander, L. J., Smith, T. P. L., Beattie, C. W., & Broom, M. F. (1997). Construction and characterization of a large insert porcine YAC library. *Mammalian Genome, 8*, 50–51.

Anderson, S. I., Lopez-Corrales, N. L., Gorick, B., & Archibald, A. L. (2000). A large-fragment porcine genomic library resource in a BAC vector. *Mammalian Genome, 11*, 811–814.

Ankeny, R. A., & Leonelli, S. (2016). Repertoires: A post-Kuhnian perspective on scientific change and collaborative research. *Studies in History and Philosophy of Science Part A, 60*, 18–28.

Archibald, A. L., Haley, C. S., Brown, J. F., Couperwhite, S., McQueen, H. A., Nicholson, D., et al. (1995). The PiGMaP consortium linkage map of the pig (*Sus scrofa*). *Mammalian Genome, 6*, 157–175.

Bonneuil, C., & Thomas, F. (2009). *Gènes, pouvoirs et profits: Recherche publique et régimes de production des savoirs de mendel aux OGM*. Éditions Quae.

Bruce, A., & Lowe, J. W. E. (2022). Pigs and Chips: The making of a biotechnology innovation ecosystem. *Science & Technology Studies*. https://doi.org/10.23987/sts.111111

Chardon, P., Vaiman, M., Kirszenbaum, M., Geffrontin, C., Renard, C., & Cohen, D. (1985). Restriction fragment length polymorphism of the major histocompatibility complex of the pig. *Immunogenetics, 21*, 161–171.

Curry, H. A. (2017). Breeding uniformity and banking diversity: The genescapes of industrial agriculture, 1935–1970. *Global Environment, 10*(1), 83–113.

Curry, H. A. (Ed.). (2019). Special issue: The conservation of plant genetic resources. *Culture, Agriculture, Food and Environment, 41*(2), 73–116.

Fahrenkrug, S. C., Rohrer, G. A., Freking, B. A., Smith, T. P. L., Oswoegawa, K., LiShu, C., et al. (2001). A porcine BAC library with tenfold genome coverage: A resource for physical and genetic map integration. *Mammalian Genome, 12*, 472–474.

García-Sancho, M., Leng, R., Viry, G., Wong, M., Vermeulen, N., Lowe, J.W.E. (2022). The Human Genome Project as a singular episode in the history of genomics. *Historical Studies in the Natural Sciences, 52*, 320–360.

García-Sancho, M., & Myelnikov, D. (2019). Between mice and sheep: Biotechnology, agricultural science and animal models in late-twentieth century Edinburgh. *Studies in History and Philosophy of Biological and Biomedical Sciences, 75*, 24–33.

García-Sancho, M., Myelnikov, D., & Lowe, J. W. E. (2017). *The invisible history of the visible sheep: How a look at the past may broaden our view of the legacy of Dolly*. Report for the UK Biotechnology and Biological Sciences Research Council (BBSRC). Retrieved December 9, 2022, from http://tinyurl.com/dollyreport

Groenen, M. A., Archibald, A. L., Uenishi, H., Tuggle, C. K., Takeuchi, Y., Rothschild, M. F., et al. (2012). Analyses of pig genomes provide insight into porcine demography and evolution. *Nature, 491*(7424), 393–398.

Haley, C., & Visscher, P. M. (1998). Strategies to utilize marker-Quantitative Trait Loci Associations. *Journal of Dairy Science, 81*(2), 85–97.

Hilgartner, S. (2017). *Reordering life: Knowledge and control in the genomics revolution*. The MIT Press.

Hogan, A. J. (2016). *Life histories of genetic disease: Patterns and prevention in postwar medical genetics*. The Johns Hopkins University Press.

Humphray, S. J., Scott, C. E., Clark, R., Marron, B, Bender, C., Camm, N., et al. (2007). A high utility integrated map of the pig genome. *Genome Biology, 8*, R139.

Kaufmann, A. (2004). Mapping the human genome at Généthon laboratory: The French Muscular Dystrophy Association and the politics of the gene. In J.-P. Gaudillière & H.-J. Rheinberger (Eds.), *From molecular genetics to genomics: The mapping cultures of twentieth-century genetics* (pp. 129–157). Routledge.

Landecker, H. (2010). *Culturing life: How cells became technologies*. Harvard University Press.

Lowe, J. W. E. (2018). Sequencing through thick and thin: Historiographical and philosophical implications. *Studies in History and Philosophy of Biological and Biomedical Sciences, 72*, 10–27.

Lowe, J. W. E. (2021). Adjusting to precarity: How and why the Roslin Institute forged a leading role for itself in international networks of pig genomics research. *The British Journal for the History of Science, 54*(4), 507–530.

Lowe, J. W. E. (2022). Humanising and dehumanising pigs in genomic and transplantation research. *History and Philosophy of the Life Sciences, 44*, 66.

Lowe, J. W. E., & Bruce, A. (2019). Genetics without genes? The centrality of genetic markers in livestock genetics and genomics. *History and Philosophy of the Life Sciences, 41*, 50.

Lowe, J. W. E., Leng, R., Viry, G., Wong, M., Vermeulen, N., & García-Sancho, M. (2022). The bricolage of pig genomics. *Historical Studies in the Natural Sciences, 52*, 401–442.

Lunney, J. K., Hob, C.-S., Wysockia, M., & Smith, D. M. (2009). Molecular genetics of the swine major histocompatibility complex, the SLA complex. *Developmental and Comparative Immunology, 33*, 362–374.

Lyon, M. F. (2002). A personal history of the mouse genome. *Annual Review of Genomics and Human Genetics, 3*, 1–16.

Meuwissen, T. H. E., Hayes, B. J., & Goddard, M. E. (2001). Prediction of total genetic value using genome-wide dense marker maps. *Genetics, 157*, 1819–1829.

Myelnikov, D. (2017). Cuts and the cutting edge: British science funding and the making of animal biotechnology in 1980s Edinburgh. *The British Journal for the History of Science, 50*(4), 701–728.

Niu, D., Wei, H. J., Lin, L., George, H., Wang, T., Lee, I. H., et al. (2017). Inactivation of porcine endogenous retrovirus in pigs using CRISPR-Cas9. *Science, 357*(6357), 1303–1307.

Paigen, K. (2003a). One hundred years of mouse genetics: An intellectual history. I. The classical period (1902–1980). *Genetics, 163,* 1–7.

Paigen, K. (2003b). One hundred years of mouse genetics: An intellectual history. II. The molecular revolution (1981–2002). *Genetics, 163,* 1227–1235.

Paszek, A. A., Schook, L. B., Louis, C. F., Mickelson, J. R., Flickenger, G. H., Murtaugh, M. A., et al. (1995). First international workshop on porcine chromosome 6. *Animal Genetics, 26,* 377–401.

Peres, S. (2016). Saving the gene pool for the future: Seed banks as archives. *Studies in History and Philosophy of Biological and Biomedical Sciences, 55,* 96–104.

Pool, R., & Waddell, K. (2002). *Exploring horizons for domestic animal genomics: Workshop summary.* National Academy Press.

Rader, K. (2004). *Making mice: Standardizing animals for American biomedical research, 1900–1955.* Princeton University Press.

Richardson, S. S., & Stevens, H. (Eds.). (2015). *Postgenomics: Perspectives on biology after the genome.* Duke University Press.

Rogel-Gaillard, C., Bourgeaux, N., Billault, A., Vaiman, M., & Chardon, P. (1999). Construction of a swine BAC library: Application to the characterization and mapping of porcine type C endoviral elements. *Cytogenetics and Cell Genetics, 85,* 205–211.

Rogel-Gaillard, C., Bourgeaux, N., Save, J. C., Renard, C., Couilin, P., Pinton, P., et al. (1997). Construction of a swine YAC library allowing an efficient recovery of unique and centromeric repeated sequences. *Mammalian Genome, 8,* 186–192.

Rohrer G. A., Alexander, L. J., Hu, Z., Smith, T. P., Keele, J. W., & Beattie, C. W. (1996). A comprehensive map of the porcine genome. *Genome Research, 6*(5), 371–391.

Rohrer, G., Beever, J. E., Rothschild, M. F., Schook, L., Gibbs, R., & Weinstock, G. (2002). *Porcine sequencing white paper: Porcine Genomic Sequencing Initiative.* Retrieved December 9, 2022, from https://core.ac.uk/download/pdf/212813224.pdf

Schook, L. B., Beever, J. E., Rogers, J., Humphray, S., Archibald, A., Chardon, P., et al. (2005). Swine Genome Sequencing Consortium (SGSC): A strategic roadmap for sequencing the pig genome. *Comparative and Functional Genomics, 6*(4), 251–255.

Szymanski, E., Vermeulen, N., & Wong, M. (2019). Yeast: One cell, one reference sequence, many genomes? *New Genetics and Society, 38*(4), 430–450.

Thorsby, E. (2009). A short history of HLA. *Tissue Antigens, 74,* 101–116.

Vermeulen, N., & Bain, M. (2014). Little cell, big science: The rise (and fall?) of yeast research. *Issues in Science and Technology, 30*(4), 38–46.

Wimsatt, W. C. (2007). *Re-engineering philosophy for limited beings: Piecewise approximations to reality.* Harvard University Press.

Yang, L., Güell, M., Niu, D., George, H., Lesha, E., Grishin, D., et al. (2015). Genome-wide inactivation of porcine endogenous retroviruses (PERVs). *Science, 350*(6264), 1101–1104.

Yerle, M., Lahbib-Mansais, Y., Mellink, C., Goureau, A., Pinton, P., Echard, G., et al. (1995). The PiGMaP consortium cytogenetic map of the domestic pig (*Sus scrofa domestica*). *Mammalian Genome, 6*, 176–186.

PART III

Contextualising and Enhancing Reference Genomes

CHAPTER 6

Making Reference Genomes Useful: Annotation

This chapter looks at the processes of annotation: the identification and adding of biologically-relevant information to the reference genome, which can then be visualised in genome browsers, with the annotations aligned against the reference sequence itself. Annotation is both a key part of the creation of a reference genome and a definitional criterion of being a designated reference genome in the RefSeq database. It is the way in which the data produced in genomics are linked with the concerns and interests of the empirical life sciences and particular problems that motivate the work of specific communities: what historian Jon Agar (2012, 2020) has termed "working worlds".

This chapter demonstrates that the establishment of ever-more automated and refined pipelines incorporating multi-dimensional data—including cross-species comparative and 'beyond the genome' data such as protein sequences—was only part of the story of the development of genome annotation. We show that the manner in which annotation has developed was affected by: the ways in which the algorithms, protocols and operations of these pipelines were configured and improved; how they related to practices of *manually* annotating genomes; and the role played by the interactions of specialist genomicists with particular research communities. These factors were also pertinent to shaping what got annotated, how, and what use was made of the resulting enriched reference resources.

© The Author(s) 2023   205
M. García-Sancho, J. Lowe, *A History of Genomics across Species,
Communities and Projects*, Medicine and Biomedical Sciences in
Modern History, https://doi.org/10.1007/978-3-031-06130-1_6

We show that different models of annotation are shaped by the relationship between reference sequence production efforts and the nature of the involvement of different communities converging around the genomes of particular species. Pig genome annotation, as a collaboration between the community of pig genomicists outlined in Chap. 5 and a well-developed annotation infrastructure at the Sanger Institute, differed in its nature and outcomes from yeast and human genomics. In yeast genomics, the community of yeast biologists was intimately involved in the reference genome production, while the initial annotation of the genome was orchestrated by the central bioinformatics coordinator, the Martinsried Institute for Protein Sequences.[1] In human genomics, two models existed: one involved the creation of high-throughput annotation pipelines at the institutions participating in the International Human Genome Sequencing Consortium (IHGSC), while the other—developed by their rival, the company Celera Genomics—was more open to input from prospective sequence users. In the former case, as with the reference genome sequencing, the medical genetics community was largely uninvolved. In the latter case, a subset of this medical genetics community was brought into the fold and contributed towards the realisation of a product—an annotated genome—distinct from that emanating from the large-scale sequencing centres leading the IHGSC effort.

One key commonality between the multiple species we have examined is the involvement of the Sanger Institute. In the previous chapter, we saw how the Sanger Institute's relationships with the different species communities varied in important and consequential respects. In this chapter, we show how the relationship of the Sanger Institute to the existing pig genetics community, already particularly close during the production of the *Sus scrofa* reference genome, was even more entangled for the annotation of the resulting sequence. This annotation used data from prior annotation and sequencing (in particular of the human genome) and availed itself of the Sanger Institute's infrastructures and procedures (pipelines) developed through human (and pre-human) sequencing projects. However, this annotation effort also had crucial input from the pig genomics community, whose members played a significant role in manually annotating the genome, confirming the automated annotations of the Sanger Institute, and contributing to an already-established panoply of comparative resources, empirical data and theoretical insights. Rather than

---

[1] Yeast genomicists often refer to annotation as *sequence analysis* or *functional analysis* of the genome. The term *annotation* is more uniformly preferred in human and pig genomics.

just being a large-scale data producer, the Sanger Institute features here as a collaborator, facilitator, trainer and provider of quality assurance, as well as the manager of various data infrastructures.

This changing role exhibited by the Sanger Institute enables us to show that the story of increasingly automated and data-intensive annotation pipelines merely corresponds to *some* of the ways in which the IHGSC institutions operated. We demonstrate that a broader multi-species approach to examining the history of annotation practices helps us to notice strategies that connect to the working worlds of the communities using the sequence data. This allows us to disclose the activities of communities that had long been generating and interpreting sequences, and to incorporate their trajectories into the history of the production of reference genomes.

6.1   Annotation: Pipelines and Jamborees

6.1.1   What Is Annotation and How Does It Contribute to the Production of a Usable Reference Genome?

Broadly speaking, annotation is the marking of features of interest in the abstract landscape of the sequences of nucleotides. Typically, representations of the genome accessible to researchers and the lay public are in the form of a browser, a window in which the user can select or deselect different features and modes of presentation of the genome to be conveyed to them (Fig. 6.1). The different selected features are aligned vertically next to a horizontal representation of the strands of the chromosome, which depicts the order of nucleotides along it if the user zooms in sufficiently. The browsers are based on database resources, perhaps incorporating several nested layers of data drawn from different sources.

The features that can be annotated include:

- Open Reading Frames (ORFs; segments between start and stop codons—specific sequences that may indicate the presence of transcribable DNA such as a gene);
- Genes (and their structure, organisation and variants);
- Repeat sequence regions, including those constituting telomeres at the ends of chromosomes and centromeres that perform a key role in the chromosome dynamics of cell division;

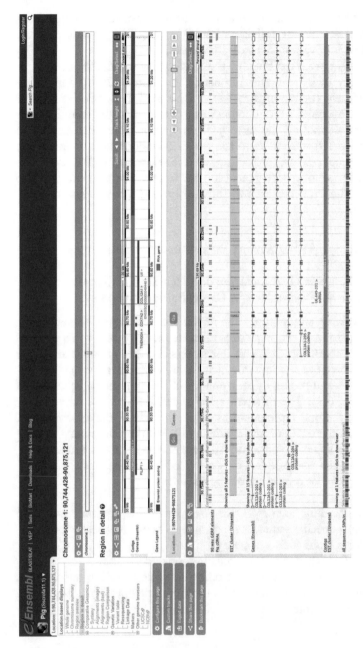

Fig. 6.1 Example of a display of a reference genome—Sscrofa11.1 for *Sus scrofa*, the pig—on a genome browser: Ensembl. Taken from Sscrofa11.1 Chromosome 1: 90,744,428-90,875,121, Ensembl (Howe et al., 2020), release 105 (https://www.ensembl.org/Sus_scrofa/Location/View?r=1:90744428-90875121;db=core, last accessed 18th December 2022)

- Pseudogenes (which appear similar to genes but do not function as such, due to mutations—these may have originally been copies of functioning genes);
- Regulatory regions that are not themselves expressed, but that affect the expression of genes.

Beyond these, many different kinds of sequence variants can also be identified and annotated, including structural variants in which stretches of nucleotides have been deleted, inserted, added, moved and inverted (Mahmoud et al., 2019).[2] Genomic variation comes in many forms, from differences in individual nucleotides, through variation in the sequence of individual coding regions, variation in the number of copies of repeat sequence in particular regions, to differences in sequence at a more gross level such as structural variants.

Two key distinctions have emerged to describe the processes and objects of annotation: manual and automated annotation; and structural and functional annotation.

Manual annotation involves the marking of genomic features using biological knowledge, such as the known sequence and location of a given gene. This way, the sequence is interpreted and contextualised using evidence from a variety of sources that may include earlier automated annotations. In automated processes, the genome assembly is first computationally analysed to identify key features such as repeat sequences and ORFs, and then existing datasets are interrogated to make predictions as to the annotation of more complex features such as protein-coding genes. These predictions are then examined further using a variety of algorithms embedded in different software to synthesise different forms of data and thus establish consensus models of the gene, which may include its structure and the existence of different forms. The data used in these automated processes include Expressed Sequence Tags (ESTs), known protein sequences and RNA sequences. These data can concern the species being annotated, as well as other species known—through prior comparative work—to be genomically close enough to the target species in order that cross-species inferences between parts of the genomes known to be equivalent can be made (Lowe, 2022).

Typically, generic pipelines have been designed and continually developed to annotate genomes, similar to the way that ones have evolved to produce and assemble sequence data (Stevens, 2013). These pipelines

[2] https://www.ncbi.nlm.nih.gov/dbvar/content/overview/ (last accessed 18th December 2022).

involve the specification of a series of sequential tasks and associated pro-tocols, though typically different options for routes along the pipeline may exist to enable projects with differing levels of resources to navigate it. While some projects may have the resources to, for example, pay for additional manual annotation to refine the automated annotation, others may not. The existence of generic pipelines, together with the use of cross-species data, shows how genomic endeavours for different species interact. The infrastructures are built to accommodate difference, but also to chan-nel it to ensure that the products of the pipelines are commensurate, even though they may serve—and be used by—different communities.

Alongside the selection of the source of DNA and the planning of the project, it is in annotation that the reference genome as a creative product of a particular configuration of actors is most manifest. The ways and the extent to which the annotation process enables new forms of genomics and genome-related research and resource development, however, depends on the details of the construction of that reference genome. Such details include the libraries used and how the genomic variation of the species was abstracted into the reference sequence. Also crucial are the relationships of particular research communities to various aspects of the process from pre-reference genomics through to annotation, as we show below.

The distinction between structural and functional annotation appears to map onto the distinction between (reference) genomics and post-(reference) genomics, which is explored further in the following chapter (Chap. 7). Structural annotation is the identification of particular features of the genome such as genes and their organisation, but also other functional and non-functional elements. Functional annotation is the connection of this structural data to other forms of data that help to make sense of the prod-ucts and role of particular genomic elements. Broadly, we discuss structural annotation more in this chapter and functional annotation in the following chapter, but in doing so, we reveal that the distinction and apparent tem-poral succession from structural to functional is not clear cut.

6.1.2   Creation of Annotation Infrastructures

Annotation practices pre-date the annotation of reference genomes, and even the invention of DNA sequencing: for instance, the annotations in Margaret Dayhoff's early DNA sequence database were modelled on those in her previously-established protein sequence database (Strasser, 2019, p. 209). In the generation, collection and curation of annotations in data-bases such as GenBank, Stephen Hilgartner (2017) and Bruno Strasser

(2019) have identified two broad periods. There was an earlier period in which database staff themselves had to collect and annotate individual sequences, by trawling the literature. Then there was the period that succeeded this, in which the producer of the sequence data was able to submit it—with pertinent annotation—directly to databases with the help of specially-designed software tools.[3] In alliance with funders and journal editors, the databases helped to increasingly transform this practice into a duty.

In the first period, in the 1980s, annotation was essentially in the form of metadata; curators would read journal articles reporting a new nucleotide sequence and annotate the sequence by indicating the source of DNA and key features within the string of nucleotides. This process was advanced by agreements forged from 1982 onwards between GenBank and the Nucleotide Sequence Data Library hosted at the European Molecular Biology Laboratory (EMBL), and later (1987) between these and the DNA Data Bank of Japan. This tripartite alliance later became formalised as the International Nucleotide Sequence Database Collaboration. They divided up the laborious tasks of going through the literature and extracting and annotating sequences between themselves. Furthermore, to get around existing compatibility problems, they strove to harmonise the format that the data was recorded in.

In spite of this, and the use of supercomputers at the US Department of Energy's National Laboratory at Los Alamos to try to automate data processing and annotation, the rapidly-increasing production of sequences led to a backlog. This encouraged GenBank to streamline the process, in part by skipping the annotation or making it more cursory (Hilgartner, 2017, pp. 157–161; Strasser, 2019, pp. 228–230). As annotation was meant to be about making the data useful and "biologically meaningful", enabling it to be picked up and re-used by researchers using the database, this was problematic (EMBL Director General Lennart Philipson, as quoted by Strasser, 2019, p. 232). The EMBL, closer to bench biology than the physicist-led GenBank (which was based at Los Alamos from 1982 to 1992), was less keen on short-cuts around or through the annotation process (Strasser, 2019). The inadequacy of the initial algorithms designed for annotating sequences at the EMBL led to the conscription of biology students and clerical staff to contribute to the effort. When this also proved insufficient, more senior biologists were cultivated, which involved

---

[3] Though, as we detail in Chap. 7, this transition does not occur so neatly for species-specific databases targeted at particular organismal communities such as the *Saccharomyces* Genome Database, or for functional annotation rather than structural annotation.

informing them about some of the basics of the operation of the database, as well as circulating new sequences that may have been of interest to them. Biological researchers at the EMBL could then work with the database staff to refine the sequences stored on the database—and their annotations—as well as helping to improve the algorithms used in automated annotation (García-Sancho, 2012, pp. 111–114).

From 1987 onwards, there was a strategic shift towards securing agreements with journals, by which they would only publish articles including DNA sequence data if they were accompanied with accession numbers, indicating that they had been submitted to a publicly-accessible database such as the DNA Data Bank of Japan, the EMBL one or GenBank. Even though these agreements and rules were variably enforced, they succeeded in encouraging more direct submission, especially when software tools making data submission easier for researchers spread. Further changes in rules and norms of submission followed in the 1990s and improvements in the way data were submitted and accessed also occurred. There was increasing adoption and ease of internet access, additional tools to interrogate the databases were developed (such as the Basic Local Alignment Search Tool—BLAST—sequence comparison software), additional databases beyond the basic sequence ones were launched, and ongoing improvements were made to the fundamental DNA sequence databases.

In 1992, GenBank came under the umbrella of the National Center for Biotechnology Information, which maintains a panoply of other reference and software resources, including the RefSeq database (Chap. 1), and ClinVar, which is explored in Chap. 7. As we showed earlier (Chap. 4), in 1994, the EMBL database moved from Heidelberg—where the EMBL headquarters are—to what is now known as the Wellcome Genome Campus in Hinxton, Cambridgeshire, to form the EMBL's European Bioinformatics Institute (EBI). The Wellcome Genome Campus is also where the Sanger Institute is based, a co-location of significance to the story of the development of annotation infrastructures, and the specific examples of annotation we detail in the following section.

For now, the relationship between the Sanger Institute and the EBI is pertinent, because of the role of these institutions in the creation of means by which the data in well-stocked nucleotide databases could be brought together and presented in a useable form for researchers. These resources, the database system AceDB and the genome browser Ensembl, were forged in the exigencies of reference genome sequencing: of the nematode worm *Caenorhabditis elegans* and the human, respectively.

AceDB, which stands for '*A C. elegans Data Base*', was originally founded in 1989 by Jean Thierry-Mieg and Richard Durbin. The former was a Centre National de la Recherche Scientifique (CNRS) researcher in France, and Durbin was in a spell at Stanford University in-between doctoral and postdoctoral work based at the Laboratory of Molecular Biology in Cambridge; he moved to the Sanger Institute in 1992 and stayed there full-time until 2017. As it developed, AceDB allowed users to access and relate different kinds of representations of the genome of *C. elegans* in an internet browser, to move between representations of the DNA sequence, and the genetic linkage and physical maps. In her historical investigation of *C. elegans* genomics and the nature of the AceDB enterprise, Soraya de Chadarevian has highlighted the infrastructuring work that is required to make maps—that have been produced in very different ways and constitute distinct representations—commensurable in databases and visualisations generated using them. The production of new kinds of maps, including the full genome sequence, was driven by specific concrete demands (e.g. of particular communities) that were often independent of those that drove the construction of preceding maps. In making different kinds of maps interoperable through this work of commensuration, the specificities of the objectives, communities, practices and historical trajectories involved in forming these resources are flattened (de Chadarevian, 2004). This eases visualisation and navigation by users, but at the cost of abstracting the underlying specificities and lineages. As we show below, this double-edged sword—of easing inter-operability at the expense of flattening specificities—persisted in other infrastructures produced at the Wellcome Genome Campus.[4]

In 1999, the same institution at which AceDB was developed—the Sanger Institute—collaborated with the EBI to launch a key platform to accelerate the IHGSC human reference sequence effort: Ensembl. The Ensembl team devised a pipeline to help assemble the reference sequence

---

[4] The adoption of the map-based approach for human genomics was due to the success of the whole-genome sequencing of *C. elegans* using a prior physical map. Maps have historically informed sequencing, and small-scale sequencing was often a key part of genome mapping. Genomics research is also inextricably entangled across species, with practices, resources and tools developed by communities working on one species regularly used and adapted for different species (de Chadarevian, 2004; Lowe, 2022; Stevens, 2013, Ch. 7). All this involves the construction of infrastructures to enable commensurability, or at least interoperability across different representations and resources. Star and Bowker (2002) is a foundational text concerning infrastructuring, while Baker and Millerand (2010) examine infrastructuring concerning data in the life sciences. For examinations of analogous trade-offs involved in the creation and mobilisation of data itself, see Leonelli (2016); Leonelli and Tempini (2020).

and present it online through a genome browser.[5] The Ensembl browser presents an abstracted view of any part of the genome one chooses to zoom-in to. It offers a variety of 'tracks' representing different annotated features of the genome that can be selected and lined up alongside the reference nucleotide sequence, which is itself arrayed horizontally (Fig. 6.1). Ensembl does not only generate these visualisations but, for verte-brate species, also produces the annotations that are included in them, through its own automated annotation pipelines. It augments this with downloaded annotation data for other key non-vertebrate species. Ensembl, therefore, exhibits the clear will in the late-1990s to automate the annotation process and to bring it 'in-house' into the small number of institutions producing sequence data.

The manual annotation of select species was conducted by the Human And Vertebrate Analysis and Annotation group (HAVANA) at the Sanger Institute. HAVANA had its origins in the Human Sequence Analysis team led by Tim Hubbard within 'Team 71', the Informatics division that was led by Durbin at the Sanger Institute. The Sanger Institute component of Ensembl led by Michele Clamp was also part of Hubbard's team. Jennifer Ashurst (later Harrow) joined this team in April 2000 and led a distinct HAVANA group within the team from 2002. At the time she joined, there were two people working on manual annotation. However, it became apparent that Ensembl's automated annotation generated too many false positives due to the quality of sequence data then available to them.[6] It did predict approximately 70% of human genes accurately, good enough for a rough-and-ready annotation of the draft genome, but not of the required quality for biomedical research or diagnostic purposes. To improve the quality of the annotation, manual annotation was required that would make use of data coming in from the automated pipelines, but also involve curatorial decisions based on biological knowledge.[7]

---

[5] Other key general genome browsers include the UCSC Genome Browser hosted by the University of California Santa Cruz and Genome Data Viewer hosted by the NCBI. There are also more specialist species or taxon-specific browsers, such as the *Saccharomyces* Genome Database.

[6] This was despite it being expressly designed to produce gene predictions of high specific-ity at the cost of high sensitivity, in other words, to try to avoid false positives even if that meant missing true positives. This is a reflection of how difficult it was to generate effective automated procedures, and from early on the Ensembl team recognised that subsequent manual curation—and evaluation and refinement of the gene structures that were the out-puts of automated annotation—was vital (Birney, Andrews, et al., 2004).

[7] Jennifer Harrow, interview conducted in Cambridge by James Lowe, October 2017.

HAVANA developed the curated Vertebrate Genome Annotation (VEGA) database and browser, which was built on Ensembl. VEGA was operational for human manual annotations from 2002, and mouse and zebrafish from 2003.[8] The browser was curated using both manual annotations conducted by the HAVANA group itself (such as for human chromosome 20) and by other groups and institutions (such as Ian Dunham's for human chromosome 22, and Genoscope and the CNRS for human chromosome 14). From early on, this annotation and curation work was accompanied by the development of protocols for manual annotation. At two 'Human Annotation Workshops' (HAWK1 and HAWK2) hosted by HAVANA in March and September 2002, participants from multiple institutions involved in manual genome annotation discussed possible standards and guidelines. A test sequence was annotated using different manual and automated methods at HAWK1, and the results of this were compared.[9] These workshops formed the basis for the manual annotation standards used in VEGA and were intended to aid commensurability across other resources and genome browsers developed at the NCBI and University of California Santa Cruz (see note 5). The Otter manual annotation system that was developed for HAVANA by Ensembl and used in VEGA was designed in accordance with the standards formulated in the HAWK workshops (Searle et al., 2004).

From 2014 to 2017, Ensembl became solely part of the EBI. HAVANA became part of Ensembl at the EBI in 2017. By then, HAVANA had branched out to work directly with some species communities on manual annotation; the pig was one of these, as we see later in this chapter.

6.2   Annotating the Yeast, Human and Pig Genomes

When we consider the annotation process across the main three species we look at, we find that the nature of it depended on: the generation and use of existing genomic resources such as maps and genome libraries; the existence of data such as that on Expressed Sequence Tags (ESTs), complementary DNA (cDNA) sequences, RNA sequences, and protein sequences; the nature of the inferential apparatus available for intra-specific and

---

[8] Sequencing the reference genome of zebrafish (*Danio rerio*), a model organism, was an initiative begun at the Sanger Institute in 2001.

[9] https://web.archive.org/web/20020825133038/http://www.sanger.ac.uk/HGP/havana/hawk.shtml (last accessed 18th December 2022).

inter-specific data analysis; the kind of community and actors involved and their interests; and the mode of organisation of genomic projects. The available data sources and inferences were marshalled to find and elucidate the fine structure of genes and other elements of the genome. A closer look at the specifics of annotation practices across these three different species allows us to complicate the relationships between automated and manual processes, as well as between structural and functional annotation.

Four basic models of annotation were identified by bioinformatician Lincoln Stein in an article published in the summer of 2001 (Stein, 2001). He associated these models with particular stages of the annotation process, in terms of its increasing complexity and recontextualisation through forming connections to other kinds of biological data and knowledge. Two of his terms—factory and cottage industry—are familiar from earlier debates concerning the proper organisation of genomics (Chaps. 2 and 3). We have interpreted his designations in the scheme displayed in Fig. 6.2.

Stein's scheme, as interpreted, highlights the importance of the establishment of mechanisms by which existing datasets and resources can be accessed and used in annotation, as well as the significance of the role of annotation itself in enabling the creation of new links to other datasets and standard references, such as the Gene Ontology. This enables the

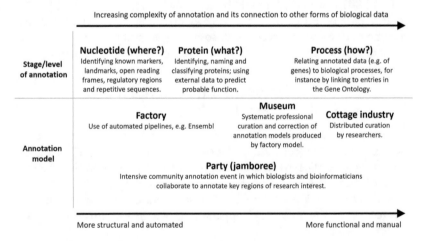

Fig. 6.2 Diagrammatic depiction of models of annotation and how they relate to different stages or levels of annotation. (Produced by both authors, based on Stein, 2001)

decontextualised reference sequence to be progressively connected to other forms of biological data and therefore recontextualised.[10] This process makes use of software and algorithms to search external databases. Crucially, it also uses maps and libraries employed in the construction of the reference genome to initially annotate the sequence. This seeds further annotation by providing reference points to aid the searching of external data, and also aids the later contextualisation of the annotated data. Stein's conception, while consisting of stages, does break down firm distinctions between manual and automated annotation, and also structural and functional annotation, as entanglements of each are implicated in any one point. Key here is that the weights of the different modes (automated/manual; structural/functional) change as the annotation process proceeds. The schematic we have drawn from Stein is a useful overview of the general trends in the annotation process, and it constitutes a helpful reference point with which to consider examples that depart from the sequential and separable stages implied by it. For instance, we may observe that the main genome browsers such as Ensembl moved towards a hybrid factory-museum model (Loveland et al., 2012).

Quite apart from the particular manifestations of sequencing, and the extent to which they may depart from Stein's ideal types, the ways in which particular communities and genomic endeavours undertake annotation is constrained by multiple factors. These include the histories, motives and resources of particular communities of genomicists. Furthermore, groups such as HAVANA developed forms of community annotation, in which they acted as facilitators—rather than the sole conductors—of the annotation process. As we detail below, these forms of community annotation involved the creation of software tools such as Otterlace/Zmap for manual annotation on the cottage industry model, and also more direct interactions with research communities, such as the one that had been working on pig genetics and genomics (Loveland et al., 2012).

6.2.1   Yeast Genome Annotation

For yeast genome sequencing, as previously noted, one finds a community of geneticists, cell biologists, biochemists and molecular biologists, often dedicated to working with standardised strains of *Saccharomyces cerevisiae*.

---

[10] Stein makes explicit mention of the entries in the Gene Ontology; such data resources also involve processes of annotation, albeit featuring different models and kinds of curatorial roles; see Leonelli (2016).

The ease of working on this unicellular eukaryote was what made it a model organism, and this engendered the virtuous cycle by which the existing weight of scientific capital—in the form of mounting knowledge, resources, tools, and mechanisms of dissemination and sharing—justified new investment in its further augmentation. When the perception began to grow that "[t]he yeast genome was becoming overstudied, and yet..., largely unexplored!"—that different research groups were working on the same genes while much of the genome was *terra incognita*—multiple laboratories across Europe, Japan, Canada and the USA rallied to participate in an unprecedented collaboration to sequence the first full eukaryotic genome (Dujon, 1996, p. 263; Chap. 2).

The structural annotation of the yeast genome reflected the hierarchical, top-down and distributed approach of the sequencing effort in Europe. Within the initiative funded by the European Commission, the centralised bioinformatics function located at the Martinsried Institute for Protein Sequences (MIPS) was married with the specific expertise of the laboratories performing sequencing, and seeking to make use of the data so generated.

MIPS, on assuring the quality of the sequences it received and assembling contiguous tracts of sequence (contigs) on the basis of them, screened the data for ORFs by identifying stretches of minimum numbers of nucleotides (from about 50 to 300, a lower number risking more false positives and a higher one more false negatives) with no stop codon. They also sought contigs with sizes below the threshold by searching for sequences that were homologous (showed sufficient similarity) to known protein sequences, based on the knowledge of the genetic code and processes of transcription and translation. Already, this analysis relied upon existing experimental knowledge of this well-studied organism, as well as the prior delineation of protein sequences and elucidation of their functions. Using sequence homologies, the MIPS team was able to classify the ORFs in terms of their putative functions (Mewes et al., 1998). Once the data had been passed on to the sequencing laboratories, the initial identification of the ORFs could be built on with a deeper analysis of these sequences. This was done either using existing biological data or materials (for example, concerning centromeric and telomeric DNA, tRNA and Ty elements for chromosome II) or by performing a variety of experiments to characterise their functional role. Following the conclusion of the reference genome sequencing, such experiments were organised and conducted in a concerted way in a successor project on functional analysis and annotation called EUROFAN, which is discussed in Chap. 7. Due to the limitations of homology analysis, with about 40% of putative genes being "orphans" either having no discovered homologues or homologues with no known function, such functional analysis would also enable the verification of the structural annotation.

Once the presumed coding regions were separated from the non-coding, the non-coding regions could be further analysed to detect sequence motifs (including promoter regions of genes) and other features such as transposable elements (Ty elements). Many of these non-coding elements were of interest to participants in the network, who could use the genomic data that they generated—and MIPS processed—to further their research. For example, Horst Feldmann at Ludwig-Maximilian University of Munich was particularly interested in Ty elements (Chap. 2) and advanced his research using the structurally annotated sequences he now had access to. These sequences had themselves been augmented using the data he had previously collected (Feldmann et al., 1994; Heumann et al., 1996; Mewes et al., 1998). While the centralised parts of this process, such as the role of MIPS, will seem analogous to some of the informatics pipelines and groups of the IHGSC discussed in the next section, the yeast biology laboratories played an important role in refining and developing the initial annotations that were made by MIPS. Unlike in human reference sequencing, in which prospective users were not involved in the processes of data production, in the yeast genome effort there was a set of users incorporated in those processes (García-Sancho, Lowe, et al., 2022).

The completion of the sequencing and sequence analysis of the different chromosomes at different times enabled innovations developed for one chromosome to be taken up by groups working on other parts of the yeast genome. For example, the methods that yeast geneticist Bernard Dujon developed for the evaluation of ORFs to identify which ones were indeed "functional genes" in the chromosome XI paper published in June 1994 were then used in the chromosome VIII paper published in September that year (Dujon et al., 1994). Chromosome XI was Europe-led, while VIII was coordinated from Washington University by Mark Johnston. While they exhibited different organisational models, as we saw in Chap. 2, there was enough of a connection for each to build on the advances of the other.

Washington University's model of annotation was also different, though in practice they used searches of public nucleotide and protein databases to identify cross-species homologies with known genes and protein sequences, as well as examining other elements such as tRNAs, much as MIPS did. For assembly and annotation, they (along with some European-led groups) used a version of AceDB: AScDB, with 'Sc' standing for *S. cerevisiae* rather than the 'ce' of *C. elegans*. AScDB had been specially adapted for yeast by Richard Durbin, young EMBL bioinformatician Erik Sonnhammer and LaDeana Hillier, the director of informatics at the Washington University Genome Sequencing Center (Johnston et al., 1994). Hillier collaborated closely with Johnston, and also worked on *C. elegans* and human genomics. With the benefit of a comparative perspective gained from interaction with the yeast, human and *C. elegans* efforts, she observed that a significant problem with "smaller numbers of groups doing the sequencing" was that "user education" could be "an issue". However, for "yeast the user education was taken care of because the sequencing was done at so many different places that everybody [...] understood the limits of the data" (Hillier, 2012, p. 7).

Dujon and Johnston gave assistance to the chromosome I team that mainly operated at McGill University. They were the next to publish—in April 1995—with Dujon helping with sequence analysis and Johnston providing the chromosome VIII sequence, which enabled some genome duplications to be identified. Later papers indicate a continuation of this cooperation around sequence analysis. These publications document a refinement of the processes, datasets and software used from the early published chromosomes onwards (Bussey et al., 1995; see also Galibert et al., 1996). This stands in contrast to the development of novel tools and the infrastructural transformations associated with human genome annotation or the adaptation of established infrastructures and processes to the particular demands of pig genomics.

For the Europe-led sub-projects, MIPS continued its role in sequence analysis. It did not see its task as restricted to identifying individual genomic elements, but also as aiding the global characterisation of the genome, by using their initial structural annotation to partition the genome into units. As a consequence, sequence comparisons could be made between these units, in order to identify gene duplications to aid future functional analysis and provide data that could be used in tracking the evolution of the *S. cerevisiae* genome. These twin approaches of targeting function and diversity that arose out of the initial work to structurally characterise the genome form an important part of the narrative of Chap. 7.

For the purposes of sequencing and annotation, yeast had clear advantages over the bulkier organisms that we consider next: humans and pigs. The yeast genome is considerably smaller in size, but also more economical, in that it contains comparatively little non-coding DNA and complex gene structures, compared with multicellular eukaryotes. As a model organism, it also had a panoply of available experimental evidence that could be used and built on to inform both automated and manual approaches to annotation. Additionally, the range and extent of functional analysis conducted by the yeast genomics community that we discuss in Chap. 7 was not possible for human and pig. This meant that distinct strategies for annotation needed to be developed for these species. For the human genome, this involved making use of the abundant ESTs and protein sequence data that had been gathered, the creation of automated and manual sequencing pipelines, and advancing the means with which to conduct analyses of homology by harnessing and further developing comparative genomic approaches.

6.2.2   Human Genome Annotation

In the three major papers describing the sequence of the entire human genome (authored by the IHGSC in 2001 and 2004, and by Celera in 2001), only the Celera paper includes details of the annotation process. For the IHGSC, the details of annotation are dealt with only in the subsequent individual papers describing the sequence of each chromosome. This reflects, we suggest, the IHGSC primary concern of getting assembled sequence out in the public domain to prevent its enclosure by some form of intellectual property. On the part of Celera, the inclusion of information about annotation evinces their commercial strategy of building the foundations for the exploitation of the genome for biomedical purposes. Even though they described aspects of their annotation process, users would still have to pay to access Celera's full annotated sequence. In this way, Celera sought to make itself an obligatory passage point for those seeking the richly-annotated data that they produced.

The first chromosome that the IHGSC sequenced was chromosome 22, by a team led by Ian Dunham at the Sanger Institute. The paper announcing this appeared in December 1999, before Ensembl and HAVANA were up and running. Tim Hubbard's sequence analysis team were involved, though, and they integrated existing data on nucleotides and protein sequences, using similarity searches (through programmes implementing the 'BLAST' algorithm developed at the NIH by Gene Myers and colleagues) and prediction programmes (Dunham et al., 1999). Like the annotation of subsequent chromosomes, an early stage was identifying repetitive sequences and 'masking' them. This meant filtering them from view so that they were not incorporated in automated analyses of the sequence data. To do this, the annotators used 'RepeatMasker', a piece of software developed and (then) hosted by the Genome Sequencing Center at Washington University. The remaining unmasked sequence was then analysed for the presence of various genomic features, such as spotting areas of the genome with a relatively high proportion of guanine and cytosine bases in order to discern the presence and location of CpG islands, in which cytosine is next to guanine. These are frequently located in the promoter regions of genes and are therefore a good indicator of the presence of genes.

At this point, the automated aspects of searches and the use of prediction programmes were interweaved with manual approaches. In large part, this was because of the calibration and verification required for each method, and the overall need to evaluate and refine the annotation process. A re-evaluation of the chromosome 22 annotation in 2003 reaffirmed the value of combining automated prediction, sequence similarity and comparative methods in annotation, but observed that the optimum configuration of them with respect to each other had not yet been found. Furthermore, at this time the ideal comparator species for similarity analysis was unclear. The authors acknowledged that while annotation processes would be improved, at that point automated approaches had significant limitations. As well as refining data categories and making use of new sources of data (e.g. new human ESTs and various kinds of data on related species), overcoming these limitations would involve manual analysis and experimentation (Collins et al., 2003).

The only other chromosome sequence published before the announcement of the completed draft of the whole genome in February 2001 was for chromosome 21, conducted by a consortium led by RIKEN (**Ri**kagaku **Ken**kyūjo, the Institute of Physical and Chemical Research) in Japan.[11] This team also conducted gene predictions and sequence similarity searches. They additionally defined criteria by which putative gene classifications were assigned to one of five categories, depending on the strength of the evidence for them being protein-coding genes. They, therefore, placed the discernment of functional elements of the genome such as protein-coding genes at the heart of their annotation effort, an orientation appropriate to the biomedical interests of many of the institutions that worked on chromosome 21. That emphasis—and the function-centred annotation—motivated and aided the paper's substantial analysis of the medical implications of their results (Hattori et al., 2000).

The biomedical interests of RIKEN's collaborators were the exception rather than the rule for most institutions involved in sequencing subsequent chromosomes within the IHGSC effort. This was reflected in the way that the sequence data was analysed in the publications announcing their completion. Advances in the analysis of sequence data were heralded, but in so doing, the potential biomedical users of the data were a secondary concern. As we now detail, these analytical advances constituted refinements and additions that augmented the annotation pipelines for each successive chromosome. The augmentations that these specialist genomicists introduced were directed towards improving the capabilities of genomics *qua* genomics, as an enterprise in itself with its own internal goals and motivations. They sought to improve their assemblies and annotations according to internal generic metrics of quality, contiguity and coverage, guided by an overall ideal of completeness. In other words, they did not primarily shape the annotation process and its products in such a way as to fulfil the requirements of any specific external community or set of users.

The first chromosome sequence published after the announcement of the draft whole sequence was chromosome 20 in December 2001; after

---

[11] Other members of the consortium were: Keio University School of Medicine in Japan and from Germany the Max Planck Institute for Molecular Genetics in Berlin, Institute for Molecular Biotechnology in Jena, and German Research Centre for Biotechnology in Braunschweig. Collaborating institutions were the National Cancer Center Research Institute and University of Tokyo (both Japan), UMR 8602 CNRS at UFR Necker Enfants-Malades and CNRS UPR 1142 at the Institute of Biology (both France), Eleanor Roosevelt Institute (USA), University of Geneva Medical School (Switzerland) and School of Pharmacy, University of London (UK): Hattori et al. (2000).

this, there was a gap in 2002 before a flurry were published across 2003 to 2006.[12] What did the progressive accretion of methods and sources of data consist of, across the five years since the completion of chromosome 20?

The chromosome 20 paper, signed only by authors from the Sanger Institute, was the first to use the Ensembl database in the analysis of the sequence; this sequence was, though, still assembled and visualised in AceDB. The genomicists were able to make use of sequence data from two vertebrates (the mouse *Mus musculus* and the pufferfish *Tetraodon nigroviridis*) in their comparative analyses rather than merely the mouse maps that the previous chromosomes had relied on (Deloukas et al., 2001).

For chromosome 14 (February 2003), a two-step annotation approach was employed by the collaboration between Genoscope, the Institute of Systems Biology in Seattle and the Washington University Genome Sequencing Center. In this, automated methods using computational predictions to formulate provisional models of the structure of genes, were refined by sequence similarity analysis. This was complemented by experimental data on gene expression using microarrays, a tool containing potentially many thousands of DNA probes that can indicate the presence or absence of specific complementary sequences. In the "manual curation" that followed, the genomicists used additional data to refine the gene models produced in the first stage and remove "suspicious data" such as partial matches that were not found to contain any significant coding sequences (Heilig et al., 2003, p. 607).

Washington University Genome Sequencing Center was also heavily involved in the completion of chromosome 7 (July 2003), as well as the Y chromosome (June 2003). These featured a significant focus on methods for the identification of pseudogenes, including K_A/K_S analysis to identify the kind and extent of selection operating on putative pseudogenes and known genes. In this type of analysis, the scientists generated reconstructed ancestral sequences to detect signatures of neutral evolution (and therefore an absence of positive or purifying selection) which would indicate the presence of a pseudogene. They then checked these inferences

---

[12] Chromosome numbers were assigned according to the observed size of the chromosomes in karyotypes. Generally speaking, this is reflected in their length, with the longest nuclear chromosome being 1, the second-longest being 2, and so forth. There are some exceptions at the shorter end: 21 is longer than 22, and 20 is longer than 19, for instance. It is easy to see why, therefore, the higher-numbered (and therefore shorter) chromosomes tended to be sequenced earlier, and the lower-numbered ones tended to be sequenced later (1 was the last to be published), though this was only a general trend.

using the available mouse sequence data (Skaletsky et al., 2003; Hillier et al., 2003).[13]

Like chromosome 20, the paper heralding the completion of chromosome 6 (in October 2003) was wholly authored by people at the Sanger Institute. Since 2001, there had been considerable developments in their annotation process. Ensembl was now more refined, and the HAVANA team was established and embarking on their extensive manual annotation. VEGA was now up and running and hosting the annotated sequence data. Built into the heart of Ensembl's automated annotated process were two sequence-matching tools: GeneWise for exploiting protein sequence data and Genomewise for using EST and cDNA data indicative of the presence of transcribed genes (Curwen et al., 2004; Birney, Clamp and Durbin, 2004). In its design, the Ensembl pipeline had been configured to integrate and more effectively deploy existing annotation methods. In addition, it was now able to make use of sequence data on the rat (*Rattus norvegicus*; an animal model), another pufferfish (*Fugu rubripes*; with a far more economical genome than other vertebrates) and zebrafish (*Danio rerio*; a model organism) as well as the mouse and *Tetraodon nigroviridis*. Using the protocols and standards forged in the HAWK meetings in 2002, the HAVANA group manually curated the gene structures generated through the Ensembl pipeline. Given their later role in facilitating community annotation of immune response genes in the pig, it is appropriate that HAVANA's first formal role in human genomics concerned chromosome 6, which contains the Major Histocompatibility Complex implicated in immune response.

---

[13] The theory behind this approach is that compared with a reconstruction of the ancestral version of the gene, a functional gene will exhibit *either* a high ratio of nonsynonymous substitutions to synonymous substitutions—reflecting positive (directional) selection—or it will show a low ratio resulting from stabilising selection. Synonymous substitutions mean that observed mutations—when compared with the ancestral version of the gene—will result in no change in the amino acid that is specified by the codon (the triplet of bases read during DNA transcription); there will therefore be no change in the function of any gene products as a result of such substitutions. A gene that has undergone positive selection has had its sequence altered in a manner that increases the fitness of its holders. Stabilising selection, by contrast, ensures that the sequence does not change—as changes would be disadvantageous to the organism. These evolutionary mechanisms can therefore be identified using this analysis. Pseudogenes can also be detected. They should exhibit a ratio of about 1, indicating that there has been no selection either way. This absence of selection is expected for nonfunctional parts of the genome such as pseudogenes.

We will return shortly to the annotation of the remaining chromosomes, focusing on the development of Ensembl and HAVANA at the Sanger Institute. For now, with the expansion of the number of creatures for which informative sequence data was available in mind, we make a brief excursion into the development of comparative genomic resources and approaches.

As we noted in earlier chapters, a comparative genomic perspective was present in genomics from its inception. Genome sequencing projects on other species were used as pilots to aid the planning of the Human Genome Project. Furthermore, the map and sequence data of those other species were used to help construct human genome maps and sequences, by applying knowledge about comparative regions between the species. Finally, it was also envisaged that establishing a rich understanding of comparative connections between human and non-human genomes would enable the more fruitful exploitation of the human resource. In one respect, this was because experimental interventions on organisms such as yeast and animal models could then be connected to and inform human biology through genomic and other omics data. In another respect, this was because of the mooted contribution of data on other species towards enriching the annotation of the human genome.

To aid human genome annotation in this way, in December 2003, the Large-Scale Sequencing Program of the US National Human Genome Research Institute (NHGRI) established two Working Groups: one on 'Annotating the Human Genome' chaired by Robert Waterston and the other on 'Comparative Genome Evolution' chaired by Laura Landweber and John Gerhart. Both groups were tasked with identifying what new sequencing could be conducted in large-scale sequencing centres to advance human genome annotation and functional analysis. The Comparative Genome Evolution group also had to identify which organisms to sequence to shed new light on human evolution and genome evolution across eukaryotes in general. Each of the groups identified three components of research, a range of organisms and appropriate sequencing strategies (including coverage to be obtained) to contribute to these components, and indicated percentages of total sequencing capacity to be allotted to each task.

The Annotation Working Group recommended that 15 non-primate mammalian genomes be shotgun sequenced at relatively low coverage in two successive sets (known as 'Bins'). They further indicated that other genome efforts already in progress, including for non-mammals such as the chicken, should proceed further so that complete high-quality sequences be produced to aid the identification of conserved sequences across mammals. The second component suggested by the Annotation Working Group was the high-quality sequencing of two primate genomes and relatively high-coverage shotgun sequencing of three others, to enable differences to be identified between these and the human genome. The third component was a recommendation to survey human genomic variation by sequencing 1000 people at very low coverage. The group additionally suggested that "a modest cDNA effort be included as a component of all genomic sequencing projects" to aid assembly and gene prediction.[14]

The Comparative Genome Evolution working group's recommendations ranged more deeply and widely across the tree of life, further extending the selection criterion employed by the Annotation Group by which some species would be preferentially sequenced due to representing key phylogenetic positions. Both groups also deployed other criteria to recommend particular organisms as candidates for sequencing, including the quality of the submissions ('white papers') sent in by the relevant communities; the role of the organism as a model; its potential biomedical significance; its economic importance; the possibility that a genome sequence for it would enable the construction of reference sequences for closely-related organisms of biological significance and the size and heterogeneity of the genome.[15]

A Coordinating Committee (chaired by William Gelbart) then evaluated the proposals, presenting a modified set of recommendations to the NHGRI's Advisory Council for approval in May 2004.[16] We consider this further in the following chapter when addressing different aspects of post-reference genome work on the human. For now, it is pertinent to note that in the documented assessment of species proposals by the Working Group on Comparative Genome Evolution, their conception of the communities working on these organisms and submitting white papers to the NHGRI was very much as groups of *users*. The evaluations that the NHGRI made of the white papers were based on the readiness of these

---

[14] "New Sequencing Targets for Genomic Sequencing: Recommendations by the Coordinating Committee", part of the documents for the Meeting of the NHGRI Research Network for Large-scale Sequencing and the NHGRI Sequencing Advisory Panel, May 16, 2004 (NHGRI History Archive 7036–021).

[15] The community of pig genomicists submitted one of these white papers (Chap. 5).

[16] "New Sequencing Targets for Genomic Sequencing: Recommendations by the Coordinating Committee", part of the documents for the Meeting of the NHGRI Research Network for Large-scale Sequencing and the NHGRI Sequencing Advisory Panel, May 16, 2004 (NHGRI History Archive 7036–021).

communities for receiving the genome. Their role was envisaged as developers of proposals for the NHGRI to judge, and as groups that needed to corral the appropriate resources to make use of what the NHGRI would end up providing for them.[17] New research goals were added for subsequent rounds of sequencing additional species, such as identifying the mammalian "core genome". The increasing apparatus and empirical basis of comparative analysis guided the number and selection of sequencing targets and the methods deployed on them.[18]

Returning to the annotation of the individual chromosomes, the remaining ones that the Sanger Institute was involved with were: 13, 9, 10, X, 17 and 1. For chromosome 13, published in April 2004, the availability of a new database for non-coding RNAs, Rfam, advanced the annotation of these, which had been deemed extremely tricky as recently as in the chromosome 6 paper published in October 2003. For chromosome 13, modifications had been made to the Ensembl pipeline to aid manual curation. With the chromosome 9 paper, published in May 2004, there was a special focus on duplications of segments of the chromosome, which were assessed using K_A/K_S analysis (see note 13). Having previously mapped Single Nucleotide Polymorphisms (single base changes; SNPs) against their sequence using data from the dbSNP database, for chromosome 9 the genomicists identified their own bank of SNPs by analysing the sequence data from overlapping portions of DNA fragments (clones). In May 2004's chromosome 10 paper, the authors continued their identification of SNPs and extended this focus at the single nucleotide level by comparing 617,071 single nucleotide sequence differences between human and chimpanzee, conducting K_A/K_S analysis on the results to ascertain the presence of sites of selection. From this paper on, there was an increasing focus on annotating alternative splice variants, which result from transcription processes that generate multiple different messenger RNA sequences from a single gene.

In the X chromosome paper published in March 2005, there was a particular focus on the evolution of the X chromosome and comparisons were made between it and the Y chromosome. The chicken (*Gallus gallus*) genome assembly was used for this analysis in addition to previously mentioned comparator species, many of which now had newer versions of their

---

[17] "Report of the Annotation of the Human Genome Working Group", dated January 3, 2005 (NHGRI History Archive 7039–005).

[18] E.g., https://www.genome.gov/Pages/Research/Sequencing/SeqProposals/2x-7x_promotion_seq.pdf (last accessed 18th December 2022).

assemblies that were used. For the April 2006 paper on chromosome 17, human sequencing was conducted at the Broad Institute; the Sanger Institute's role focused more on the sequencing of mouse chromosome 11 as part of the Mouse Genome Sequencing Project.[19] The paper was mostly dedicated to a comparative analysis of the two chromosomes and a reconstructed ancestral chromosome, with the authors focusing on an assessment of the different changes to the chromosomes that occurred in the distinct evolutionary lineages.

The final chromosome to be published, in May 2006, was 1. In the paper, the genomicists aligned the chromosomal sequence to the now-standard array of comparator species (minus the chicken) to identify regions of evolutionary conservation. This paper also represented a culmination of the increasing focus on SNPs from 2004 onwards. These SNPs were used to identify and map genomic diversity within species, identify recombination at a higher resolution than previously possible, detect signals of selection, and as a resource to augment the utility of the reference genome (Dunham et al., 2004; Humphray et al., 2004; Deloukas et al., 2004; Ross et al., 2005; Zody et al., 2006; Gregory et al., 2006).[20] The comparative approaches and cataloguing of diversity were conducted to ease the process of developing genomic resources, by feeding into and augmenting the pipelines of the IHGSC participants. The intended use of the resources so produced, however, was generic rather than tailored to specific user communities.

Compared to the IHGSC effort discussed above, Celera's approach was quite distinct, giving potential communities of users of genomic data a more active and participatory role than in the IHGSC and NHGRI's annotation strategies. As noted above, Celera's 2001 paper discussed annotation far more than the contemporary IHGSC one. It was an automated annotation that it chronicled, though, in a discussion of their Otto gene prediction system. This software was designed to weigh different forms of data constituting evidence for particular annotations, namely cDNAs and ESTs. The weighting was based on Celera's previous

[19] The Broad Institute was opened in 2004, the result of collaboration between the Whitehead Institute, Harvard University and hospitals affiliated with Harvard.

[20] The other chromosomes were handled by the Stanford Human Genome Center and the US Department of Energy (19, 5, 16), Washington University (2 and 4), the Broad Institute (18, 8, 15, 11, 17; 18 with RIKEN, 11 primarily RIKEN with the Broad Institute, and 17 with the Sanger Institute), and Baylor College of Medicine (12 and 3; 3 with BGI, formerly known as the Beijing Genomics Institute).

experience of the manual annotation of the *Drosophila* genome. This approach therefore reaffirmed and reflected the process of genomic discovery promoted by Venter in the early-1990s, especially the crucial importance it conferred to protein-coding regions of the genome, as revealed by EST and cDNA sequence data. While the paper reported some computational validation of Otto's results, it acknowledged that the "[e]xtensive manual annotation to establish precise characterization of gene structure" that was still deemed necessary lay in the future (Venter et al., 2001, p. 1317).

As their automated annotation took inspiration from prior work on *Drosophila,* so did their manual annotation, by using the jamboree model. *Drosophila* genomics was not the only inspiration, however. A challenge that Celera faced was the absence of information about the means and decision-making procedures by which the public project's annotations were made. Therefore, to develop their own annotation capabilities, they needed to obtain institutional knowledge of how the sausage was made. To that end, they recruited Peter Li from Johns Hopkins University, who had worked on the GDB Human Genome Database and the Online Mendelian Inheritance in Man (OMIM) catalogue while there, and as a result was acutely aware of the details of the annotation process. The OMIM connection, deepened by the use of data from it in the annotation of Celera's gene sets, was just as significant as the model of *Drosophila* genomics to the way that Celera manually annotated the human genome. OMIM used curators who were experts on particular diseases, with their knowledge of the relevant genetics feeding into the published data. The need for biological expertise to contribute towards the annotation—and more broadly, the contextualisation of the data that Celera was generating—was keenly felt by the company. Due to its particular sequencing strategy, it had invested considerably in computational infrastructure and expertise for the purposes of assembly rather than in acquiring biological knowledge. But because of the need to generate rich and translationally-relevant data to be incorporated into proprietary databases (such as The Celera Discovery System™), drawing on this kind of expertise was essential.

A variety of academics were therefore invited to participate in a human annotation jamboree that took place in April 2001, two months after the publication of the draft reference sequence. This jamboree built on the previous one that Celera had held on the *Drosophila* genome and involved some of the OMIM curators (García-Sancho, Leng, et al., 2022). The human genome jamboree presented an opportunity for participation on the part of medical geneticists who had been largely uninvolved in the

IHGSC effort. They would contribute their expertise, in concert with the computational experts at Celera, and in turn were given access to the latest proprietary data on their area of interest, as well as the fruits of their collaboration with Celera. Following the publication of their sequence in *Science* in 2001, Celera kept further improvements to their assembly behind a paywall for their clients, who were primarily pharmaceutical and biotechnology companies rather than academics. At the jamboree, though, the academics could assess the sequence assemblies in regions on which they had expertise, contributing information that would not just refine the gene structures predicted by Otto, but also inform improvements to the overall automated annotation pipeline.

The involvement with medical geneticists did not end there. A further Chromosome 7 Annotation Project was initiated, prompted by a suggestion by medical geneticist Stephen Scherer to Richard Mural, the head of the Annotation Team at Celera. The result was a higher quality resequenced chromosome 7 that better connected to biomedical and clinical research due to the expertise and physical mapping data provided by medical geneticists. This provided the medical genetics community with a useful resource, as well as aiding Celera in its strategic reorientation towards identifying diagnostic and therapeutic targets.[21]

The ways in which genomes are improved and connected to other forms of data are explored further in the next chapter. For now, we note that the institutional imperatives of the IHGSC and Celera shaped the design of their respective annotation processes. Annotation, therefore, emerged in ways that reflected the trajectories, networks and goals of practitioners; Celera was more open to the medical genetics community, while the IHGSC was more self-contained.

In the following section, we consider the annotation of the pig genome, an effort in which existing pig genomicists interacted closely with teams at different stages of the sequencing and analysis pipeline established at the Sanger Institute. This reflected the model of interaction between medical geneticists and Celera more than the way that annotation unfolded within the IHGSC human reference genome sequencing. Furthermore, the relationship between the existing community of researchers working on the pig and the Sanger Institute helped to shift

---

[21] Peter Li, interview conducted over Skype by both authors, September 2020. See also Kerlavage et al. (2002).

some of the Sanger Institute's operations towards a model closer to the community annotation advanced by Celera.

6.2.3   Pig Genome Annotation

As it came after the sequencing of other genomes at the Sanger Institute, by the time the pig genome was sequenced, the annotation process used an established pipeline derived from procedures that had been deployed and refined in previous initiatives, in particular the sequencing and anno- tation of *Homo sapiens*. Like in sequencing and assembly, the pig project adopted and used repertoires established through the experience of proj- ects on other species, while adding distinctive twists on these.

For the sequencing itself, the community of pig genomicists through the Swine Genome Sequencing Consortium (SGSC) had contracted with the Sanger Institute rather than the project being initiated from within the IHGSC (Chap. 5). This contractual relationship did not, however, imply a hands-off approach by the community; it was intimately involved in guiding the strategic—and in some cases operational—direction of the project. Part of this direction meant indicating to the Sanger Institute where they should target sequencing efforts, so they could focus on par- ticular areas associated with genes of interest to individual research groups. This was reflective of a desire to make genome data useable as promptly as possible. As a result, even while the sequencing was still underway the com- munity pursued annotation, the identification of SNPs and the creation of a SNP chip that captured agriculturally-relevant genetic variation.

We discuss the creation of the SNP chip in the following chapter. Here we detail the annotation effort. Just over £1.1 million of funding was secured from the UK Biotechnology and Biological Sciences Research Council (BBSRC) for 2007–2010 by the Roslin Institute (with Alan Archibald as Principal Investigator and Andrew Law as co-investigator), the EBI (Ewan Birney as Principal Investigator) and the Sanger Institute (Tim Hubbard as Principal Investigator and Jane Rogers as co- investigator).[22] These grants funded four posts, one each in Hubbard and

---

[22] https://gtr.ukri.org/projects?ref=BB%2FE010520%2F1#/tabOverview (last accessed 18th December 2022); https://gtr.ukri.org/projects?ref=BB%2FE010520%2F2#/ tabOverview (last accessed 18th December 2022); https://gtr.ukri.org/ projects?ref=BB%2FE010768%2F1#/tabOverview (last accessed 18th December 2022); https://gtr.ukri.org/project/6AB44634-8225-4645-8935-CC9977F581BD#/ tabOverview (last accessed 18th December 2022).

Rogers' teams at the Sanger Institute, one in Archibald's group at the Roslin Institute and one supervised by Birney at the EBI. Two of these positions (with Hubbard and Birney) were in the Ensembl teams at the EBI and Sanger Institute. As noted above, the annotation effort began while the sequencing itself was still being conducted. Like in human genome sequencing, the pig genome was scanned using algorithms to predict the presence of genomic features. Pig protein and RNA sequence data were obtained from specific databases, and data on pig cDNA and ESTs were also downloaded from GenBank. Many of the cDNAs and ESTs had been generated by the Animal Genome Research Program at the National Institute of Agrobiological Sciences in Japan, and the Japan Institute of Association for Techno-innovation in Agriculture, Forestry and Fisheries (Groenen et al., 2012 and Supplementary Information; Lowe, 2018). These resources were generated in part using samples from cloned offspring of TJ Tabasco (Schook et al., 2005; Uenishi et al., 2012).[23]

A key feature of the automated annotation in the Swine Genome Sequencing Project (SGSP) was the integration into the Ensembl pipeline of multiple forms of data already generated by the community from prior projects. These data concerned maps, Quantitative Trait Loci, and clones, in addition to the cDNA and ESTs mentioned above. The community provided Ensembl with these rich resources to enable the annotated reference sequence to be connected with—and immediately contextualised by—other forms of data and information produced by pig geneticists. This enabled functional inferences to be made concerning parts of the genome, but also inferential pathways to be constructed between the pig genome and other porcine biological data, and also between the pig genome and the genomes of other species. With the means to generate comparisons with other mammalian genomes being a key product of the grant work, this connectivity was intended to boost the pig as a comparative model, with data and the results of experiments intended to travel along the connections forged within the species, but also then to be able to travel beyond the species. Crucially, this wider horizon was accompanied by a desire to embrace the varied research needs of the community of pig researchers in the annotation, through the addition of tracks comprising other forms of data to the Ensembl browser. This was

---

[23] This Japanese effort also used tissues from crossbred pigs derived from Landrace, Large White and Duroc breeds, and ones from a Chinese Meishan pig, two Landrace pigs, a Berkshire pig and a miniature pig (Uenishi et al., 2012).

effected through Ensembl's Distributed Annotation System, and pig geneticists who were interested in adding these tracks for the forms of data valuable to them were invited to contact Archibald, who was in regular liaison with teams at the Sanger Institute and the EBI.[24]

There were therefore multiple kinds of community involvement in even the automated annotation of the pig genome. The community helped to define the nature of the annotation, taking advantage of the clone-based sequencing to squeeze as much use out of the products of sequencing and assembly as possible, through integrating assembly and annotation as well as incorporating data and resources already developed by the community into the pipeline, or through the Distributed Annotation System. This was particularly important, as the resource limitations of the overall genome project entailed a trade-off between comprehensiveness and utility, with the community opting for a more rough-and-ready but more immediately exploitable resource, above aspirations for completeness.

This meant that the drawbacks of automated annotation, well-appreciated by the Ensembl team for the more refined human genome, were even greater for the pig genome. As Jennifer Harrow reported to us, the algorithms at the heart of Ensembl were only as good as the assemblies they were working on, and for the pig these were incomplete and of lower quality than for the human. Manual curation of the data by the biologically-trained members of the HAVANA team was therefore more critical for improving and developing the initial assemblies of the pig genome produced by the Ensembl pipeline, than it was for human or mouse.[25]

As with human genome sequencing, the annotated sequences produced through the Ensembl pipeline were published in the Ensembl database, while additional manual annotation was published on the HAVANA-led VEGA database, built on the Ensembl database.[26] HAVANA worked closely with some of the members of the pig genomics community, such as Christopher Tuggle at Iowa State University. James Reecy, an animal geneticist in Tuggle's group, spent his faculty leave (equivalent to a sabbatical) with them from September 2007 to August 2008. Like many pig geneticists, Reecy worked on multiple livestock species, in his case primar-

[24] "PIG TALES: Newsletter of the International Swine Genome Sequencing Consortium (SGSC) Pig Genome Sequence Project", 2nd Quarter 2007—Volume 1 Issue 3. On the Distributed Annotation System, see: Dowell et al. (2001).

[25] Jennifer Harrow, interview conducted in Cambridge by James Lowe, October 2017.

[26] For more on VEGA, see Harrow et al. (2014).

ily cattle. Reecy was interested in developing skills in manual annotation and areas of programming, and HAVANA had put together the most comprehensive approach to manual annotation in the world at the time. He was able to pursue this because of the close interactions between the pig genomics community and leading figures at the Sanger Institute, which we saw in Chap. 5. During his visit, Reecy met with Jane Rogers, Tim Hubbard and Richard Durbin, as well as Jennifer Harrow and Jane Loveland of HAVANA, discussing what he could offer in situ at the Sanger Institute. Aided by his demonstration that an animal geneticist could pick up the techniques of manual annotation, Reecy's advocacy of community involvement in annotation met a receptive audience in the HAVANA team.

As a result, HAVANA decided to dedicate more attention to manual annotation than they had been contracted to do and in so doing developed new means of manually annotating a genome.[27] This new model took two forms. HAVANA consulted with the SGSC members on an informal basis for guidance on what precise parts of the genome they wanted special attention paid to. This was a continuation of the targeted approach to sequencing and meant that the annotation could be preferentially refined in particular regions of interest to researchers. In the process, information was fed back to the assembly team if a problem was detected in the course of the manual curation.[28] As the annotation started while the reference genome was being assembled,[29] this allowed it to feed into the assembly (and even inform the amendment of algorithms in automated assembly pipelines), as well as adding value to the eventual sequence.

Additionally, HAVANA shifted its mode of operation, developing new capabilities in education, training and engagement to increasingly function as community annotation facilitators, providing the pig geneticists with the tools, training and assistance so that they could annotate the genome themselves. This began with a training programme hosted at the

[27] This illustrates the importance of the initial choice of the Sanger Institute to host the sequencing of the pig genome, even if this was not made with the eventual model of annotation in mind. The Human Genome Sequencing Center at Baylor College of Medicine, the other candidate to sequence the pig genome, as it had the cattle, was comparatively quite small. The kind of manual annotation employed for the pig and the development of community annotation would therefore have been less likely to occur there.

[28] Jennifer Harrow, interview conducted in Cambridge by James Lowe, October 2017; Kerstin Howe, interview conducted at Wellcome Genome Campus (Hinxton, Cambridgeshire) by James Lowe, October 2017.

[29] Craig Beattie, interview conducted over Skype by James Lowe, March 2017.

Sanger Institute in July 2008. While this event was labelled as a "jamboree", it differed from the *Drosophila* and human jamborees organised by Celera. Rather than just annotating the genomes in situ, the Sanger Institute event was intended to equip the researchers to go back to their own institutions and conduct annotation on regions of the genome pertinent to their existing research projects there. Abridged guidelines were created for pig annotation, due to the need to do the annotation quickly because of resource constraints, but also to economically document the key processes and procedures for these amateur annotators scattered around the world. Conference calls were used to share problems, observations and advice, but a manual was still needed for the HAVANA facilitators to refer to, and for the manual annotators to consult in their own offices and labs between meetings (see Fig. 6.3).

This community annotation effort was aided by the availability of the Otterlace/ZMap system combining a relational database and graphical interface for the manual annotators to use (Loveland et al., 2012; Dawson et al., 2013). In turn, HAVANA used their close working relationship with the pig genomicists to develop their tools and annotation processes.

The initial step in the manual annotation process was the computational alignment of multiple forms of data from the pig—and other species such as human and mouse—onto the *S. scrofa* genome assembly. A crucial feature of the Otterlace/ZMap manual annotation system used by HAVANA and VEGA was that it enabled annotation of an ongoing assembly rather than just individual clones, which was all that previous curation tools had allowed users to annotate (Searle et al., 2004). This functionality was helpful to pig genomicists, who wanted to promptly exploit and further augment the sequences so assembled. It meshed with the more significant role that manual procedures had in the annotation of the *S. scrofa* reference genome. The combination of the automated pipeline with the bespoke manual sequencing distributed in laboratories across the world constituted a combination of Stein's factory and cottage industry models, and was therefore different to the case of Ensembl discussed above (Lowe, 2018).[30]

This initial curation created a visualisation that displayed the sequence data along with another layer of information indicating evidence for the possible presence of genes. With this, anyone with an account could log in

---

[30] As we discuss later, the manual annotation of the X and Y chromosomes was performed by the Sanger Institute itself.

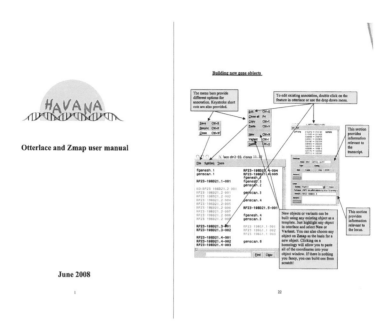

Fig. 6.3 Cover and selected page of a manual produced by the HAVANA team for use by manual annotators of the pig genome community. From personal papers of Alan Archibald, "Pig Sequencing" folder, obtained 17th May 2017. Reproduced with permission, courtesy of Alan Archibald and the Human and Vertebrate Annotation group at the Wellcome Trust Sanger Institute. For a larger version of this figure that can be zoomed in and out, see https://www.pure.ed.ac.uk/ws/portalfiles/portal/314800096/higheres_fig_6_3.pdf

to the Otterlace/Zmap system and start to annotate a chosen gene. The annotator could weigh the different forms of evidence presented to them, and amend the model of the gene according to that evidence and any specific knowledge of the gene that they have. They would then be able to submit it for inspection by a HAVANA team member, who could then work on it further to finish off the annotations to the required standard.[31]

In the earlier annotation of the human genome, as well as for well-funded model organisms such as the mouse, HAVANA had generally performed manual annotation wholly in-house. Its role was quite different for the pig, instead conducting education and training to enable researchers

---

[31] Jennifer Harrow, interview conducted in Cambridge by James Lowe, October 2017.

to themselves manually annotate genes, with the HAVANA team then performing quality control on the results. The only other species that HAVANA was providing community annotation support for at the time was cattle (*Bos taurus*). There were, though, weaker interactions between HAVANA and the cattle genomics community, partly because its greater funding meant that a close relationship was less necessary, but also because of the less-established links that the Sanger Institute had enjoyed with members of this community compared to pig genomicists (Chap. 5).

Parallel to HAVANA's tasks, the pig genomics community itself helped to organise the manual annotation activity. As bioinformatics coordinator, Reecy led the community side of the work and provided training on manual annotation in the USA and China. In Scotland, training was also provided by the Roslin Institute. With Reecy, Iowa State University colleague Zhi-Liang Hu set up a website listing the genes and gene families that were candidates for manual annotation, and individual researchers were invited to indicate which they intended to annotate. This has been described as an "adopt-a-gene type approach" by Reecy, building on the targeting strategy in the sequencing phase.[32] The community did not have the resources to manually curate the whole genome to a high standard. They needed to maximise the utility of the genome for their particular research purposes, and for this, selectivity and distribution of the sequencing were appropriate. The value of the genome was therefore not primarily assessed in terms of generic metrics, even if data on the number of genes annotated still constituted a useful barometer of progress. The key was the utility of what had been done, not the extent of it; such concerns with completeness were more of a priority for the IHGSC. The pig community assessed the *S. scrofa* genome in terms of its use as a research tool for their own purposes. They were themselves deeply imbued with an awareness of what was required in the domains of agricultural or other forms of translation that they worked towards.[33]

---

[32] The website (still live as of 18th December 2022) is: https://www.animalgenome.org/cgi-bin/host/ssc/gene2bacs. It was actively updated from November 2009 to September 2010. We thank Zhi-Liang Hu for kindly providing us with the information on this, following an initial lead provided to us by James Reecy. Most of the adopted genes were taken by the Immune Response Annotation Group (see below) and Cathy Ernst's research group. Zhi-Liang Hu defines himself as a bioinformaticist: someone who programmes new tools as well as using them (personal communication with James Lowe, January 2022).

[33] James Reecy, interview conducted over Zoom by both authors, May 2021.

For this manual annotation, particular groups were established based on common research and translation interests. Some of these focused on resolutely structural elements such as repetitive sequences, while others operated in areas where the line between structural and functional was blurred. Examples of the latter were the groups that aimed to annotate genes and analyse genomic regions relating to olfaction, immune response, and retroviral insertions into pig DNA such as Porcine Endogenous Retroviruses (PERVs). The range of interests of the pig genome community was reflected in these groups. In addition to the interests listed above, genomicists working on domestication and the relationships between the sequenced domesticated pig and European and Asian wild boar contributed analyses to the publication heralding the reference sequence (Groenen et al., 2012).

It was the involvement of the pig genomics community in annotation processes that helped to blur the line between structural and functional annotation. This is illustrated by the most developed of the annotation groups, which became the Immune Response Annotation Group (IRAG) and continued its activities well beyond the initial analysis of the reference genome. IRAG comprised 51 researchers based in thirteen institutions in China, France, India, Italy, Japan, UK and USA. There had been considerable work on immune response prior to genomic research, as we showed in Chap. 5. Further, a high-quality manually annotated sequence of the pig's MHC (the Swine Leucocyte Antigen complex, or SLA) was published in 2006, as a result of work by Laboratoire Mixte CEA-INRA de Radiobiologie Appliquée (CEA-INRA), Genoscope, Tokai University in Japan, and the Sanger Institute. The HAVANA team and the CEA-INRA group (in particular, Christine Renard) performed the manual annotation of the SLA region (Renard et al., 2006). It did not therefore need to be developed further in the subsequent 'immunome' project.

This ambitious 'immunome' project and group arose out of discussions between researchers at CEA-INRA and Iowa State University, in particular Claire Rogel-Gaillard at the former and Christopher Tuggle at the latter. They each had straightforward motivations for establishing this effort, since they both worked on the immunogenetics of the pig. We have already encountered Rogel-Gaillard, part of the team at CEA-INRA (and later, just INRA) that had adopted genomic approaches to investigating immune response. This had involved studying the dense polymorphic regions containing genes implicated in it from the 1980s, as well as investigating PERVs in the late-1990s, which had implications for the prospective xenotransplantation of pig organs and tissues into humans. Together with Patrick Chardon, she had led the development of the YAC and BAC libraries of pig DNA to aid those research efforts (Chap. 5). Her research interests had increasingly been directed towards studying the genetics of immune response variability in terms of pig health and resilience against disease. Tuggle's research had trended in a similar direction, though from a different origin: his work in the 1990s was at the heart of the mapping endeavour to try to identify (and then exploit) genes and Quantitative Trait Loci primarily involved in livestock production traits in pigs.[34]

From this nucleus, a call for interested parties was issued, and once the participants were confirmed, the group set about seeking data from databases and the literature to identify a list of genes to annotate.[35] Once this list was agreed and the rules for annotation established, particular sets of genes were assigned to individual teams. The approach embodied the advantages and disadvantages of distributed, targeted community annotation, as while expertise could be applied to particular regions by researchers, this meant that some regions went unadopted, for instance those with lower sequence quality that were difficult to annotate as a result or ones that simply did not contain genes of interest.[36]

Reecy provided training for the group's annotators in a workshop, but beyond that people worked in their own offices and labs, using Otterlace. Annotators would be able to see the analysis for their particular region, with the data tracks (for example the RNAs aligned to it) depicted. They would also be able to use the software tools to tweak the predictions made at the Sanger Institute.[37] The work was coordinated, and credit negotiated, in regular conference calls, using the Webex videoconferencing application to share screens. Jennifer Harrow had overall oversight at the HAVANA end, which included making the decisions about which annotations to exclude. She guided Jane Loveland in the day-to-day management, coordinating annotation between different groups, showing

---

[34] Tuggle took over from Max Rothschild, his frequent co-author and superior in the Iowa State University Department of Animal Science in the 1990s, as the National Swine Genome Coordinator for the US Department of Agriculture.

[35] Claire Rogel-Gaillard, interviews conducted over Skype by James Lowe, May 2017. See also Dawson et al. (2013). In particular, the group searched for annotations in the Gene Ontology, using "immune system process", GO:0002376, as the inclusion criterion.

[36] Jane Loveland, interview conducted at Wellcome Genome Campus (Hinxton, Cambridgeshire) by James Lowe, October 2017.

[37] Christopher Tuggle, interview conducted over Skype by James Lowe, March 2017.

annotators how to use tools and access data, conducting quality control on the annotations and giving feedback. The motivation for HAVANA was to enable communities to take on as much of the task of annotation themselves as possible, both as a general aim and a particular solution for the resource-poor pig genomics community.[38] While the HAVANA team primarily supplied support for the informatics aspects of the manual annotation, on the community side a trio of coordinators—Rogel-Gaillard, Tuggle and Harry Dawson—guided the effort with a view to making the resulting annotated sequence as valuable as possible for those who would make use of it. Dawson, based at the USDA's Beltsville facility in Maryland, monitored which genes were being annotated, following up on any genes that remained unannotated. He also conducted cross-species comparative analyses based on the annotation data he compiled from the whole project.[39] Dawson had led the development of the Porcine Immunology and Nutrition (PIN) Database at Beltsville, which was launched in 2005 containing data on 2600 annotated pig genes, with gene expression data linked to information on gene function. The database (now known as the Porcine Translational Research Database) was configured to enable users to identify genetic pathways related to genes of interest and to connect to human and mouse databases for comparative purposes, as well as to other pig genomic databases (Dawson et al., 2007).[40]

Because the annotation began with a panel of genes, rather than simply annotating the assembly that was there, genes missing from the assembly could be identified, and therefore areas of the assembly that needed further work could be pinpointed. Indeed, having conducted the annotation using version (build) 9 of the swine genome, the results of the annotation fed into the newer and improved version 10.2. The annotators refined the models of 1369 genes and elucidated 3472 transcripts from these, around a third of which were inferred using only data from other species. They extended the analysis concerning genes under positive selection undertaken in the 2012 *Nature* paper announcing the reference sequence. And finally, the group used transcriptomic data derived from experiments to

[38] Jennifer Harrow, interview conducted in Cambridge by James Lowe, October 2017; Jane Loveland, interview conducted at Wellcome Genome Campus (Hinxton, Cambridgeshire) by James Lowe, October 2017.

[39] Claire Rogel-Gaillard, interviews conducted over Skype by James Lowe, May 2017. See also Dawson et al. (2013).

[40] https://web.archive.org/web/20220928072749/https://www.ars.usda.gov/news-events/news/research-news/2005/pig-gene-database-supports-human-nutrition-immunity-studies/ (last accessed 18th December 2022).

discern the role of some of the genes involved in immune response, identify networks of co-expression of genes and to annotate accordingly (Dawson et al., 2013).

This work had direct translational impact motivating it, and this gave the group clear indications on how to target their focus and structure the division of labour within the project. To achieve the translational ends of the researchers involved, the methods and approaches employed in the project were comparative, and explorations of function were knitted together with examinations of diversity and evolution.[41] For example, inferences that the researchers made about the evolution of genes accompanied functionally-oriented transcriptomic studies. Genes identified for their putative function enabled both the functional and structural annotation of the genome to be improved. And these in turn fed into the refined assembly of the genome itself.

Concerning the improvement of the reference genome as a community-generated resource, we close with an account of the sequencing and annotation of the pig's X and Y chromosomes. This project filled the gap left by the SGSP, which had excluded the sex chromosomes due to the complexities involved in their sequencing. The sequencing of the sex chromosomes therefore finally completed a reference sequence for the whole of the nuclear genome of *S. scrofa*. This project also shows how the existing community of pig genomicists were able to broker and contribute to a collaboration between the Sanger Institute and an external group of researchers who had been working on these sex chromosomes for both biomedical and agriculturally-oriented purposes.

This project involved the EBI and the Sanger Institute, was funded with a BBSRC grant, and used infrastructure and work that was supported by the European Commission and the Wellcome Trust, much like previous work we have described. It did not involve any of the 'usual suspects' from the pig community as a collaborative partner, however, but a group based in the Department of Pathology at the University of Cambridge who had been consistently investigating the sex chromosomes of the pig since the turn of the century.[42] Their research had a dual aspect, being

---

[41] In Chap. 7, we term such research on diversity and evolution as 'systematic' and examine the different ways in which explorations of these topics relate to functional studies across yeast, human and pig.

[42] https://gtr.ukri.org/projects?ref=BB%2FF021372%2F1#/tabOverview (last accessed 18th December 2022).

motivated by biomedical objectives, as well as being supported by a major pig breeding firm, the Pig Improvement Company (PIC), due to the implications of the genetics of sperm development and male fertility for breeding purposes.[43] The Cambridge University-led arm of the sequencing and annotation of the pig X and Y chromosomes was also conducted in collaboration with PIC. A key figure in the mapping of individual genes relating to sperm fertility was Andy Day. His funding came from PIC, who he had worked for since leaving university in 1995 and continued to be employed by until 2006. Day's research at the University of Cambridge used comparative approaches to exploit the more plentiful and refined data and resources concerning the human genome to aid in the mapping of specific genes in the pig (Day et al., 2003; Kollers et al., 2006). One of his collaborators, Claire Quilter, approached human–pig comparative genomics from a medical genetic angle: she worked on the role of the Y chromosome in male infertility and Turner syndrome, a condition that affects women and involves the lack of all or part of an X chromosome.[44]

In the early-2000s, Quilter had been the lead author of a paper that surveyed porcine sex chromosomes, identifying and mapping 19 genes onto them. For this, she made use of the PigEBAC library developed by the Roslin Institute and the UK Human Genome Mapping Project Resource Centre. This work explored the evolutionary consequences of this mapping data, in part by comparing the order of genes determined on the porcine Y chromosome with the corresponding order of those genes on the human and mouse Y chromosomes (Quilter et al., 2002). As well as representing a convergence of biomedical and agriculturally-inclined research, it also presaged the entanglement of comparative, evolutionary and functional studies that would be further realised in the work conducted with the Sanger Institute, and also the relationship between systematic and functional genomics explored in Chap. 7.

The X and Y chromosomes were an interesting challenge for the HAVANA team, due to the high level of conservation in X chromosomes and the tricky genomics of the Y chromosome. Y chromosomes contain a

---

[43] As with many of the institutions mentioned in this book, we have affixed one name for an institution that changed names and did not have a straightforward institutional history. The Pig Improvement Company was founded in 1962, was bought by Dalgety plc in 1970, which became the PIC International Group in 1998, and then Sygen International Group in 2001. Genus, a cattle breeder, bought Sygen in 2005. 'PIC' remains a brand for the pig breeding side of their business; for more, see Bruce and Lowe (2022).

[44] https://www.researchgate.net/profile/Claire-Quilter (last accessed 18th December 2022).

lot of repetitive sequences and degenerated genes due to its near-complete isolation from recombination with the X chromosome during meiosis.[45] In the original reference genome operation by the SGSP, some limited sequencing of the Y chromosome had been conducted using clones from the DNA libraries derived from males. However, only 11 clones were sequenced—in a draft rather than finished condition—and a limited number of scaffolds containing positioned contigs were placed on the chromosome: hardly an assembly (Groenen et al., 2012, Supplementary Information).

On the sequencing side, the X and Y chromosomes project began under the leadership of Jane Rogers. When she left the Sanger Institute, it was taken over by Chris Tyler-Smith, a human evolutionary geneticist. The sex chromosome sequencing project began in 2009. Both sides of the project were funded by the BBSRC for three years, with the Sanger Institute being alloted £1,369,161 to Cambridge's £349,639.[46] The endeavour would contribute an improved assembly and annotation of the X chromosome and the first assembly and annotation of the Y chromosome.

Beyond the original pig genome sequencing, the X and Y work benefited from a change in mapping techniques and improvements to sequencing techniques.[47] Optical mapping was used to build a new assembly of X. To conduct this, Kerstin Howe—who led the team that analysed, validated and improved genome assemblies such as the pig one—worked alongside David C. Schwartz, who pioneered the method for eukaryotes.[48] Optical mapping does not require the use of library clones and the technique obviates the need for reconstruction of the order of the clones. It was therefore useful in correcting problematic repetitive regions that are difficult to resolve using clone-based mapping. The new optical-based map enabled the corrected assembly to be produced, which was then improved further, for example with targeted sequencing to close gaps and resolve assembly problems. This improved assembly in turn enabled an improved annotation, with 690 protein-coding genes annotated, a

---

[45] Jane Loveland, interview conducted at Wellcome Genome Campus (Hinxton, Cambridgeshire) by James Lowe, October 2017. See also Skinner et al. (2016).

[46] https://gtr.ukri.org/projects?ref=BB%2FF02195X%2F1#/tabOverview (last accessed 18th December 2022); https://gtr.ukri.org/projects?ref=BB%2FF021372%2F1#/tabOverview (last accessed 18th December 2022).

[47] Jennifer Harrow, interview conducted in Cambridge by James Lowe, October 2017.

[48] Kerstin Howe, interview conducted at Wellcome Genome Campus (Hinxton, Cambridgeshire) by James Lowe, October 2017.

considerable advance over the 422 in the original (for Sscrofa10.2), with increased numbers of non-coding genes and pseudogenes identified as well. As with the SGSP, there was close interaction between the annotation and assembly teams at the Sanger Institute.

For the Y chromosome, a bespoke library was created using DNA from a Duroc boar (the same breed as the originator of the CHORI-242 clones from which the bulk of the reference sequence was derived) donated by Genus, the company that incorporated PIC. At the Sanger Institute, a fingerprint contig map was produced using this library to create a map of overlapping clones which formed the basis of a minimum tiling path to guide the sequencing and assembly. They used and combined the outputs of multiple sequencing platforms, and then improved it further as with the X chromosome, to bring the sequence towards 'Finished' standard. This updated assembly was validated using PacBio long-read technology, which affirmed the high quality of the new assembly, using the same clone library as the original sequencing conducted by the SGSP.

For both the X and Y chromosomes, annotation involved the alignment of various EST, messenger RNA, and protein sequence data against the sequence. This was performed through the Otter annotation pipeline, and it then underwent manual curation by the HAVANA team, using the Otterlace/Zmap tools according to the procedures developed for both human genome annotation through GENCODE (Chap. 7) and the immunome project (Skinner et al., 2016 and Supplementary Information).[49] The Y chromosome assembly subsequently became incorporated into the updated Sscrofa11.1 assembly, which became the reference genome (at 'representative genome' level in RefSeq) for the pig in 2017 (Warr et al., 2020).

Cambridge University's side of the project involved identifying shared regions between the two chromosomes to aid in the sequencing of them and in tracing their evolutionary history, identifying functional genes and non-coding sequences on the Y chromosome, and locating and analysing a gene—*HSFY*—found in cows to study chromosomal evolution across pigs and closely-related species. The insights gained from this project were explicitly designed to inform the sequencing and assembly of the chromosomes using the knowledge gained about their structure and the location

---

[49] On the PacBio validation: Jane Loveland, interview conducted at Wellcome Genome Campus (Hinxton, Cambridgeshire) by James Lowe, October 2017.

of repetitive sequences, but also to guide the exploitation of the data.[50] This research was therefore a good example of the functional and systematic synergies that are explored further in the following chapter.

It also shows how the specific genetic expertise of a group of researchers newly admitted to the community of pig genomicists, fed into and informed the highly-developed pipelines and expertise at the Sanger Institute. Here, the Sanger Institute did not conduct this work merely at its own initiative or at the behest of the Wellcome Trust or an international collaboration like the IHGSC. It also was not merely contracted to perform the work, as per the original relationship with the pig genomicists. Instead, building on the relationships developed through pig genome sequencing, which intensified as attention was directed towards annotation and the development of a new community-oriented model of it, the X and Y project constituted a more horizontal peer-to-peer collaboration from the start. This collaboration involved the highly-refined infrastructures and personnel of a large-scale genome centre. It incorporated a community of pig genomicists with a core of operators such as Alan Archibald who married a drive towards the development of genomic resources intended for wide use with a sensitivity to particular uses to which they could be put. And finally, it included an existing set of researchers seeking to conduct sequencing and annotation pertaining directly to their ongoing interests.

The X and Y project instantiates deep entanglements between different models of sequencing and annotation. It challenges strict demarcations and distinctions, and also the linearities indicated by presumed separations between stages, whether in particular projects or pertaining to the wider development of genomics. Who would dare reduce this X and Y project—or any part of it—to a singular form of annotation along the lines of Stein's ideal types, or even to any of the strategies pursued in prior genomics projects such as the genome centre model of the IHGSC, or the distributed model of the European Commission-funded Yeast Genome Sequencing Project? Instead, as the progression of pig genomics illustrates, aspects of these models were mobilised and combined, mediated by the historical trajectories of the actors coming together to form particular projects.

---

[50] https://gtr.ukri.org/projects?ref=BB%2FF021372%2F1#/tabOverview (last accessed 18th December 2022).

6.3   ANNOTATION STRATEGIES AND LINEAGES OF GENOMICS

In examining the different models of reference genome annotation for yeast, human and the pig, this chapter has begun to explore the development and use of genomic resources beyond the determination of the nucleotide sequence of the reference genome. This broader perspective expands the range of narratives that historians can mobilise to capture genomics as an ongoing and multifaceted endeavour, moulded in distinct ways by different communities.

The yeast genome annotation followed the distributed-but-hierarchical model of the European Commission's sequencing project, with a key role for MIPS as the bioinformatics coordinator. The centralisation through MIPS reflected the division of labour of the sequencing across multiple, often small, laboratories and the need for a genome-wide perspective for some forms of genome analysis that the consortium wanted to perform. In this model, we see a strict separation of structural from functional annotation.

The human reference genome, on the IHGSC side, involved the development of the Ensembl pipeline and HAVANA to automatically and then manually annotate the sequence data. IHGSC institutions progressively added new sources of data and methods for the annotation of various elements in the human genome, such as protein-coding genes. Compared with Celera's approach, this involved far less interaction with wider communities of researchers, and instead a concentration on developing pipelines and repertoires to improve the quality and extent of annotation, without directing or targeting it towards particular users. The aims and operations were therefore internal to a community of specialist genomicists, institutions and operatives, who sought to improve the output as measured by general metrics and guided by an ideal of completeness.

This, as we have seen, was not a fixed or essential characteristic of the genome centres, the key institution in the IHGSC model. In the case of the Sanger Institute, for example, the relationship of some of its departments and key personnel to a well-coordinated pig genome community effected a change in the way this institution worked. As a result, the model and results of the annotation of the pig genome were quite distinct from the human annotation that preceded it.

Some of this was driven by resource constraints that limited the quality of the pig genome assembly in some respects, making manual curation more crucial in correcting the automated predictions. As funding would only go so far in paying for in-house manual curation, the community would need to take up the slack. The extent they were able to do this owed much to the community's own history of coming together to coordinate the work of identifying genetic markers, compiling and integrating genetic, cytogenetic and physical maps, and creating databases and materials (such as genome libraries and radiation hybrid panels). They pursued the creation of genomic resources because they knew what kinds of data they needed to advance their own research. Together, they advanced their overall endeavour of improving the genomic reference resources concerning the pig, secured pots of money from various sources to do so, and then worked out how to stretch what they had as far as they could. This accommodated but also drew upon the heterogeneous but often overlapping interests held across the pig genome community. For their members, like those forming the yeast genomics community, genomics has constituted a nexus around which multiple different interests could draw upon the resources generated through it, with those interests and motivations also shaping the creation of those resources in distinctive ways.

Indeed, a reference genome is a creative and dynamic product. The selection of the materials that are used in its creation and the decisions made in sequencing and assembly reaffirm that. It matters what libraries are used, what methods are used in sequencing and assembly, and what is or is not targeted for special treatment to refine sequence quality. This is even more the case for annotation. Annotation is affected by the prior steps, but in turn, what is annotated can feed back to further develop the assembly. It will also affect what the genome can be used for. The model of distributed community annotation—involving individuals, laboratories and groupings of researchers interested in genes with particular hypothesised functions—guided the annotation of the pig genome towards those regions deemed useful for proximate research purposes. In terms of the allotting of work, there was a similarity with the yeast genome sequencing network, though for the pig it was less hierarchical and comprehensive, and more discretionary.

The activities of the SGSP more generally, and IRAG and the X and Y chromosome sequencing more specifically, involved a wider set of actors, approaches and interests than the IHGSC. IRAG involved members of an existing community of pig genomicists that dated back to at least the 1990s. The project to sequence the X and Y chromosomes, though, showed how that community still had the ability to form new connections.

While the scale, speed and automation of sequencing operations had all increased at the Sanger Institute, this did not intensify the tendency we observed in the IHGSC effort: the narrowing of participation and the concentration of operations in-house (Chap. 4). Indeed, the Sanger Institute, and in particular the HAVANA group, opened out to and engaged with a specific external community to develop new genomic resources, tools and expertise through the assembly and annotation activities of the SGSP, IRAG and the X-Y project. That community shaped the direction of various aspects of the sequencing process, in so doing affecting the nature of the product. In turn, the Sanger Institute, at a time in which it was adjusting to the period following the 'completion' of the human reference sequence and each chromosome in turn, itself changed the way it worked.

In considering how the Sanger Institute and the pig genomics community shaped their emerging community annotation strategy and practices, we observe that the cottage industry model (Stein, 2001) needed to be implemented and combined with factory-style approaches. These genomicists, therefore, deployed modes of annotation regarded as characteristic of earlier 'pre-genomic' stages, in conjunction with the concentrated factory style that came to dominate the sequencing of the human reference genome. This challenge to the idea of progression through distinct and separate models and stages of activity, is an important historiographical consequence of our account of pig genome annotation.

As well as helping to re-shape the way that HAVANA operated, the work of pig genome annotation fed into the processes of assembly, automated annotation and indeed manual annotation itself. This was enabled by the temporality of annotation that existed in the pig genome project, with manual annotation occurring alongside ongoing assembly. The manual annotation was therefore able to help correct the assembly as well as contributing to the improvement of automated prediction algorithms. The pig genome community conceived the genome they were helping to produce as provisional and incomplete; their attitude was one of satisficing (on satisficing, see Wimsatt, 2007).

Of course, as we see at the outset of the following chapter, reference genomes are never complete; they are always subject to changes intended to improve their quality and utility. But the pig genome community did not hold an ideal of completeness or comprehensiveness to be paramount in the creation of the first reference assemblies. In one respect, they shared this attitude with Celera. For Celera, the very provisionality of their human sequence was its selling point; it was important that the publicly-available data it had released in 2001 quickly became outmoded, and that it was widely known to be so. This was to make access to the continually-improved genome and associated data that they held behind a paywall more valuable to potential subscribers. It was this commercial strategy, along with the model of OMIM and their experiences with *Drosophila* sequencing and annotation, that encouraged Celera to forge collaborations with medical geneticists who had been peripheral to the IHGSC.

We have shown that distinctions between manual and automated annotation, annotation and assembly, and functional and structural annotation should all be qualified. In the next chapter, we demonstrate something analogous as we explore the changing relationship between the functional and systematic genomic research that followed the initial sequencing and annotation of the reference genomes of our three species.

REFERENCES

Agar, J. (2012). *Science in the twentieth century and beyond.* Polity Press.

Agar, J. (2020). What is science for? The Lighthill report on artificial intelligence reinterpreted. *The British Journal for the History of Science, 53*(3), 289–310.

Baker, K. S., & Millerand, F. (2010). Infrastructuring ecology: Challenges in achieving data sharing. In J. N. Parker, N. Vermeulen, & B. Penders (Eds.), *Collaboration in the new life sciences* (pp. 111–138). Routledge.

Birney, E., Andrews, T. D., Bevan, P., Caccamo, M., Chen, Y., Clarke, L., et al. (2004). An overview of Ensembl. *Genome Research, 14,* 925–928.

Birney, E., Clamp, M., & Durbin, R. (2004). GeneWise and Genomewise. *Genome Research, 14,* 988–995.

Bruce, A., & Lowe, J. W. E. (2022). Pigs and Chips: The making of a biotechnology innovation ecosystem. *Science & Technology Studies.* https://doi.org/10.23987/sts.111111

Bussey, H., Kaback, D. B., Zhong, W.-W., Vo, D. T., Clark, M. W., Fortin, N., et al. (1995). The nucleotide sequence of chromosome I from *Saccharomyces cerevisiae. Proceedings of the National Academy of Sciences of the United States of America, 92*(9), 3809–3813.

Collins, J. E., Goward, M. E., Cole, C. G., Smink, L. J., Huckle, E. J., Knowles, S., et al. (2003). Reevaluating human gene annotation: A second-generation analysis of chromosome 22. *Genome Research, 13,* 27–36.

Curwen, V., Eyras, E., Andrews, T. D., Clarke, L., Mongin, E., Searle, S. M. J., & Clamp, M. (2004). The Ensembl automatic gene annotation system. *Genome Research, 14,* 942–950.

Dawson, H. D., Guidry, C. A., Vangimalla, V., & Urban, J. F., Jr. (2007). The Beltsville Human Nutrition Research Center's porcine immunology and nutrition resource database. *The FASEB Journal, 21*(5), A377.

Dawson, H. D., Loveland, J. E., Pascal, G., Gilbert, J. G. R., Uenishi, H., Mann, K. M., et al. (2013). Structural and functional annotation of the porcine immunome. *BMC Genomics, 14*, 332.

Day, A. E., Quilter, C. R., Sargent, C. A., & Mileham, A. J. (2003). Chromosomal mapping, sequence and transcription analysis of the porcine fertilin beta gene (*ADAM2*). *Animal Genetics, 34*, 375–378.

de Chadarevian, S. (2004). Mapping the worm's genome. Tools, networks, patronage. In J.-P. Gaudillière & H.-J. Rheinberger (Eds.), *From molecular genetics to genomics: The mapping cultures of twentieth-century genetics* (pp. 95–110). Routledge.

Deloukas, P., Matthews, L. H., Ashurst, J., Burton, J., Gilbert, J. G., Jones, M., et al. (2001). The DNA sequence and comparative analysis of human chromosome 20. *Nature, 414*, 865–871.

Deloukas, P., Earthrowl, M. E., Grafham, D. V., Rubenfield, M., French, L., Steward, C. A., et al. (2004). The DNA sequence and comparative analysis of human chromosome 10. *Nature, 429*, 375–381.

Dowell, R. D., Jokerst, R. M., Day, A., Eddy, S. R., & Stein, L. (2001). The Distributed Annotation System. *BMC Bioinformatics, 2*, 7.

Dujon, B. (1996). The yeast genome project: What did we learn? *Trends in Genetics, 12*(7), 263–270.

Dujon, B., Alexandraki, D., André, B., Ansorge, W., Baladron, V., Ballesta, J. P., et al. (1994). Complete DNA sequence of yeast chromosome XI. *Nature, 369*, 371–378.

Dunham, A., Matthews, L. H., Burton, J., Ashurst, J. L., Howe, K. L., Ashcroft, K. J., et al. (2004). The DNA sequence and analysis of human chromosome 13. *Nature, 428*, 522–528.

Dunham, I., Shimizu, N., Roe, B. A., Chissoe, S., Hunt, A. R., Collins, J. E., et al. (1999). The DNA sequence of human chromosome 22. *Nature, 402*, 489–495.

Feldmann, H., Aigle, M., Aljinovic, G., André, B., Baclet, M. C., Barthe, C., et al. (1994). Complete DNA sequence of yeast chromosome II. *The EMBO Journal, 13*(24), 5795–5809.

Galibert, F., Alexandraki, D., Baur, A., Boles, E., Chalwatzis, N., Chuat, J. C., et al. (1996). Complete nucleotide sequence of *Saccharomyces cerevisiae* chromosome X. *The EMBO Journal, 15*(9), 2031–2049.

García-Sancho, M. (2012). *Biology, computing, and the history of molecular sequencing: From proteins to DNA, 1945–2000.* Palgrave Macmillan.

García-Sancho, M., Leng, R., Viry, G., Wong, M., Vermeulen, N., & Lowe, J. W. E. (2022). The Human Genome Project as a singular episode in the history of genomics. *Historical Studies in the Natural Sciences, 52*(3), 320–360.

García-Sancho, M., Lowe, J. W. E., Viry, G., Leng, R., Wong, M., & Vermeulen, N. (2022). Yeast sequencing: 'Network' genomics and institutional bridges. *Historical Studies in the Natural Sciences, 52*(3), 361–400.

Gregory, S. G., Barlow, K. F., McLay, K. E., Kaul, R., Swarbreck, D., Dunham, A., et al. (2006). The DNA sequence and biological annotation of human chromosome 1. *Nature, 441*, 315–321.

Groenen, M. A., Archibald, A. L., Uenishi, H., Tuggle, C. K., Takeuchi, Y., Rothschild, M. F., et al. (2012). Analyses of pig genomes provide insight into porcine demography and evolution. *Nature, 491*, 393–398.

Harrow, J. L., Steward, C. A., Frankish, A., Gilbert, J. G., Gonzalez, J. M., Loveland, J. E., et al. (2014). The Vertebrate Genome Annotation browser: 10 years on. *Nucleic Acids Research, 42*, D771–D779.

Hattori, M., Fujiyama, A., Taylor, T. D., Watanabe, H., Yada, T., Park, H. S., et al. (2000). Chromosome 21 mapping and sequencing consortium. The DNA sequence of human chromosome 21. *Nature, 405*, 311–319.

Heilig, R., Eckenberg, R., Petit, J. L., Fonknechten, N., Da Silva, C., Cattolico, L., et al. (2003). The DNA sequence and analysis of human chromosome 14. *Nature, 421*, 601–607.

Heumann, K., Harris, C., & Mewes, H. W. (1996). A top-down approach to whole genome visualization. *ISMB-96 proceedings*, 98–108. Retrieved December 18, 2022, from https://citeseerx.ist.psu.edu/document?repid=rep1&type=pdf&doi=21e9d845b94393d0d8371452cd8c2e61cb6a9581

Hilgartner, S. (2017). *Reordering life: Knowledge and control in the genomics revolution*. The MIT Press.

Hillier, L. (2012). Interview conducted over telephone by Kathryn Maxson, Robert Cook-Deegan, 5 April 2012. Retrieved December 18, 2022, from https://dukespace.lib.duke.edu/dspace/bitstream/handle/10161/7701/2012%2005%20April%20LaDeana%20Hillier%20interview.pdf?sequence=1&isAllowed=y

Hillier, L. W., Fulton, R. S., Fulton, L. A., Graves, T. A., Pepin, K. H., Wagner-McPherson, C., et al. (2003). The DNA sequence of human chromosome 7. *Nature, 424*, 157–164.

Howe, K. L., Achuthan, P., Allen, J., Allen, J., Alvarez-Jarreta, J., Amode, M. R., et al. (2020). Ensembl 2021. *Nucleic Acids Research, 49*(D1), D884–D891.

Humphray, S. J., Oliver, K., Hunt, A. R., Plumb, R. W., Loveland, J. E., Howe, K. L., et al. (2004). DNA sequence and analysis of human chromosome 9. *Nature, 429*, 369–374.

Johnston, M., Andrews, S., Brinkman, R., Cooper, J., Ding, H., Dover, J., et al. (1994). Complete nucleotide sequence of *Saccharomyces cerevisiae* chromosome VIII. *Science, 265*, 2077–2082.

Kerlavage, A., Bonazzi, V., di Tommaso, M., Lawrence, C., Li, P., Mayberry, F., et al. (2002). The Celera Discovery System™. *Nucleic Acids Research, 30*(1), 129–136.

Kollers, S., Day, A., & Rocha, D. (2006). Characterization of the porcine *FSCN3* gene: cDNA cloning, genomic structure, mapping and polymorphisms. *Cytogenetic and Genome Research, 115,* 189–192.

Leonelli, S. (2016). *Data-centric biology: A philosophical study.* The University of Chicago Press.

Leonelli, S., & Tempini, N. (Eds.). (2020). *Data journeys in the sciences.* Springer Nature.

Loveland, J. E., Gilbert, J. G. R., Griffiths, E., & Harrow, J. L. (2012). Community gene annotation in practice. *Database,* 2012, bas009.

Lowe, J. W. E. (2018). Sequencing through thick and thin: Historiographical and philosophical implications. *Studies in History and Philosophy of Biological and Biomedical Sciences, 72,* 10–27.

Lowe, J. W. E. (2022). Humanising and dehumanising pigs in genomic and transplantation research. *History and Philosophy of the Life Sciences, 44,* 66.

Mahmoud, M., Gobet, N., Cruz-Dávalos, D. I., Mounier, N., Dessimoz, C., & Sedlazeck, F. J. (2019). Structural variant calling: The long and the short of it. *Genome Biology, 20,* 246.

Mewes, H.-W., Frishman, D., Zollner, A., & Heumann, K. (1998). The bioinformatics of the yeast genome. In A. J. P. Brown & M. Tuite (Eds.), *Methods in microbiology. Volume 26: Yeast gene analysis* (pp. 33–51). Academic Press.

Quilter, C. R., Blott, S. C., Mileham, A. J., Affara, N. A., Sargent, C. A., & Griffin, D. K. (2002). A mapping and evolutionary study of porcine sex chromosome genes. *Mammalian Genome, 13,* 588–594.

Renard, C., Hart, E., Sehra, H., Beasley, H., Coggill, P., Howe, K., et al. (2006). The genomic sequence and analysis of the swine major histocompatibility complex. *Genomics, 88,* 96–110.

Ross, M. T., Grafham, D. V., Coffey, A. J., Scherer, S., McLay, K., Muzny, D., et al. (2005). The DNA sequence of the human X chromosome. *Nature, 434,* 325–337.

Schook, L. B., Beever, J. E., Rogers, J., Humphray, S., Archibald, A., Chardon, P., et al. (2005). Swine Genome Sequencing Consortium (SGSC): A strategic roadmap for sequencing the pig genome. *Comparative and Functional Genomics, 6*(4), 251–255.

Searle, S. M. J., Gilbert, J., Iyer, V., & Clamp, M. (2004). The Otter annotation system. *Genome Research, 14,* 963–970.

Skaletsky, H., Kuroda-Kawaguchi, T., Minx, P. J., Cordum, H. S., Hillier, L., Brown, L. G., et al. (2003). The male-specific region of the human Y chromosome is a mosaic of discrete sequence classes. *Nature, 423,* 825–837.

Skinner, B. M., Sargent, C. A., Churcher, C., Hunt, T., Herrero, J., Loveland, J. E., et al. (2016). The pig X and Y chromosomes: Structure, sequence, and evolution. *Genome Research, 26,* 130–139.

Star, S. L., & Bowker, G. C. (2002). How to infrastructure? In L. A. Lievrouw & S. Livingstone (Eds.), *The handbook of new media: Social shaping and consequences of ICTs* (pp. 151–162). Sage.

Stein, L. (2001). Genome annotation: From sequence to biology. *Nature Reviews Genetics, 2,* 493–503.

Stevens, H. (2013). *Life out of sequence: A data-driven history of bioinformatics.* The University of Chicago Press.

Strasser, B. J. (2019). *Collecting experiments: Making big data biology.* The University of Chicago Press.

Uenishi, H., Morozumi, T., Toki, D., Eguchi-Ogawa, T., Rund, L. A., & Schook, L. B. (2012). Large-scale sequencing based on full-length-enriched cDNA libraries in pigs: Contribution to annotation of the pig genome draft sequence. *BMC Genomics, 13,* 581.

Venter, J. C., Adams, M. D., Myers, E. W., Li, P. W., Mural, R. J., Sutton, G. G., et al. (2001). The sequence of the human genome. *Science, 291*(5507), 1304–1351.

Warr, A., Affara, N., Aken, B., Beiki, H., Bickhart, D. M., Billis, K., et al. (2020). An improved pig reference genome sequence to enable pig genetics and genomics research. *GigaScience, 9*(6), giaa051.

Wimsatt, W. C. (2007). *Re-engineering philosophy for limited beings: Piecewise approximations to reality.* Harvard University Press.

Zody, M. C., Garber, M., Adams, D. J., Sharpe, T., Harrow, J., Lupski, J. R., et al. (2006). DNA sequence of human chromosome 17 and analysis of rearrangement in the human lineage. *Nature, 440,* 1045–1049.

CHAPTER 7

Improving and Going Beyond Reference Genomes

Throughout this book, we have mapped the participation of different scientific communities in genomic endeavours across three species—yeast, human and pig—and the distinct processes, epistemic goals and domains of application that informed the creation of annotated reference genomes. In this chapter, we examine how the existence of reference genomes enabled the creation of increasing amounts of additional genomic data, as well as other kinds of biological data. This involved the generation of new *reference resources* intended to represent forms of variation within a species that were either missing or insufficiently incorporated into its reference genome. Such reference resources could include new maps or sequences, novel ways to relate or align freshly identified variation to the reference genome, or tools to capture and document variants.

We examine two main currents of post-reference genome data production, collection and analysis: functional and systematic studies. By functional analysis, we mean the investigation of the effects of variation in genes and genomes, in terms of alterations to biological processes and therefore differences in phenotypes: the measurable presentation of

© The Author(s) 2023
M. García-Sancho, J. Lowe, *A History of Genomics across Species, Communities and Projects*, Medicine and Biomedical Sciences in Modern History, https://doi.org/10.1007/978-3-031-06130-1_7

traits in organisms.[1] By systematics, we mean the exploration of the patterns and specific details of genomic variation both within a species and between it and related species. We conclude by considering the implications of the increasing interrelationship of the functional and systematic modes in research pertaining to each of the three species.

Before a reference genome exists, and after its creation, communities tend to focus their interest on intra- and inter-species variability. Constructing a reference genome involves abstracting away—to a greater or lesser extent—variation to create a single canonical reference standard. Although the nature of reference genomes varies across species, this abstraction is often raised as a source of concern by the many different communities that may use them but who were not involved in their construction. This concern, we contend, arises from a tension between the presumed representativeness of reference genomes and their role as standards, as stipulated bases of reference and comparison.[2] This tension may seem to place—and in some cases, conflate—conflicting demands on reference genomes. Yet, the two demands are linked. Reference genomes, through their role as standards, enable researchers to gain a greater appreciation of the range of biologically-meaningful variation present across a species, and so help seed critiques of their representativeness. Furthermore, through their very role as a scaffold on which other representations of variation can be constructed and connected, reference genomes have enabled the development of data, tools and representations to facilitate more bespoke functional and systematic explorations of the biology of the species. Such developments have also encouraged visions of the refinement or even replacement of the reference genome as a central object in genomic and allied research.

In this chapter, we once again consider genomics research on all three of the species we have concentrated on in this book, in the context of understanding the nature of genomics after the reference genome. This is often referred to as 'postgenomics' (Richardson & Stevens, 2015). In Sect. 7.1, we contend that this label, and some of the meanings that have been attributed to it, reflects and reinforces a misleading picture of the history of

---

[1] The term 'functional genomics' itself has been traced back to the mid-1990s, when large-scale sequencing projects—especially the determination of the human reference genome—started to accelerate (Guttinger, 2019). 'Functional analysis' was used earlier than this, for example in connection with yeast genome sequencing (Grivell & Planta, 1990).

[2] This echoes the debate concerning the use of model organisms in the biological sciences, e.g., Jessica Bolker (2012) critiquing them on the basis of their unrepresentativeness in multiple respects, and Ankeny and Leonelli (2011; Leonelli & Ankeny, 2013) arguing that this should not eclipse their key infrastructural and comparative role across biology as a whole.

genomics that has arisen through a disproportionate focus on the elucidation of the human reference genome. In this view, a concern with relating multiple kinds of genomic and non-genomic data (what we refer to as *multidimensionality*) and the biological contextualisation of that data is *post*genomic. As we have seen in the preceding chapters, however, these facets were evident in pre-reference genome research and even featured alongside the generation of the reference genome for yeast and especially the pig. Even within *Homo sapiens*, the compilation of genomic data went hand-in-hand—rather than preceding—an aspiration to capture variation and connect this with other biological and medical problems outside the concerted effort of the International Human Genome Sequencing Consortium (IHGSC).

To begin to make that case, in Sect. 7.2 we consider the extent to which reference genomics is an ongoing project, both for species that already have a reference genome, and for species still lacking one. This continued reference genomics does not just constitute a tidying up exercise or involve incremental improvement. To accept that would presume that there is some final standard of completion for reference genomes. Further, it would imply that reference genomes are something to be discovered, rather than constituting creative products. As we have shown in previous chapters, reference genomes are abstractions from the variation found in nature, in which decisions made about data infrastructures, mapping, library construction and use, sequencing method, assembly and annotation are pertinent to shaping the final product. Who was involved in these processes, and when, are therefore matters of deep significance.

In Sect. 7.3, we develop our argument by examining two modes of genomic research—functional and systematic studies—and the relationships between them, by first inspecting an example of pre-reference genome work: pig genetic diversity projects that ran from the mid-1990s to the mid-2000s. These were efforts that explicitly aimed at apprehending the diverse genetic resources that might be tapped for breeding programmes. They also neatly aligned with interests in the domestication, evolution, phylogeny and natural history of pigs that were held by many researchers who primarily worked on pig genetics for agricultural purposes. They therefore instantiate an early entanglement between functional and systematic work and show how some of the supposedly 'postgenomic' concerns with variation and multi-dimensionality operated before the creation of reference genomes.

In Sect. 7.4, we compare the manifestation of these functional and systematic modes of genomic research across yeast, human and pig. We show that genomics research across these three species exhibits different forms of entanglement (or lack thereof, at times) between systematic and functional approaches. This crucially affects the ways in which reference genomes are used and how new reference resources are developed and connected to each other. We document this through particular examples of work conducted after the release of the reference genomes of each species:

- **Yeast**. EUROFAN, Génolevures and allied projects that followed the completion of the reference genome of *Saccharomyces cerevisiae*. EUROFAN sought to systematically produce mutants for particular genes and combinations of genes in *S. cerevisiae*, and so generate mutant stock collections as a standard reference intended for wider circulation. Génolevures was a network that sequenced and comparatively analysed the genomes of multiple yeast species, and in so doing explored their evolutionary dynamics and generated the comparative means for further developing functional analyses.

- **Human**. ENCODE, the post-reference sequence project aiming to catalogue the functional elements of the human genome, and GENCODE as a sub-project of this. We also examine attempts to map and make sense of human genomic diversity, the establishment of reference sequences for particular populations, as well as the creation of ClinVar: a database of genomic variants associated with clinical interpretations of their possible implication in disease.

- **Pig**. The Functional Annotation of Animal Genomes network (FAANG), which has grouped the pig genomics community with genomicists working on other farm animals. We also examine research on pig genomic diversity across breeds, particularly that related to tracking and understanding patterns of evolution, domestication and dispersal. Finally, we recount the creation of a SNP chip or microarray to test for the presence or absence of particular Single Nucleotide Polymorphisms (SNPs). The advent of this chip enabled further functional and systematic studies, as well as the development of novel resources and the inferential means by which researchers could connect new and existing resources. This eased the connection of genomic resources to particular modes of research and domains of application or 'working worlds' (Agar, 2020).

In yeast, we see first a pursuit of functional analysis to build on and enrich the reference genome as a resource, followed by systematic studies. Leading yeast genomicists, pursuing their own lines of molecular biological and biochemical research, increasingly realised the synergies between these two modes. The relationship between functional and systematic research in yeast reflected the 'do-it-yourself' approach of the yeast biologists who made up the genomics community, and the nature of yeast as a model organism.

In human genomics, we see continuity from the way that the reference genome effort was organised, with grand concerted efforts led by many of the same institutions encompassing the IHGSC. We focus on ENCODE and GENCODE and compare these with some contemporary systematic studies that examined human genomic diversity, such as the production of reference sequences for particular populations. We conclude the discussion of human post-reference genomics by looking at a relatively new initiative, ClinVar, which aims to connect the infrastructures, norms and practices of large-scale genomics with those of the medical geneticists who were peripheral to the IHGSC, and who had instead developed their own separate and parallel data infrastructures.

For the pig, however, after the discussion of the functional and systematic motivations and consequences of pig genetic diversity research in Sect. 7.3, it will not come as a surprise that the post-reference genome distinction between functional and systematic modes has been far fuzzier than for yeast and human; there have been multiple crossovers between the modes and an early appreciation of their synergies by the community. After examining how the pig genomics community immediately set about functionally and systematically exploiting the reference genome they had helped to create, we examine their collaboration with the sequencing technology company Illumina to produce a SNP chip. The SNP chip is an excellent illustration of the significance of the involvement of a particular community in the creation of genomic resources, in particular in shaping the generation of new reference resources. This, however, introduces constraints into these resources as much as it engenders capabilities or affordances.

We conclude (in Sect. 7.5) by observing that the coming together of functional and systematic modes of genomic research instantiates a particular stage in the development of what we term a *web of reference*. Over time, webs of reference feature ever-denser webs of connectedness between distinct representations of the variation of, for example, a particular species. Such representations include reference sequences (e.g. of the species or sub-species populations), genome maps and resources such as SNP chips. Connections between such representations are progressively forged by data linkages and the process of identifying and validating inferential and comparative relationships between them. This is enabled by the creation of reference resources that seed the web, with new nodes representing new forms of data scaffolded on and linking to existing ones. These webs of reference are especially dense within particular species, but they can—and indeed have and often must—be connected to genomic reference resources beyond them. The way in which these webs develop depends on prior genomic research and reference resource creation (including that for other species) and the involvement of specific communities of genomicists in those efforts.

7.1   POSTGENOMICS OR POST-REFERENCE GENOMICS?

This chapter explores the surplus of data concerning genomic variation that has been generated in the wake of the elucidation and publication of reference genomes. We refer to this as 'post-reference genomics', to indicate the differences between our treatment of this work with what is usually connoted by the term 'postgenomics'. 'Postgenomics' has often implicitly referred to genome-related research that followed the determination of the human reference sequence. It therefore ignores the ongoing 'reference genomics' of the human—beyond the conclusion of the IHGSC endeavour—as well that being conducted for other species. Many species, of course, still do not have a reference genome, while others—as we have shown throughout the book—had their reference sequences produced in a substantially different manner to the human one.

There has been some debate on what postgenomics means, beyond the chronology of simply following the initial publication of the human reference genome. Some scholars have suggested that it was conceived by IHGSC scientists to market their post-reference sequence research agenda, with parallels drawn between this agenda and the contemporary rise of the notion of *translation*: an imperative to transform research results and data into medical outcomes (Stevens & Richardson, 2015). While obtaining further grant funding may have been a significant driver of the framing of postgenomics as something distinct and new, other accounts have sought to characterise postgenomics as a more substantial endeavour. A common theme in this latter school of thought is that post-genomics constitutes research that aims and aimed to relate other forms of biological data—to integrate additional dimensions—to DNA sequence data, and therefore begin to properly capture the complexity of biological processes. Here, the technologies, methods, infrastructures and data generated through genomics have been used as a platform for further biological research. In this version, postgenomics comprises a recognition of complexity and a non-deterministic, non-reductionist, interactionist and holistic vision of the organism.[3]

This perspective on organismal complexity was first outlined at a 1998 conference held at the Max Planck Institute for the History of Science. As by any definition, this conference was held before 'postgenomics' came into being in some form, postgenomics was envisaged in conjectural and promissory ways. At the conference, the biologist Richard Strohman outlined four phases of genomics: the first two being "monogenetic and polygenetic determinism", then "a shift in emphasis from DNA to proteins" and then "functional genomics". Following this was a burgeoning fifth stage, presumably postgenomics, but not labelled as such, which is "concerned with non-linear, adaptive, properties of complex dynamic systems". This was an early statement of the idea that genomics pertains to the linear and deterministic while postgenomics opens out to nonlinear and nondeterministic facets of biology, but it differed from some of the accounts of later scholars by including aspects of this extra- and multi-dimensionality in genomics itself, rather than this being characteristic of postgenomics (Thieffry & Sarkar, 1999, p. 226).

Adrian Mackenzie has also evaluated genomics in terms of dimensionality. He identifies the period roughly between 1990 and 2015 as "the 'primitive accumulation' phase of genomics" that "has yielded not only a highly accessible stock of sequence data but sequences that can be mapped onto, annotated, tagged, and generally augmented by many other forms of data". The single dimensionality of sequence data produced in this phase of genomics is something to be augmented with—and related to— other forms of data; it is the challenge of dealing with dimensionality that characterises "post-HGP [Human Genome Project]" biology (Mackenzie, 2015, pp. 79 and 91). In this conception, postgenomics is defined in terms of both the use of existing sequence data and associated infrastructures, and the establishment of connections between genomic data and other forms of 'omic' data (Stevens, 2015). Here, postgenomics involves the results and modes of research of genomics being brought together with other types of biological traditions and outputs, a process that is characterised by the advent of new forms of labour, for example the figure of the curator (Ankeny & Leonelli, 2015).

---

[3] Rheinberger and Müller-Wille (2017) would say, rather, that if this was the case, postgenomics constitutes a *rediscovery* of this holistic vision.

The designation of something called postgenomics as an endeavour to contextualise sequence data, indicates that genomics became conceptualised in terms of sequence production, rather than involving both sequence production and use, and featuring a range of different ways in which production and use were related and combined. This production-centred interpretation tallies with an approach to genomics that foregrounded an increase in the efficiency and speed of data production, with the pressure of this drive helping to manifest and reify a strict division between producers (submitters) and users (downloaders). However, as we have shown elsewhere (García-Sancho & Lowe, 2022) and further illustrate throughout the book, a sharp division only existed within the IHGSC effort; other approaches to genomics featured different configurations and entanglements between sequence production and use. Additionally, contextualisation of sequence data has been pursued both before and during the production of a reference genome, as much as afterwards. We can, therefore, conclude that contextualisation is not a defining attribute of postgenomics: rather, following the advent of a reference genome, existing forms of contextualisation are altered and new ones are established.

What do multi-dimensionality, augmentation, integration and contextualisation mean in the post-reference genome world? They relate to ideas of completeness, comprehensiveness and the capturing of a whole or a totality. In that 1998 conference previously mentioned, biologist and scientific administrator Ernst-Ludwig Winnacker, the founder of major yeast and human sequencing centre Genzentrum (Chap. 2), said that postgenomics should be about "an understanding of the whole" (Thieffry & Sarkar, 1999, p. 223).

Historian Hallam Stevens has articulated how genomics itself seeks wholeness and comprehensiveness. Drawing on his detailed study of the Broad Institute, and mostly informed by human genomics, he presents genomics as a special form of data-driven bioscience. The nature of data in genomics makes it amenable to the adoption and development of bioinformatics and information technology-based approaches more generally.[4] Stevens' interpretation of genomics is that the investigation of the particu-

---

[4] The Broad Institute was formed out of a partnership of the Whitehead Institute's Center for Genome Research and several Harvard University-affiliated institutions. Stevens (2013) focuses on this major sequence producer, as well as AceDB as an example of database technology. As we discuss in Chap. 6, this database was designed to present data in a user-friendly manner, with the assumption that once produced and released in AceDB, it would be the user who would add dimensionality to the sequence data.

lar is replaced by a sensibility that aims to characterise the totality. Totality and generality are key. For example, he points to the "Added value" generated by having completely sequenced genomes (Stevens, 2013, p. 161, quoting Bork et al., 1998).

Stevens stresses the dialectic of sequence data production and the development and incorporation of informatics infrastructures and approaches. Successive different structures of databases are indicative of shifts from pre-genomics to genomics to postgenomics. Genomics is about producing databases; the reference genome and a particular way of storing and presenting data—relational databases—are mutually constitutive. Distinctions between pre-genomics, genomics and postgenomics are therefore made on the basis of the structure of databases and the place and role of DNA sequence data within them. When researchers increasingly wanted to relate DNA sequence data to other forms of data (e.g. various omics data), this necessitated a shift from one kind of database structure to another. The relational databases that were able to capture well the single dimension of DNA sequence data catalogued in strings of As, Ts, Cs and Gs, therefore gave way to more complex *networked* databases (Stevens, 2013).

These interpretations of genomics are usually based on specific institutions and infrastructures, often in the orbit of the IHGSC. In such expositions, the effort of producing a reference genome is detached from prior genomic research, parallel genomic research (for example, by medical geneticists), and work following it. Accompanying this separation is the projection of distinct and exclusive attributes to pre-genomic, genomic and postgenomic research.

As an alternative, we propose the designations of *pre-reference genomics*, *reference genomics* and *post-reference genomics*. This periodisation scheme is based on the availability (or otherwise) of an object—the reference genome—and the relationship of particular communities to it. It does not presume that each stage will exhibit specific essential characteristics. Our approach emphasises the historicity and specificity of reference genomes and helps us to discern a more fluid interconnectedness between stages. In the rest of the chapter, we illustrate this by comparing post-reference genome research on yeast, human and pig.

7.2   IMPROVING GENOMES

Reference genomes are not static: they are amended over time, with updated versions evaluated and validated using metrics that enable direct comparisons to be drawn between the new and the old. Even when a reference genome is considered to be 'complete'—as the human reference genome was famously deemed in 2004—it still subsequently undergoes revisions that are intended to improve it according to existing and novel benchmarks. In what follows, we examine revisions of the human, yeast and pig reference genomes and how metrics and judgements of quality changed according to evolving and distinct objectives for the three species.

We have seen that Celera Genomics saw their full human sequence as provisional and in need of constant improvement and enrichment. This was in order that their corporate effort would be seen to offer sufficiently more value than the publicly-available data to justify paying a subscription to access it. Indeed, as we indicated in Chap. 6, Celera kept developing its whole-genome sequence: new additions that were incorporated after the initial public release in 2001 were only accessible with a paid subscription.

The working draft of the IHGSC sequence (release name: hg3) was published on the University of California Santa Cruz's (UCSC) website on 7th July 2000. At this stage, though, it was just the sequence data that could be downloaded, with a UCSC browser to visualise it still being in the works. This version had significant gaps and ambiguous positioning of sequenced fragments. The major draft published in February 2001 could also be downloaded from the UCSC website. In the *Nature* paper accompanying its release, it was estimated that the draft encompassed 96% of the euchromatic regions, the parts of DNA open to transcription.[5] Much as with the addition of the pig Y chromosome sequence to the new *Sus scrofa* reference genome assembly in 2017 (Chap. 6), future reference assemblies of the human genome would incorporate data from several sources.

The quality of the human and other reference genomes has been assessed in a number of ways: in terms of coverage, contiguity and accuracy.

Coverage is a metric we have already encountered; it is a function of the depth of sequencing, roughly how many 'reads' or particular determined nucleotides are present on average across the genome. It is expressed in

---

[5] 92% of human DNA is euchromatic. Together with data from other public databases, the draft was thought to encompass 94% of the entire human genome. On the UCSC Genome Browser, see: https://genome.ucsc.edu/goldenPath/history.html (last accessed 19th December 2022).

terms of number-X, with the number designating the average amount of reads across the genome. However, there may be heterogeneity in the coverage of different regions of the genome. This outcome can be inadvertent, due to the clones captured in library production not evenly representing all areas of the genome, or be because of the exigencies of assembling regions with different genomic properties. Or it may be deliberate, due to the kind of targeting we saw in swine genome sequencing at the Sanger Institute.

Contiguity is the extent to which the building blocks of an assembly, such as contigs or scaffolds, are connected together. A contig is a continuous sequence in which the statistical confidence level in the order of the nucleotides exceeds a stipulated threshold, while a scaffold is a section of sequence that incorporates more than one contig, together with gaps of unknown sequence. The measured level of contiguity affects the classification of the level of a sequence assembly in the GenBank database. The designation of being a "complete genome" requires that all chromosomes should have been sequenced without gaps within them. Then there is a "chromosome" level of assembly: to qualify for this level, a sequence must encompass at least one chromosome, ideally with a complete contiguous sequence; if gaps remain, there need to be multiple scaffolds assigned to different locations across the chromosome. The other two levels are "scaffold" and "contig", pertaining to the definitions of those objects.[6] Note that these are ways of assessing genome assemblies. They do not necessarily determine whether an assembly is designated as a 'reference genome' or the lesser category of 'representative genome' by the RefSeq database (Chap. 1, note 3), both of which are incorporated in the notion of reference genome we deploy across this book. As with improvements to mapping procedures or the evaluation of new genome libraries (Chap. 5), completeness can also be ascertained by searching for known genes or markers in the assembly, and enumerating those found and not found.

As well as these designations, there are metrics that are used to assess the contiguity of assemblies in a more fine-grained way. The most significant are the enumeration of the gaps (and the different kinds of gaps) and the estimated sequence length they represent, and also the calculation of N50 and L50 figures. The L50 figure is the smallest number of contigs whose total sequence lengths add up to at least 50% of the total length of the assembly. The N50 figure is the length of the shortest contig that constitutes part of the smallest set of contigs that together add up to at

---

[6] https://www.ncbi.nlm.nih.gov/assembly/help/ (last accessed 19th December 2022).

least 50% of the total length of the assembly. The L50 figure will therefore be expressed as a simple integer, while the N50 figure will be expressed in terms of numbers of nucleotides. These figures pertain to the length of the assemblies, rather than the presumed length of the actual chromosomes or whole genomes that are being assessed. For assemblies of the same length, the quality is presumed to be higher if the N50 figure is larger and the L50 figure smaller. The original draft human reference sequence, published in 2001, contained N50 figures for individual chromosomes and the genome as a whole. Gaps were counted across the assembly. These metrics enabled areas for improvement to be identified and analysed, but also provided a benchmark against which further improvements could be assessed.

Finally, there are measures of accuracy, which is the extent to which an assembly—and the parts thereof—is 'correct'. This can relate to different aspects, such as the order and orientation of sequenced clones in the assembly, or pertain to the 'base calls'—the assignment of the identity of individual nucleotides in each position in a DNA molecule—at the sequence level. This is, of course, trickier to execute than the other measures of quality, as it requires not just the measurement of the properties of the assembly and the construction of comparable metrics, but also necessitates assessment against a recognised standard. In the 2001 human reference sequence paper authored by the IHGSC, the accuracy of the assembly was evaluated by comparing it against an ordering of parts of the genome as dictated by sequence data derived from the ends of the cloned fragments in the DNA libraries used in the sequencing. This resulted in the identification of clones that did not overlap with others. These non-overlapping clones had been sought, as their presence could indicate misplacement of fragments in the assembly; they were subjected to closer investigation, resulting in "about 150" of the 421 "singletons" being attributed to misassembly.

Sequence quality at the level of nucleotides was evaluated in terms of the 'PHRAP score' for each one. The IHGSC used PHRAP and PHRED, software packages that were developed by Phil Green (both of them) and Brent Ewing (PHRED) at the University of Washington in Seattle. Together, they were—and are still—used for base calling. The software analyses the fluorescent peaks in the sequence read-out. It estimates error probabilities for each base call based on figures obtained from the read-out data and generates consensus sequences with error-probability estimates (Ewing et al., 1998;

Ewing & Green, 1998).[7] The resulting PHRAP scores indicate the probability that an individual base call is incorrect, and therefore the overall accuracy of the sequencing. A score of 10 denotes an accuracy of 90% and that there is a 10% chance that any given base is wrong. A score of 20 means an accuracy of 99% (and a 1% chance of a given base being wrong), 30 means 99.9% (0.1% chance of a given base being incorrect), and so on. The 1998 Second International Strategy Meeting on Human Genome Sequencing held in Bermuda promulgated sequence quality standards that included an error rate of less than 1 in 10,000 (e.g. 99.99% accurate, a PHRAP score of at least 40) and a directive that the error rates derived from PHRAP and PHRED be included in sequence annotations.[8]

Following the initial 2001 publication and online availability of a draft sequence, further assemblies were made available on the internet through GenBank, the DNA Data Bank of Japan and the European Nucleotide Archive (ENA), the last of which encompassed the databases housed at the European Bioinformatics Institute (EBI) from 2007 onwards. From December 2001, these assemblies were released using the name of the National Center for Biotechnology Information (NCBI), the institution into which Genbank was incorporated. NCBI Build 28 was the first release labelled in this way.[9] Then, in April 2003, the first assembly that constituted a human reference sequence was published, known as NCBI Build 33.[10]

The 2004 IHGSC paper on the 'finished' euchromatic sequence was working from a subsequent assembly, NCBI Build 35 (International Human Genome Sequencing Consortium, 2004). In their analysis of this build, the authors compared it against the 2001 version using some of the measures indicated above, but also pursued some deeper analysis of the quality of the new sequence. The new assembly had 341 gaps, compared to 147,821 in 2001. The N50 for the 2004 sequence was 38,500 kilobases, a dramatic improvement from the 81 kilobases determined for the

---

[7] The error-probabilities were validated in the same issue in which these papers were published (Richterich, 1998). See also: https://www.codoncode.com/productsservices/phrap.htm (last accessed 19th December 2022).

[8] https://web.ornl.gov/sci/techresources/Human_Genome/research/bermuda.shtml#2 (last accessed 19th December 2022). See also Felsenfeld et al. (1999).

[9] UCSC retained their own naming system for subsequent human genome assembly releases. https://genome.ucsc.edu/FAQ/FAQreleases.html (last accessed 19th December 2022).

[10] https://www.ncbi.nlm.nih.gov/Web/Newsltr/Spring03/human.html (last accessed 19th December 2022).

2001 version. To further examine the completeness of the 2004 assembly, the consortium looked for 17,458 known human cDNA sequences in it and found that the "vast majority (99.74%) could be confidently aligned to the current genome sequence over virtually their complete length with high sequence identity".

The 2004 paper assessed the accuracy of sequencing by inspecting discrepancies between nucleotides in the overlapping regions of 4356 clones from the same Bacterial Artificial Chromosome (BAC) library. This required some appreciation of the rate of polymorphism (genetic variation) across humans, as a difference in a single nucleotide could be due to this inter-individual or inter-group variation rather than constituting an error. While later, we see how an appreciation of genomic variation and diversity was vital to making functional use of genomic data, here we see how such an understanding, however tentative, played a part in fundamental analyses of the quality of a reference sequence itself.

Alongside these assessments, the IHGSC members evaluated whether junctions "between consecutive finished large-insert clones" that they had used "to construct the genome sequence" were spanned by another set of fosmid clones derived from a library that they created for this purpose (International Human Genome Sequencing Consortium, 2004, p. 936). With approximately 99% of the euchromatic sequence deemed to be of the requisite finished quality, the attention of the sequencers turned to the recalcitrant 1% and the heterochromatic regions, which would require new methods and materials to resolve, rather than merely a continuing scale-up of sequence production. Next-generation sequencing methods, including long-read technologies that sequence larger stretches of DNA and therefore reduce the number of problematic gaps or misassemblies, have assisted in this (e.g. Nurk et al., 2022). Furthermore, fundamental research pertaining to particular problematic regions has generated data and information that has enabled the creators of successive assemblies to amend and improve these refractory areas.

In addition to improvements to a single canonical reference sequence, attempts were increasingly made to ensure that the reference genome was more reflective of the variation manifested by the target organism. For instance, this was realised by creating the possibility of depicting alternate loci, contigs and scaffolds that differ from the reference sequence in databases and visualisations. An example of a new presentational mode that conveys different kinds of variants alongside the reference sequence is the *pangenome graph*, showing where these variants diverge from the standard and how common their departures from the reference version are (Khamsi, 2022).

In order to move towards a model of reference assemblies that incorporated variation, and to manage and conduct this ongoing work, the Genome Reference Consortium was established in 2007 by the Sanger Institute, the McDonnell Genome Institute (the new name of the genome centre at Washington University), the EBI, and NCBI. They initially focused on three species: human, mouse and zebrafish, the latter two because of their role as model organisms and due to existing investments in creating gene knock-out collections for these species (Church et al., 2011). Since then, rat and chicken—also model organisms—have been added, and The Zebrafish Model Organism Database and the Rat Genome Database have joined the consortium.[11]

Pig and yeast are notably absent from the Genome Reference Consortium. In the case of yeast, ongoing improvements to the sequence and annotation of the reference genome—first released in 1996—are performed by the *Saccharomyces* Genome Database at Stanford University, with both the sequence and annotation treated as "a working hypothesis" subject to continual revision (Fisk et al., 2006). A major revised new version of the yeast reference genome was completed in 2011, using a colony derived from the AB972 sub-strain of S288C. Linda Riles had used AB972 to construct the genome libraries for the original sequencing of the yeast genome. The new sequence reads were aligned to the existing reference genome, with low quality mismatches discarded and manual assembly and editing of the genome conducted, which involved checks of the literature for particular sequences and annotations.

While the comparison with the older reference affirmed the quality of that earlier standard, the new assembly made numerous corrections to it. The authors of the paper announcing it had sufficient confidence in it to suggest that the reference sequence was now comprehensive and accurate enough so that in future revisions greater weight would be given to incorporating variation rather than fixing errors. They also suggested that

---

[11] The publication of subsequent assemblies constituting reference genomes has increasingly reflected the nomenclature of software releases, with periodic new versions (e.g., 2.0 and 3.0) being corrected and augmented with more regular patches (e.g., 2.3 and 3.1). Once published, these new reference assemblies are picked up and developed by the major genome browsers—such as the UCSC one, the NCBI, and Ensembl—and form the basis for further annotation. The advent of a new major release requires the commensuration of the new assembly to the old, by the mapping of coordinates—and therefore features—between them.

having worked towards and largely achieved a highly veridical representation of a single strain, the focus of yeast reference genomics should shift towards creating the most *useful* representation of the organism. One of the stated implications of this was the need to develop a *pangenome* including annotated sequences representing different *S. cerevisiae* laboratory strains and wild specimens, using some of the copious data being generated on these, as well as on related species (see Sect. 7.4; Engel et al., 2014).

In pig genomics, the first major revision after the completion of the reference genome (represented by the 2011 Sscrofa10.2 assembly) was released in 2017. The impetus for producing a new reference genome was provided by a team led by Tim Smith at the US Department of Agriculture's Meat Animal Research Center (USDA MARC). They sequenced a boar from a population whose breed ancestry was estimated to be half Landrace, quarter Duroc and quarter Yorkshire. Smith was using Pacific Biosciences long-read sequencing technology, which held the promise of greater contiguity of sequence and fewer potential issues with assembly. However, when others in the pig genome community found out about Smith's endeavour, the error rate for this technology made them sceptical of its worth.

They did, however, work with Smith and his team to produce a new reference genome. Together, they hit upon the strategy of using Pacific Biosciences long-read technology in conjunction with more reliable Illumina short-read technology. This, combined with the improved chemistry of the newer versions of the Pacific Biosciences technology, helped them to produce a high-quality assembly that formed the basis for Sscrofa11, which became the designated reference genome Sscrofa11.1 when the Y-chromosome data from the X+Y project (Chap. 6) was incorporated. Alan Archibald at the Roslin Institute used money acquired from the UK Biotechnology and Biological Sciences Research Council to fund a large part of this effort, paying Pacific Biosciences for an initial assembly that the community could then work on further. He was fortunate that the contractor Pacific Biosciences had engaged to do this had fallen behind schedule, meaning that Pacific Biosciences took it in-house and conducted the work themselves, ensuring that the project benefitted from the latest chemistry and the best expertise on deploying their technology.[12]

[12] Interview with Alan Archibald, conducted by James Lowe, Roslin Institute, November 2016.

The USDA assembly—resulting from Smith's original work—was submitted separately, though it was compared with the new reference sequence in the eventual paper reporting its completion. Multiple metrics—such as the number of gaps between scaffolds, the coverage and the N50—demonstrated the superiority of Sscrofa11.1 to Sscrofa10.2, and this higher quality ensured a better automated annotation through the Ensembl pipeline, including a doubling of the number of gene transcripts identified (Warr et al., 2020).

It is worth observing here, though, that interpretations of the quality of assemblies are not straightforward. For example, the 2011 assembly Sscrofa10 has a higher number of scaffolds, gaps between scaffolds, and 'worse' N50 and L50 figures for scaffolds and contigs than 2010s Sscrofa9.2. This does not mean that the assembly is of a lower quality, but that additional chromosomes (such as the Y chromosome) and extranuclear DNA had been included in the assembly. The Y chromosome notoriously contains many repetitive sequences that are consequently difficult to assemble.

This example shows that reference assemblies can constitute—and therefore represent—different objects, even within the same species. Furthermore, for the pig, in addition to the reference assemblies of the Swine Genome Sequencing Consortium, there is the USDA MARC assembly. There have also been other assemblies published for different breeds of pig (including Chinese breeds by the company Novogene) and the minipig used for biomedical research (sequenced by GlaxoSmithKline and BGI-Shenzhen, formerly the Beijing Genomics Institute), as well as other sequences concerning a variety of breeds and populations of pigs. These more specific references, with some recognised in formal designations and database entries and others not, are examined later in the chapter.

The discussion above shows that reference genomes are not monolithic, static objects. They are continually improved, impelled towards an ever-receding horizon of completeness. But parallel to this continual improvement of the standard reference sequence, genome assemblies have also ramified, as we see with the compiling of genomes for distinct breeds of pig. Additionally, for human and yeast, new aims that guide the evaluation of reference genomes in ways that go beyond the quality metrics of old (e.g. N50) have emerged, especially concerning the variation that the reference sequences instantiate. However, this concern with variation and variants is not something that arises after the reference genome, as the story of the IHGSC and the supposed emergence of a postgenomic era may suggest: it was already present beforehand.

7.3   Functional and Systematic Genomics Before Reference Genomes

Pre-reference genomics occurred in different eras for each of the species: up to the mid-1990s in the case of yeast, until the late-2000s for pig and preceding the turn of the millennium for the human. These distinct time-frames are pertinent because none of the developments in genomics for these species or any others have occurred in a vacuum: particularities of each were mediated by the adaptation and adoption of tools, methods and data produced for other species, and the comparative inferential apparatus that was constructed to enable such translations.

For yeast (Chap. 2), we saw that comprehensive genetic linkage maps were produced well before the initiative to sequence the genome started. Extensive physical maps were produced by Maynard Olson in the 1980s, building on Robert Mortimer's earlier genetic linkage maps, and then later physical mapping was conducted by the groups in charge of the sequencing to aid this undertaking for each chromosome. In the case of yeast, the dominant focus of the community was on one laboratory strain that had already had much of its variation abstracted from it in the process of its construction as a model organism.

For human, as discussed in Chap. 3, a great deal of data was generated on variation through the medical genetics community, which extensively catalogued variants of particular genes and associated these with clinical cases of specific diseases, such as for cystic fibrosis. Significant hospital-based human DNA sequencing took place, such as at the John Radcliffe Hospital in Oxford, Guy's Hospital in London or the University of Toronto Hospital for Sick Kids. Yet, because of the notable absences of these medical genetics groups from the IHGSC membership, these maps and sequences were only marginally accounted for in the production of the reference sequence.

For the pig, mapping projects generated considerable amounts of data concerning the variation of particular genetic markers, which were discerned through crosses of different breeds suspected to be genetically distinct owing to the geographical disparity of their origins and their morphological differences. The familiarity of these geneticists with the kinds of markers used in these studies enabled a subset of them to pursue the European Commission (EC) funded projects PigBioDiv 1 and 2 (1998–2000 and 2003–2006, respectively) to characterise the genetic diversity of pig breeds first within Europe, and then across Europe and

China (Ollivier, 2009). These projects, as well as prior studies of pig genetic diversity that had been conducted from the mid-1990s, represented an integration of functional and systematic approaches and concerns.

Many researchers in the pig breeding community have had research interests connected to the variation and diversity of both domesticated pigs and their wild cousins. As a result, these topics were even included in early genome mapping initiatives. A pilot study of genetic diversity across twelve rare and commercial breeds of pig formed part of the EC's PiGMaP II programme (1994–1996).[13] PiGMaP II's organisation reflected the collaborative division of labour approach of the PiGMaP projects more broadly, with various groups supplying DNA from, and pedigree information concerning, animals from specific breeds they had access to. Meanwhile, researchers from Wageningen University and INRA Castanet-Tolosan (a station near Toulouse) selected a panel of 27 microsatellite markers—repetitive sequences of variable length—on the basis of their level of polymorphism, distribution across the genome, and practical ability to use in genomic studies. This panel of 27 microsatellites was subsequently adopted by the Food and Agriculture Organization of the United Nations (FAO) for studying pig genetic diversity. Max Rothschild, in his capacity as the pig genome coordinator for the USDA's Cooperative State Research, Education, and Extension Service, ensured that the appropriate PCR primers for these markers were produced and distributed among the community. In addition to the use of the microsatellites that were themselves a key product of the PiGMaP collaboration, minisatellites and DNA fingerprinting for detecting genetic variation and diversity were also trialled in PiGMaP II.[14]

Beyond PiGMaP, in addition to some of the other projects discussed in Chap. 5, the community sought to further develop their work on pig biodiversity. An initial follow-up was the 'European gene banking project for pig genetic resources' that ran from 1996 to 1998, which assessed nineteen breeds of pig using eighteen of the standard set of 27

---

[13] The breeds were: Basque, Gascon, German Landrace, Great Yorkshire, Limousin, Piétrain, Porc Blanc de l'Ouest, Schwäbisch-Hällisches Schwein, Sortbroget, Dansk Landrace, Swedish Landrace and Wild Boar; "The pig gene mapping project (PiGMaP)—identifying trait genes" final report, March 1997; in "EC PiGMaPII—Final Report" folder, personal papers of Alan Archibald, obtained 15th May 2017.

[14] "The pig gene mapping project (PiGMaP)—identifying trait genes" final report, March 1997; in "EC PiGMaPII—Final Report" folder, personal papers of Alan Archibald, obtained 15th May 2017.

microsatellites together with the blood group variants and biochemical polymorphisms that had been traditionally employed in studies of variation (Ollivier, 2009).

A major development in the elucidation of pig genetic diversity was the advent of the EC-funded demonstration project, 'Characterization of genetic variation in the European pig to facilitate the maintenance and exploitation of biodiversity', which officially ran from October 1998 to September 2000 and was retrospectively referred to as PigBioDiv1.[15] It was led from the Jouy-en-Josas station of the French Institut National de la Recherche Agronomique (INRA) with quantitative geneticist Louis Ollivier as the coordinator. The participation of Graham Plastow of the Pig Improvement Company (PIC) reflected interest in the project by the breeding sector. On the FAO side, the involvement of Ricardo Cardellino and Pal Hajas showed that those with a longer-term and strategic view of the future of livestock also held this work to be important.[16]

The aim of PigBioDiv1 was to create a means to maintain and track genetic variation. This was motivated by the breeding sector's assumption that additional sources of genetic variation were needed in order to enable the further improvement of their commercial breeding lines,[17] to ensure the sustainability of livestock agriculture, and to respond to changing consumer and regulatory demands that might entail new breeding goals. This approach was stimulated by, and aimed to address, a growing policy concern with the conservation of "animal genetic resources" to safeguard global food security. The FAO were central to this drive and published "The Global Strategy for the Management of Farm Animal Genetic Resources" in 1999 to that end (Food and Agriculture Organization, 1999). The concept of "genetic resources", which has been traced back to the 1970s, was adopted by the FAO in 1983 and formed part of the framework of the UN Convention on

---

[15] https://cordis.europa.eu/project/id/BIO4980188 (last accessed 19th December 2022).

[16] https://web.archive.org/web/20070817113534/http://www.projects.roslin.ac.uk/pigbiodiv/contact.html (last accessed 19th December 2022).

[17] A contention that was challenged by some quantitative geneticists working close to animal breeding, such as William G. Hill at the University of Edinburgh, who argued that breeding populations were not in fact short of variation, and that identifying and accessing potentially beneficial genetic variation in non-commercial populations presented significant problems that would make it less preferable to other approaches to breed improvement (Hill, 1999).

Biological Diversity in 1992. It has been criticised for foregrounding an instrumental value of biodiversity (Deplazes-Zemp, 2018), and this is certainly true in the case of the PigBioDiv projects.

Following the widespread adoption of microsatellites in pig genome mapping and the pilot diversity project, and in the light of FAO recommendations for using them in examining genetic diversity, these highly polymorphic markers formed the basis of both the PigBioDiv1 and PigBioDiv2 (February 2003 to January 2006) projects (see Table 7.1).

Sharing a view expressed by other participants, Chris Haley—a quantitative geneticist involved in the PigBioDiv projects—has observed that this work was based on the assumption that genetic diversity reflected functional diversity. Yet, microsatellites were known to be non-functional parts of the genome.[18] It was this property, however, that enabled them to be so polymorphic, and therefore useful in mapping and tracking diversity. Furthermore, despite being non-functional parts of the genome, microsatellites still had applications in functional research. Indeed, markers such as these can and have been used in animal breeding, where it is not strictly necessary to find a causative gene, but merely something—like a microsatellite—that is statistically associated with one or many genes that may themselves be implicated in phenotypic variation for traits of interest (Lowe & Bruce, 2019). As we shall see later in this chapter, SNPs generated by the pig genomics community and compiled into a SNP chip were used in this way, but were also be applied in more systematic studies of pig genetics concerned with variation and diversity.

The importance of the particular historicity of the pig genomics community, and its involvement in multiple different projects of data collection and resource generation, cannot be underestimated here. The creation of the means to identify and map markers, and exploit the data and mapping relations so generated, relied on a coming together of molecular and quantitative geneticists. In some cases, this occurred within institutions (such as at the Roslin Institute with Chris Haley and Alan Archibald, partly driven by the immediate history of that institution, see Myelnikov, 2017; Lowe, 2021) or within the overall cooperative division of labour that had been forged. This community has been able to work with populations of livestock with well-recorded pedigrees, manipulate breeding in

---

[18] Interview with Chris Haley, conducted by James Lowe and Ann Bruce in Edinburgh, December 2017.

Table 7.1 Summary of the participating institutions, breeds studied, genetic markers used and some of the results and outputs of the PigBioDiv1 and PigBioDiv2 projects. Based on multiple sources, including Ollivier, 2009, Megens et al., 2008, https://cordis.europa.eu/project/id/QLK5-CT-2002-01059 and https://web.archive.org/web/20070817113503/http://www.projects.roslin.ac.uk/pigbiodiv/index.html (both accessed 19th December 2022)

	PigBioDiv1 (1998–2000)	*PigBioDiv2 (2003–2006)*
Participating institutions (*Italics: coordinator*, **Bold: also participated in PiGMaP**)	*INRA Jouy-en-Josas* **INRA Castanet-Tolosan** **Roslin Institute** Pig Improvement Company **Wageningen University** FAO	*Sygen International (PIC)* China Agricultural University Huazhong Agricultural University **INRA** Jiangxi Agricultural University **Roslin Institute** **The Swedish University of Agricultural Sciences** **Wageningen University**
Breeds studied	58 European populations (from INRA, Roslin, PIC, University of Trás-os-Montes and Alto Douro, and PiGMaP); and Chinese Meishan (from INRA, Roslin and PIC).	52 from PigBioDiv1; 46 additional Chinese breeds/populations.
Markers used	Microsatellites and Amplified Fragment Length Polymorphisms	Microsatellites, mitochondrial and Y-chromosome genes, trait genes and SNPs
Select results and outputs	Pattern of diversity identified. DNA from project stored in banks. Data from project held on new EC Pig Diversity Database.	Creation of DNA banks for the Chinese breeds. Extensive genotype data made available on the EC Pig Diversity Database. Identification of markers for breed assignment. Identification and explanation of different patterns of variation and diversity in Europe and China.

those populations, and produce data, tools and techniques intended to aid the improvement of selective breeding practices.[19] For this community, associated as they have been with the pragmatic and instrumental concerns

---

[19] Note that these tools and techniques are not merely molecular biological or biochemical in nature, or even deemed part of classical genetics. Statistical, quantitative and computational approaches and methods have been just as central to innovation in this area: Lowe and Bruce (2019).

of breeding, genetic variation has constituted a potential resource that breeders could exploit to improve populations in the ways they desired. The pig geneticists therefore developed a different disposition to the one that prevailed in medical genetics, a discipline that has been chiefly concerned with *deleterious* variants, or the one in yeast biology wherein the use of a standardised model strain with variation abstracted away has been a crucial basis of research. This helps to explain why systematic and functional studies were less entangled early on in yeast and human genomics compared to research on the pig.

As well as aiding this functionally-oriented research, the instrumental discernment of pig genetic diversity has also contributed to the identification of Quantitative Trait Loci (QTL), sites of genomic variation associated with phenotypic variation. This is unsurprising, given that the mapping of the pig genome from the early-1990s onwards involved the crossing of breeds that were assumed to be genetically distinct, and that this work was itself directed towards developing the methods to home in on QTL. The generation and exploitation of diversity was implicated in this more direct form of functionally-oriented research from the beginning of pig genomics. In the words of the summary of PigBioDiv2 on the European Union's CORDIS website, through this research, "the discerning customer can not only demand tasty meat but can help to power the academic drive for conservation".[20]

This research also added to the data and knowledge concerning other systematic aspects of the pig: phylogenetic relationships, evolutionary history, processes of domestication, and more recent histories of genetic exchange and relationships between breeds. One of the key challenges and contributions of the project was in measuring diversity. They adapted an approach to measure diversity devised by the economist Martin Weitzman, which involved measuring the genetic distance between pairs of populations using the marker data.[21] The genetic distances were then used to cluster the populations and infer phylogenetic trees and relationships between them. It therefore provided insights into the relationships between different populations, including between European and Chinese ones, and between the patterns of variation prevailing in those two

---

[20] https://cordis.europa.eu/article/id/85133-diversity-database-helps-conserve-rare-pig-breeds (last accessed 19th December 2022).

[21] This magpie-like approach to methods, techniques and resources outside of their own field reflects the bricolaged nature of pig genomics, as explored in Lowe et al. (2022).

regions.[22] They attributed these patterns to historical flows of genes that resulted from different modes of domestication and ways of organising livestock farming and breeding (Megens et al., 2008; SanCristobal et al., 2006).

In the next section, we show that the close relationship between these two modes of functional and systematic research—and the continuity of researchers and institutions—persisted through the production of the pig reference genome and into the aftermath of its completion. In yeast and human, with few exceptions, these modes of research were considerably less entangled in the immediate aftermath of the production of the reference genome.[23]

7.4   After the Reference Genome

7.4.1   Yeast: Successive Endeavours

EUROFAN—the European Functional Analysis Network—was always considered to be the next step after the Yeast Genome Sequencing Project (YGSP) by the community of *S. cerevisiae* genomicists. Although individual laboratories functionally interpreted and made use of some of the data from the sequencing project in their research, more concerted large-scale functional analysis was postponed until after the completion of the reference sequence. A high-quality reference would be needed in order to effect the targeted gene deletions that formed the centrepiece of EUROFAN.[24] Like the pig biodiversity projects examined above, EUROFAN benefitted from initial pilot programmes. In the case of yeast, researchers used these pilots to develop modes of gene disruption and methods of phenotypic assay for functional analysis.

---

[22] For example, they found that drift rather than mutations was primarily responsible for divergences within European populations, but that mutations were more salient in differences between European populations and Meishan pigs (SanCristobal et al., 2006).

[23] Though evolutionary analysis of genomic variation was performed for the compilation of the yeast and human reference sequences themselves (see Chap. 6, Sects. 6.2.1 and 6.2.2).

[24] There was some overlap between the projects, however, with EUROFAN beginning in January 1996 and the sequencing of the yeast genome being completed in April 1996. Preparations for a follow-up project to EUROFAN—EUROFAN 2—advanced in 1996 as well, with the application submitted in October of that year and the project beginning the following year: Peter Philippsen, personal communication with James Lowe and Miguel García-Sancho, February 2022.

The EUROFAN participants used the same yeast strain as in the sequencing project: S288C. S288C is a laboratory strain and has therefore had invariance in and between its colonies strictly enforced. As a result, for functional analysis it was necessary to *create* variation, so researchers could uncover the functions of genes in the reference genome. This was done by producing a new resource, a library of mutants. EUROFAN was conceived as a continuation of the annotation of the well-established and comprehensive reference genome, and indeed recapitulated the hierarchical but dispersed nature of the prior effort to sequence the yeast reference genome, especially the EC-funded portion of it. A division of labour was instituted, between:

• The overall coordination of the project;
• Liaison with the Yeast Industrial Platform (Chap. 2);
• An informatics strand—based at the Martinsried Institute for Protein Sequences (MIPS)—to manage and assess the quality of submitted data and develop a database and computational tools for data analysis;
• The creation of the mutants;
• The storage, curation and distribution of the mutant collection;
• Various kinds and stages of functional analysis occurring at the bench.

Like the YGSP, EUROFAN therefore involved a wide variety of institutions. There was considerable continuity between the participants in the YGSP and EUROFAN, and consequently it involved a set of laboratories working on the cell biology, molecular biology and biochemistry of yeast. This approach reflected a continued perception of the value of these large-scale networked projects for the research endeavours of these laboratories, and the advantages of coordinating such laboratories in a network for further genomic analysis.[25] In this way, the model of functional analysis was conceived as a means of contributing material towards the further investigation of genes, rather than it being intended to transform the basis of

[25] https://cordis.europa.eu/project/id/BIO4950080 (last accessed 19th December 2022). The project to produce a reference genome helped to unite various yeast communities (biochemists, geneticists, cell biologists, molecular biologists), which constituted a key difference with the human genome, but was something that it had in common with pig genome research (Chaps. 2, 3, 4, 5). On the discursive use of the notion of "the yeast genome" to establish and maintain a community and link it to a deeper history of yeast genetics research, see Szymanski et al. (2019).

"normal" yeast biology.[26] Indeed, all but two of the 21 participating laboratories in EUROFAN had also been involved in the YGSP; roughly a quarter of the members of that prior effort took part in EUROFAN.[27] The creation of a curated resource in the form of a mutant collection as well as the ongoing annotation of the reference genome was attractive to the EC, but also meshed explicitly with the imperative to add more resources to the toolkit of yeast as a eukaryotic model organism.

The project was labelled as systematic (in the adjectival sense) rather than comprehensive. This is because only some of the sequences that were potentially thought to contain protein-coding genes were investigated. The work included an assessment of Open Reading Frames (ORFs) identified in the reference genome sequencing. ORFs are DNA sequences between the start and stop codons that begin and terminate the initial transcription of DNA into messenger RNA. A workflow determined which of these ORFs would undergo successive forms of "increasingly specific" functional analysis. As a result, only a portion of the total of ORFs and genes identified through the initial sequencing and structural annotation of the reference genome were fully functionally characterised.

The functional analysis commenced with deletions of specific ORFs through the design of constructs—known as 'gene replacement cassettes'—and their insertion into yeast DNA. These gene replacement cassettes contained a gene (*kanMX*) that conferred resistance to the fungicidal chemical geneticin. The application of the said antifungal agent—geneticin—thus yielded only the yeast that had integrated the cassette into its DNA and therefore had suffered the deletion of the ORF. This method was developed in the midst of the YGSP by Peter Philippsen—the coordinator of the sequencing of chromosome XIV of *S. cerevisiae*—and Achim Wach at Biozentrum (Wach et al., 1994). By observing and measuring the impact of the successful deletion of a specific ORF on the organism, researchers could infer the functional role that it played in yeast, for instance whether the deleted ORF was part of a protein-coding gene.

---

[26] The benefits to 'normal' basic biology laboratories were emphasised in several overviews of EUROFAN (Dujon, 1998; Oliver, 1996). However, these laboratories did not feel the need to establish a domain of yeast genomics separate from their day-to-day yeast biology. In human genomics, by contrast, the promoters of large-scale genome centres sought to differentiate their endeavour from laboratory biology (Hilgartner, 2017).

[27] 82 institutions have been listed as participating in the European Yeast Genome Network (Parolini, 2018), based on the affiliations listed in 1997's *The yeast genome directory*.

By 1996, researchers at the European Molecular Biology Laboratory (EMBL) had finished comparing ORF sequence data to protein sequences held in public databases. On the basis of sequence similarities, they made functional predictions for over half of all identified yeast genes (Bassett Jr et al., 1996). EUROFAN concentrated on the genes for which functional predictions of this kind were not possible. These were the so-called 'orphans': "novel genes discovered from systematic sequencing whose predicted products fail to show significant similarity when compared to other organisms, or only show similarity to proteins of unknown functions" (Dujon, 1998, p. 617). Functionally characterising these kinds of genes in EUROFAN would be particularly useful, considering yeast's role as a model organism and in biotechnology. As a model organism, it would constitute a richer platform for inferring the functional implications of homologous sequences found in the less well-characterised genomes of other species. For biotechnology, the genes with novel functions that were identified could be expressed within yeast itself to yield potentially valuable products or be inserted into other organisms by transgenic techniques.

EUROFAN, as well as filling in the orphan gaps left after the EMBL's analysis, aimed to observe gene effects and functions in ways that were missed by what leading yeast biologist Stephen Oliver described as the "function-first" approach of "classical genetics", which relied on the detection of some observative heritable variation or change to infer the presence and function of a gene. Instead, in EUROFAN they deleted known genes to produce mutants, and then measured the quantitative effects of this, for instance on growth rates of the cells through competitive growth experiments, or the biochemical effects as assessed through measurement of metabolite concentrations (Oliver, 1997).[28]

EUROFAN created mutants based on the deletion of 758 ORFs and then proceeded towards analysis of the deletants, which was led first by Peter Philippsen and then by Steve Oliver. In addition to this, parallel projects led by YGSP participants created mutant strains of smaller numbers of ORFs and Bernard Dujon's laboratory committed "mass murder" by deleting multiple ORFs at a time and then characterising the mutant phenotypes arising from these (Goffeau, 2000).

Nevertheless, the desire to identify all of the genes in *S. cerevisiae* and characterise all ORF deletants remained. Funds to realise this came

---

[28] In this way, the EUROFAN approach differed from that of the medical geneticists, for whom the 'function-first' approach was integral.

through a collaboration between two of the leading US figures in the original sequencing project: Mark Johnston at Washington University and Ron Davis of Stanford University. Johnston got a grant from the National Institutes of Health (NIH) for the period 1997 to 2000 for 'Generation of the Complete Set of Yeast Gene Disruptions', an initiative to create a comprehensive catalogue of S288C deletion strains, affecting all its genes. Davis also obtained a grant from the NIH to provide the tens of thousands of oligonucleotides—synthetic DNA sequences—that were needed for the production of the deletion cassettes (Giaever & Nislow, 2014).

This work, running from 1998 to 2002 and hosted at Stanford, became the *Saccharomyces* Genome Deletion Project, now Yeast Deletion Project, a consortium that involved many of the leading actors in European yeast genomics as well as the North Americans, including Howard Bussey at McGill in Canada (Giaever et al., 2002; Winzeler et al., 1999).[29] It was complementary to, and in many respects a development of, EUROFAN. The consortium analysed the deletion strains thus produced under several growth conditions[30] and sent the strains—containing DNA barcodes to enable linkage of material and data resources—to be preserved and distributed by repositories such as ATCC (the American Type Culture Collection) and EUROSCARF (the European Saccharomyces Cerevisiae Archive for Functional Analysis).[31]

All this functional annotation was captured by databases set up specifically for yeast biologists to be able to exploit the data deluge being generated by these projects. The *Saccharomyces* Genome Database (SGD) was founded in 1993 and first made available through the internet in 1994. It

[29] https://web.archive.org/web/20210427053303/www-sequence.stanford.edu/group/yeast_deletion_project/deletions3.html (last accessed 19th December 2022).

[30] As noted in Bassett Jr et al. (1996, p. 764), "As there are an infinite number of possible growth conditions, it would be impossible for any systematic effort to analyze any given deletion strain comprehensively", which negates the possibility of true comprehensiveness or completeness, though "the public availability of these strains for future in-depth analyses by yeast labs specializing in the study of a particular class or family of genes would represent a powerful resource". Therefore, while the territory could never be completely explored, the means now existed for any laboratory to 'visit' any part of it they desired to.

[31] DNA barcodes use sequences of genes or parts of genes that are known to be specific to particular species to determine and label species membership for a given organism. On DNA barcoding and its multiple uses, see: Hollingsworth et al. (2016) and https://transgene.sps.ed.ac.uk/blog/investigating-barcoding-life (last accessed 19th December 2022). EUROSCARF is a service run by Scientific Research and Development GmbH, a company based in Oberursel near Frankfurt-am-Main.

is primarily funded by the NIH—through the National Human Genome Research Institute (NHGRI)—and is hosted at Stanford University. SGD curators compile and integrate data on *S. cerevisiae* with the aim of presenting functionally annotated genomic data to yeast biologists in a usable form, providing them with a variety of tools that allow them to interrogate functional relationships and interactions (Dwight et al., 2004).[32] The Comprehensive Yeast Genome Database (CYGD) was established at MIPS and intended to be a development of the prior work conducted at MIPS and the European sequencing and functional annotation consortia. Expert curators manually annotated the yeast genome, using data from EUROFAN and other allied projects. Its main objectives were two-fold: to develop an informatics infrastructure to analyse and annotate complex interactions in the yeast cell and later to link data being generated on other species of yeast to *S. cerevisiae*, using comparative genomic approaches to improve the annotation of *S. cerevisiae* using this data (Güldener et al., 2005).[33]

The functional efforts that populated these databases involved the creation of variation in a compendious fashion using a single well-characterised strain of yeast on the basis of a high-quality reference genome. This was a key difference with pig genomics, in which there was a long tradition of investigating variation before the reference genome was produced. For the human, there was also this tradition of investigating variation through medical genetics, but it became disconnected from the IHGSC effort to produce a reference sequence of the whole human genome.

In yeast, the functional analysis of this variation was intended to improve the value of the reference sequence by producing data to help annotate it. More broadly, it was pursued to generate and provide data and physical

---

[32] We may observe here the utility of local, specialist databases, though there is significant overlap in the data contained in them and in more global ones such as GenBank. The main reason for having data in a local or specialised database is to adapt its presentation and analytic tools to the requirements and preferences of a specific group of users. We speculate that this may be why the annotation of yeast sequences in the yeast databases are more community-based and less automated than more general-purpose databases such as GenBank or the ENA. On the importance of similar local databases in biomedicine, see Cambrosio et al. (2020).

[33] CYGD received most of the funds for its establishment from the EC from 2000 to 2004, and also received support from the German federal government, the German Research Foundation and the government of the Brussels Region of Belgium: https://cordis.europa.eu/project/id/QLRI-CT-1999-01333 (last accessed 19th December 2022).

resources (the mutant strains), which could be used by the wider yeast research community for their own purposes, thereby improving the value of the species as a model organism. As with the YGSP, the creation of reference resources, both bioinformatic and material, was accompanied by the generation of implementable knowledge about the genome of the species that could inform the further study of wider aspects of its biology.

The creation of these reference resources also enabled the production of reference sequences for other strains of *S. cerevisiae* and related species. This led to a florescence of comparative and evolutionary-focused studies on *S. cerevisiae* and other types of yeast. One leading example is a network in which six French laboratories associated with the Centre National de la Recherche Scientifique (CNRS)[34] worked with the French national sequencing centre Genoscope. This network, Génolevures, was a programme of comparative genomics research concerning the 'Hemiascomycetous' budding yeasts, a group that includes *S. cerevisiae*.[35] In the first round of this initiative, Genoscope sequenced the genomes of thirteen species in this group at a low coverage of between 0.2 and 0.4X. The participating laboratories then analysed this sequence data with reference to *S. cerevisiae*, which served as a comparator, an "internal standard" according to Horst Feldmann's description (Feldmann, 2000). This comparative approach facilitated the manual annotation of the thirteen new genomes, and, in turn, enabled the identification of 50 new genes for improving the annotation of the *S. cerevisiae* (S288C) reference genome. From 2000, all sequence and comparative data were stored in the

[34] The CNRS units were at Institut Pasteur, INRA Agro-Paris-Tech, University of Paris-Sud (University of Paris XI), University of Lyon I, University of Bordeaux II, and University of Strasbourg. All but Lyon participated in the YGSP (Souciet, 2011; Goffeau et al., 1997; Parolini, 2018).

[35] This is another example of a large-scale sequencing centre working with the yeast genomics community, on the community's terms, in a similar manner to that seen in pig genomics. In this respect, it is quite distinct from human post-reference genome projects, in which either large-scale sequencing centres dominated, with their work augmented by smaller institutions and laboratories on the terms of the project defined by the IHGSC, rather than those smaller-scale actors setting the agenda. This may be due to the fact that, whereas the direction of yeast and pig genomics was shaped by people working with these organisms (e.g., André Goffeau and Alan Archibald), the direction of human genomics was shaped by James Watson and John Sulston, outsiders to the human and medical genetics communities.

Génolevures database, which has since been succeeded by three more specialised databases to hold the results produced by the consortium.[36]

In 2002, Genoscope agreed to sequence the reference genomes of four species at a much higher 10X coverage: *Kluyveromyces lactis, Debaryomyces hansenii, Yarrowia lipolytica* and *Candida glabrata*, the first three of which were analysed in the initial Génolevures project, the last of which is a human pathogen closely related to *S. cerevisiae* (Souciet, 2011). The comparison between the genomes involved a study of evolutionary conservation and divergence, which allowed researchers to identify and then investigate a variety of evolutionary changes that occurred in and between each of the phylogenetic branches—the lineages—that the species represented. This formed the basis for further investigations in the systematic mode, including the sequencing of additional species. Intriguingly, the comparative genomics that constituted—and was enabled by—Génolevures also allowed researchers to unveil manifold differences in gene content between the related species. These data were useful for further investigation into the physiological differences between them, and therefore advanced functional analysis as well (Bolotin-Fukuhara et al., 2005; Souciet, 2011).

This connection between the functional and systematic modes of yeast genomics was recognised by leading members of the community. For example, the next major grant that Mark Johnston secured following the 1997 to 2000 creation of deletion strains was another from the NIH: 'Comparative DNA sequence analysis of the yeast genome' running from 2001 to 2005. Using BLAST programmes (see Chap. 6) to compare nucleotide and protein sequence data between *S. cerevisiae* and other members of the *Saccharomyces* genus, Johnston and collaborators at the Washington University Genome Sequencing Center were able to estimate genetic distances between the species. This information, they supposed, would indicate which pairings would produce the most valuable comparative data. From these comparisons, they were able to identify various genomic elements, such as potential protein-coding genes and functional non-coding sequences (Cliften et al., 2001).

Most of the collaborators on that work then pursued a comparative study of the genomes of *Saccharomyces* species: *S. cerevisiae* itself, three others with genetic distances indicative of enough evolutionary distance to

---

[36] http://gryc.inra.fr/ (GRYC: Genome Resources for Yeast Chromosomes), http://fungipath.i2bc.paris-saclay.fr/ (FUNGIpath) and http://phylomedb.org/ (PhylomeDB)—all last accessed 19th December 2022.

ensure divergence of non-functional sequences, and two more distantly related species. The objective of this was to identify signals of conserved "phylogenetic footprints" in the sequence that would indicate the presence of functional parts of the genome, including those that had been previously difficult to find, such as non-coding regulatory elements. The results enabled the further improvement of the annotation of the S. cerevisiae reference genome, and also included predictions of functional sequences that could be experimentally tested (Cliften et al., 2003).[37]

Throughout, this work was accompanied by Johnston's ongoing molecular biological research programme on glucose sensing and signalling in the yeast cell. He became involved in Génolevures in the late-2000s (The Génolevures Consortium, 2009), contributing further to de novo and improved sequencing and annotation of the members of the Saccharomyces genus. The increasingly dense comparative relations and data so established helped forge synergies between reference genomics, functional analysis of the genome, molecular biological research and systematic studies. Indeed, this had developed to the extent that the status of "model genus" was claimed for the Saccharomyces sensu stricto genus encompassing S. cerevisiae and close relatives, due to the magnitude of data and experimental resources available across and within it (Scannell et al., 2011).

This dynamic was explicitly articulated in the yeast genomics community. They were aware of the limitations of relying solely on a reference sequence of a highly-standardised laboratory strain that was phenotypically atypical. They believed that more reference sequences were required, within the S. cerevisiae species itself and for related species. They appreciated that the data and knowledge of genome variation and evolution that they wrought from these could be used for functional analyses and inform the improvement of the reference resources that they were based on. Ed Louis, who we encountered providing advice on telomeres and chromosomal evolution during the YGSP (Chap. 2), conveyed this in terms of a virtuous cycle (Fig. 7.1). In this cycle, additional data on genomic variation allows researchers to increase their knowledge concerning conservation across genomes. This helps them to improve annotations. Better annotations allow a refinement of the localisation of features such as synteny breakpoints: regions in-between two stretches of conserved sequence

---

[37] Such research was not restricted to the yeast genomics community; there was a similar effort conducted by the Whitehead Institute and MIT, motivated by improving the basis of cross-species comparative genomics (Kellis et al., 2003).

Refinement of synteny breaks via annotated gene order

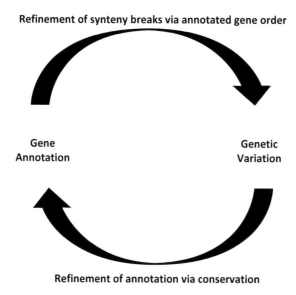

**Gene
Annotation**   **Genetic
Variation**

Refinement of annotation via conservation

Fig. 7.1 Illustration of the synergistic relationship between systematic and functional modes of post-reference genome research, as depicted by Ed Louis (2011, p. 32). Reprinted by permission from Springer Nature Customer Service Centre GmbH: Humana Press, *Yeast Systems Biology. Methods in Molecular Biology (Methods and Protocols)* by Castrillo J., and Oliver S. (eds), 2011, https://link.springer.com/book/10.1007/978-1-61779-173-4

of a particular kind. And these, in turn, allow fresh appreciation of structural variation (Louis, 2011).[38]

Ian Roberts of the National Collection of Yeast Cultures at the Institute of Food Research (Norwich, UK) and Stephen Oliver characterised research on the vast genomic and physiological diversity of yeasts and (functionally-oriented) systems biology as the "yin and yang" of biotechnological innovation involving these creatures, therefore emphasising the complementary and co-constitutive nature of these modes (Roberts & Oliver, 2011). As well as aiding manual improvements to annotations, the data and resources concerning diversity across yeast strains and species and their comparative relationships have also been harnessed to power automated annotation pipelines (Dunne & Kelly, 2017; Proux-Wéra et al., 2012).

---

[38] Analogously, we also described the international coordination of efforts around the sequencing and annotation of the yeast genome as a virtuous cycle that builds on—and facilitates—research and investments in this model organism (Chap. 6, Sect. 6.2.1).

In yeast, then, there was a passage from creating the reference genome, to pursuing functional analysis of that resource, to then producing data on other strains and related species, and using this to seed comparative and systematic research. The particular interpretation of comprehensiveness for these researchers was not restricted to a 'complete' reference genome but was far richer and heterogeneous. It involved the establishment of relations between a variety of different forms of data and the creation of tools to make use of them. This reflected the desire of the yeast genomicists themselves to make use of the resources; they therefore had knowledge of what was needed for research purposes, and how the data, resources and tools could be deployed and contextualised. All this also reflected the disposition of people who were aware of what their stewardship of a model organism entailed.

Major drivers of the yeast genome research agenda, such as Stephen Oliver and Mark Johnston, were able to appreciate and leverage the synergies that could be created between the functional and systematic modes of research, because they were engaged in both. Thus, the continuity of participants across these different successive phases of yeast genome research eased and motivated their ultimate integration. It was something of a different tale than in pig genomics, where, as we showed above, systematic and functional forms of analysis had been entwined since the pre-reference genome stage. In human genomics, our next object of analysis, the functional and systematic modes were more like twin tracks, than successive or permanently-entwined endeavours.

7.4.2   Human: Twin Tracks

As we have seen, in the sequencing of the human reference genome, the intended user communities were progressively detached from involvement in the production and annotation processes. However, in Chaps. 2 and 3 we showed how laboratories based in hospitals or medical schools had been conducting their own sequencing and making novel contributions by identifying genes and gene variants associated with particular pathological manifestations since before the start of whole-genome sequencing efforts. This programme of variant-focused and medically-oriented sequencing continued throughout the 1990s and beyond, with more and more mutations of particular genes catalogued and analysed, and more genes and key pathological variants associated with particular diseases or conditions. In some cases, research collaborations combined this approach with the sequencing of larger genomic regions: in the early-2000s, researchers at the Toronto Hospital for Sick Children (SickKids) joined forces with other medical genetics groups and Celera to sequence, analyse and extensively annotate human chromosome 7 (García-Sancho, Leng, et al., 2022; Scherer et al., 2003).

Several databases have been established to manage and present data on gene variants concerning human pathogenicity. These include Online Mendelian Inheritance in Man and the subscription access Human Gene Mutation Database (HGMD), while other databases have been created by particular communities focused on specific diseases or genes. The HGMD was founded in 1996, at the University of Cardiff in Wales. Its model is to scan biomedical literature and curate entries on 'disease-causing mutations', 'possible-disease-associated polymorphisms' and 'functional polymorphisms', according to the judgement of the curators assessing multiple lines of genomic, clinical and experimental evidence. Since 2000, HGMD has collaborated with commercial actors: the up-to-date version with enriched annotations and features is available on subscription from them, while a more basic free public version is also made available containing data that is at least three years old. Celera was the first commercial collaborator and included the extensive HGMD data in its Discovery System™ until 2005. From 2006 to 2015, the German bioinformatics company BIOBASE then developed HGMD Professional, a web application accessible upon purchase of a license, to hold this premium data. In 2014, BIOBASE was purchased by the German biotechnology company QIAGEN, which had participated in the sequencing of the yeast genome.[39]

Specialist disease-centred databases, such as the Toronto-based cystic fibrosis mutation database and network (Chap. 3), constitute resources and tools that are curated by the community of medical genetics of clinicians themselves, rather than being provided top-down by the NCBI or any other specialist genomics organisation. In this respect, these specialist databases are similar to some of the ones that arose out of yeast and pig genomics initiatives. They are, however, more long-lasting than many of the pig ones, more fragmented than the yeast ones, and more specialised than both. The more concentrated and global databases of yeast genomics, and the more ephemeral ones of pig genomics, result from different funding and support regimes, but also reflect the role of genomic resources

---

[39] http://www.hgmd.cf.ac.uk/ac/index.php (last accessed 19th December 2022). See also García-Sancho, Lowe, et al. (2022); Stenson et al. (2020).

in each community. Yeast, as a model organism, requires comprehensiveness and the inclusion of a multitude of different forms of data in one or a few repositories that exhibit some form of persistence and longevity. The pig community, however, corrals certain kinds of genomic data that are appropriate to the research and translational problems that need to be solved at a certain point in time, with such prioritisation trumping completeness (and permanence). For medical genetics, on the other hand, the community is much larger and divided by disease categories. The pig genomics community is not as partitioned by a focus on particular traits (even if some pig researchers have investigated some traits more than others) nor is the yeast one divided into silos investigating specific kinds of molecular mechanisms or processes.[40]

We return to the medical genetics track shortly. For now, we observe that it constituted a particular form of entanglement between functional and systematics research, which looked both at variation within genes, and variation between individuals, with this data linked to functional information drawn from a variety of sources. These sources even included evolutionary ones, insofar as they provided informative evidence used by curators, such as those at HGMD. Now, though, we consider a separate track that followed the publication of the human reference sequence by the IHGSC in 2004. In this track, distinct annotation efforts were conducted, on the analogy of EUROFAN, but in a quite different form. As during the determination of the human reference sequence, the medical genetics and IHGSC-based tracks remained largely separate throughout the 2000s until recent attempts at rapprochement, including the establishment of a centralised repository of clinically-relevant genomic data in 2013. This is why we refer to them as twin tracks: they developed simultaneously but maintained separated trajectories for a significant period of time.

---

[40]We would hypothesise that, in part, this is because disease states are more independent of each other than molecular mechanisms and processes in a cell are, and therefore the study of a particular disease can be more detached from research concerning other diseases. There is not necessarily a hard-and-fast distinction between the genetics and physiology of different traits in livestock animals, with immune response genes being implicated in physiological processes involved in other traits, for example. Selection for lean meat content has given rise to Porcine Stress Syndrome, simultaneously a disease, welfare and meat quality problem. Members of the pig genomics community tend to have to diversify their activities as well, to take advantage of different pots of money available to them as much as possible.

We have already encountered the comparative genome sequencing effort across the tree of life sponsored by the NHGRI in Chap. 6, in which two working groups provided recommendations to a Coordinating Committee that then amended and submitted them to the NHGRI Advisory Council. The aim of this was to generate data on non-human primates, mammals and selected other species to inform human genome annotation. Here, it is instructive to note two significant changes to the recommendations of the Working Group on Annotating the Human Genome made by the Coordinating Committee. One was to propose even lower coverage sequencing for non-primate mammals, effectively down-grading this component to a pilot project. The rationale for this was that there was insufficient knowledge of mammalian genome evolution at that point to be able to definitively identify particular species as ideal candidates for the deeper shotgun sequencing originally recommended. Instead, they argued that a shallower study should provide sufficient grounds for identifying candidates for deeper sequencing or *de novo* sequencing. Thus, systematic knowledge needed further development before it could begin to yield data from which a comparative inferential apparatus could generate homologies and hypotheses for searching the human genome for functional elements.

The other change was to postpone working on a survey of human genome variation. In spite of the Committee identifying this element as a "high priority", it baulked at committing significant resources to what amounted to a "resequencing project", and recommended instead to wait and see whether resequencing costs declined sufficiently over the coming years.[41] A 'Workshop on Characterizing Human Genetic Variation' was held in August 2004 to discuss possible ways forward, with further proposals for studying human genomic variation developed within the NHGRI in 2005, alongside collaboration with the ongoing HapMap project.[42] That, and other initiatives surveying human genomic variation are discussed later in this section. For now, it is worth noting that this

[41] "New Sequencing Targets for Genomic Sequencing: Recommendations by the Coordinating Committee", part of the documents for the Meeting of the NHGRI Research Network for Large-scale Sequencing and the NHGRI Sequencing Advisory Panel, May 16, 2004 (NHGRI History Archive 7036-021).

[42] "Report of the Annotation of the Human Genome Working Group", dated January 3, 2005 (NHGRI History Archive 7039-005); https://www.genome.gov/13514604/executive-summary-workshop-on-characterizing-human-genetic-variation (last accessed 19th December 2022).

systematic exploration of human genomic variation became decoupled from the effort to develop resources for human genome annotation.

Operating parallel to the ongoing efforts to develop a comparative approach to human annotation was ENCODE, the Encyclopedia of DNA Elements, an ongoing project that was conceived as a follow-up to the IHGSC effort. ENCODE was launched in September 2003 by the NHGRI, five months after the 'completion' of the euchromatic human genome sequence in April 2003. It has passed through successive phases and associated consortia since then (see Table 7.2 for the main participants in the Pilot Phase) but continued to work towards the overarching goal of building "a comprehensive parts list of functional elements in the human genome, including elements that act at the protein and RNA levels, and regulatory elements that control cells and circumstances in which a gene is active".[43] The rationale behind this effort was that a "comprehensive encyclopedia of all of these features is needed to fully utilize the sequence to better understand human biology, to predict potential disease risks, and to stimulate the development of new therapies to prevent and treat these diseases". It was therefore conceived as a bridge from the structural data-set 'completed' in 2003, to the ability to make use of it.[44]

ENCODE therefore aimed at, and presumed the possibility of achieving, completeness. Despite constituting an essay in functional genomics, it involved considerable structural annotation, as it involved identifying and annotating genes and other key functional elements such as regulatory regions. Its methods, though, have extended beyond those that are applied in both automated and manual annotation pipelines. The search for regulatory elements that affect the expression of genes entailed the development of a panoply of other approaches, including a return to 'wet lab' experimentation analogous to the functional analysis activities in the laboratories participating in EUROFAN. This involved the use of techniques that aimed to identify signs of activity in the genome, for example biochemical signatures of particular chromatin structures (the way DNA is packed) that enable access to the DNA so it can be transcribed (Kellis et al., 2014).

---

[43] https://www.encodeproject.org/help/project-overview/ (last accessed 19th December 2022).

[44] https://www.genome.gov/Funded-Programs-Projects/ENCODE-Project-ENCyclopedia-Of-DNA-Elements/pilot (last accessed 19th December 2022).

Table 7.2 Table of the main participants in the ENCODE Pilot Project. This project was dominated by the members of the International Human Genome Sequencing Consortium (Chap. 4, Table 4.1), although it also included institutions that did not participate in that reference sequencing effort. Elaborated by both authors from: https://www.genome.gov/Pages/Research/ENCODE/Pilot_Participants_Projects.pdf (last accessed 19th December 2022)

Core groups involved in ENCODE Pilot Phase

Group leader (institution)
Ian Dunham (Wellcome Trust Sanger Institute)
Anindya Dutta (University of Virginia)
Thomas Gingeras (Affymetrix, Inc.)
Roderic Guigo (Municipal Institute of Medical Research, Barcelona)
Richard Myers (Stanford University)
Bing Ren (Ludwig Institute for Cancer Research)
Michael Snyder (Yale University)
George Stamatoyannopoulos (University of Washington)

Additional Pilot Phase participant institutions
National Human Genome Research Institute
Duke University
Children's Hospital Oakland Research Institute
NIH Intramural Sequencing Center
NimbleGen Systems, Inc.
Pennsylvania State University
University of California, Santa Cruz
British Columbia Cancer Agency Genome Sciences Centre
Broad Institute
The Institute for Genomic Research
National Center for Biotechnology Information, National Library of Medicine
Boston University

One of the main outcomes of ENCODE has been the increasing realisation that what constitutes a functional element is relational and context-dependent. The move towards once again conducting genomics research in biological laboratories reflects this shift, as capturing what elements of the genome become functional in particular circumstances "requires a diverse experimental landscape" (Guttinger, 2019). While the establishment of the ENCODE project came out of the IHGSC effort, its investigation of the biology of the human genome has triggered the involvement of a broader range of experimental laboratories, due to ENCODE being concerned with living biological function and not merely constituting a data gathering exercise.

During the pilot phase of ENCODE, the GENCODE consortium was created to produce reference annotations of the human genome. From its inception, it was led by the Sanger Institute, and involved participants from several institutions including the EBI. GENCODE incorporated data from a variety of automated prediction pipelines and experimental data into the Ensembl pipeline and HAVANA manual curation (Chap. 6). The desire to demarcate truly functional regions from non-functional ones meant that genes had to be distinguished from pseudogenes, and that the significance of non-coding regions needed to be assessed. In both cases, inspired by research indicating the salience of regulatory regions in complex developmental processes, the identification and annotation of transcripts assumed great significance in the project and formed the basis for the manual annotation. They used transcriptomic data from EST and messenger RNA sequences and protein sequences obtained from GenBank and Uniprot, using BLAST to align these against the sequences of the original BAC clones used in human reference genome sequencing. The data arising from these efforts led to a mounting appreciation of the prevalence of alternative splicing across the genome, wherein there may be multiple products of a single gene.[45] Reflecting on their findings, the GENCODE team emphasised that the way in which a reference annotation is constructed "is extremely important for any downstream analysis such as conservation, variation, and assessing functionality of a sequence" (Harrow et al., 2012, p. 1760; see also Kokocinski et al., 2010).

Alongside this, efforts to catalogue the extent and diversity of human genomic variation were already underway. More so than in pig genomics, and far more so than for yeast genomics, this research has concentrated on variation *within* the target species. Even before the production of the human reference genome, there was a concerted project to map human genetic diversity. Although it received some support through the Human Genome Organisation (HUGO) and the NIH, the Human Genome Diversity Project founded in 1991 was unconnected with the

---

[45] Demarcating functional and non-functional elements in the genome was a task that became thornier as GENCODE—and ENCODE—went on. It has been the source of much of the controversy arising around ENCODE. Guttinger & Dupré (2016) provide a summary of the contestation around ENCODE that refers to much of the key literature on the topic.

IHGSC effort, and indeed also from medical genetics, being largely an initiative of researchers interested in human evolution and anthropology (M'Charek, 2005; Reardon, 2004). This concern with intra-specific human variation and diversity, and the connection of this with the study of the inheritance of traits—particularly disease traits—pre-dated the determination of the reference sequence. This work heavily relied on the use of genetic markers such as the Restriction Fragment Length Polymorphisms developed in the 1980s (Chap. 1). The advent of micro-array technology—SNP chips—made a qualitative difference to this line of inquiry and the relationship between research on human genetic diversity and the identification of particular genes with functional and pathological roles. Rather than recapitulate the research that has examined this (e.g. Rajagopalan & Fujimura, 2018), we instead explore the creation and impact of a SNP chip in pig genomics in the next section and relate this to the work on microarrays in human and medical genetics when appropriate.

In the 2000s, arising from the centralised and top-down world of the IHGSC, were the International HapMap Project (2002–2010) and the 1000 Genomes Project (2008–2015), which drew samples from populations around the world to identify common variants: those with at least 1% prevalence in any given population. For HapMap, SNPs were generated and selections of them were made according to the project criteria. Ten centres were used to genotype—genetically assess—the samples that were collected, with over 60% of this genotyping done at either RIKEN (**Ri**kagaku **Ken**kyūjo, the Institute of Physical and Chemical Research) in Japan or the G5 institutions: Sanger Institute, Whitehead Institute/Broad Institute, Baylor College of Medicine, Washington University in St Louis and the US Department of Energy's Joint Genome Institute. The resulting haplotype map identified sets of human genome variants that tended to be inherited together. The mapping was conceived as a "short-cut" to identifying candidate genes and aiding *association studies* to ascertain the genomic variants implicated in disease (The International HapMap Consortium, 2003). Like the follow-up 1000 Genomes Project, which sequenced whole genomes to capture genomic variation rather just sequencing parts of them, it was a top-down initiative that sought to provide a dataset to be picked up and exploited by a presumed external user community.

The efforts to cultivate and inform a user community, while directed towards helping researchers realise the value of the resource, demonstrated how separated producers and users were during the conception and realisation of such projects.[46] Furthermore, though they intended the data produced to be useful for what we describe as systematic studies, they were not conceived or generated for those purposes, but for the anticipated potential biomedical use of the data. To the extent that the data was analysed by the project for systematic purposes (e.g. The 1000 Genomes Project Consortium, 2010), it was presented as a separate application of the results. This systematic information was not articulated as being informative or indicative for functional studies, in the synergistic manner understood by the yeast genomics community by this time.

As with the reference genome produced by the IHGSC, however, we can interpret the fruits of these top-down projects in terms of the ways that they have been used as a means to create genomic resources more tailored to particular research needs. Consider the effort to produce "a regional reference genome" by a consortium of Danish researchers, "to improve interpretation of clinical genetics" in that country, enhance the power of association studies (examining the relationship between genomic and phenotypic variation) and aid precision medicine research (Maretty et al., 2017, pp. 87 and 91). This team produced 150 high-quality *de novo* assemblies, which they validated by aligning them against the then-current human reference genome assembly. They identified multiple forms of variants, aided by the reference panel produced by the 1000 Genomes Project and data from the NCBI's Single Nucleotide Polymorphism database (dbSNP), which had itself been considerably enriched through the efforts of the International HapMap Project and the 1000 Genomes Project. The Danish team were therefore able to use the infrastructure and resources developed from these top-down projects, not directly to produce research that could be translated into clinical outcomes, but to construct their own local, targeted resources in the form of a local reference

---

[46] https://web.archive.org/web/20210728170223/http://1000gconference.sph.umich.edu/ (last accessed 19th December 2022).

genome and a catalogue of variation pertinent to the populations they work with (Maretty et al., 2017).[47]

The relationship between large-scale data infrastructures and more local and specific ones focusing on concrete objects, communities or research areas has recently strengthened. One manifestation of this shift has been the establishment of ClinVar and ClinGen. These represent an attempt at liaison between the separate tracks of human genome research and medical genetics research. ClinVar and ClinGen capture forms of variation and processes of evidential evaluation of their functional and pathological significance that are found in medical genetics and clinical research. It therefore promises a form of synergy involving the alignment of different modes of data practices, methods, analytical approaches and community norms.

ClinGen and ClinVar were established by the NIH in 2013, with the aim of providing open-access data on variants, tied to clinical interpretations of them. ClinGen is the overall programme that works in partnership with ClinVar, the database that is run by NCBI. Both continue to be funded by the NIH. Their founding was based on the concern that such clinically-relevant genetic data was being kept locally, either by individual researchers and laboratories, in disease-specific databases available to members of a particular community, or hidden behind a paywall like the most recent and rich data contained in HGMD. Furthermore, the different architectures of such databases and treatments of data were thought to stymie clinical interpretation.

The answer was a centralised repository, with uniform data standards and clear processes of curation and attribution of labels to individual variants indicating their potential clinical (or otherwise functional) significance. However, to make this work, it would be necessary for the submission of data to ClinVar to be contextualised with its putative medical significance. Rather than being stripped of all but a few items of contextualisation in the form of metadata—as sequence data to GenBank and

---

[47] However, such population-specific references may not be generated for all geographical areas and populations that may want them or need to make use of them. For instance, it has been observed that, as of 2022, "[l]ess than 2% of [human] genomes analysed in the two decades" after the conclusion of the Human Genome Project "are from African individuals, even though Africa harbours more human genetic diversity than any other continent." This is not merely a problem of representation, but also a comparative lack of local sequencing and informatics capacity, as well as reliable infrastructures in these underrepresented regions (Ebenezer et al., 2022).

other similar databases is—this data on sequence variants needs to travel with clinical interpretations made by the submitters and the various kinds of evidence used in them. ClinVar serves as a repository for this information, with agreements or disagreements in interpretation assessed by the user researchers, rather than being solved by the database itself.[48]

Where ClinVar takes a more active role, is in the convening of expert panels to curate interpretations for particular genes. Applications can be made to the ClinGen Steering Committee for approval of the formation of an expert panel. The interpretations of these bodies then outrank virtually all other levels of "review status" (Landrum & Kattman, 2018).[49] One of these expert panels was called CFTR2, a group that worked—and still works—on the *CFTR* cystic fibrosis gene. Most of its members belong either to the Johns Hopkins University or the Toronto Hospital for Sick Children, reflecting a parallel route of research stretching back to the 1980s (Chap. 3) that was long separated from the established mainstream of human genomics research and infrastructure.[50]

ClinGen also takes an active role in aggregating and curating genomic and health data from various sources and feeding this into ClinVar (Rehm et al., 2018). ClinGen and ClinVar constitute the platform for a convergence between the once wholly distinct tracks associated with the IHGSC enterprise and medical genetics. While these remain separate in day-to-day practice, the creation of a data infrastructure to draw upon the findings and expertise of clinicians and researchers—including those working in medical genetics—enables them to participate in a more concerted and unified whole-genome effort. This also provides human genomicists outside the medical genetics community—including at specialist genome centres—with access to information about variation and its clinical effects that is essential for the medical translation of sequence data.

7.4.3   Pigs: A Fuzzier Distinction

As with the production of a reference genome, by the time the pig genomics community was in a position to develop their own concerted functional annotation effort, they were able to benefit from the protocols, methods,

---

[48] ClinVar limits itself to identifying one type of possible conflict between interpretations.

[49] https://www.ncbi.nlm.nih.gov/clinvar/docs/review_status/ (last accessed 19th December 2022).

[50] https://cftr2.org/about_cftr (last accessed 19th December 2022).

data and experience of human functional annotation. This legacy enabled them to devise a pared down approach more appropriate to the levels of funding they enjoyed. There were, however, aspects of functional annotation that drew on the particular history of this community, the uses that they envisaged for the data and the affordances provided by their particular subject organisms.

An initial call for the concerted annotation of non-model organism animals was made by Alan Archibald, Ewan Birney of the EBI and Paul Flicek (who had primarily worked on mouse genomics), at the International Society for Animal Genetics conference in Cairns (Australia) in July 2012.[51] This alliance reflected the ongoing connections between the pig genome community and the EBI. However, it was at the annual Plant and Animal Genome (PAG) conference, in San Diego in January 2014, that genomicists working on a variety of farm animals started developing the basis for an international multi-species collaboration to advance functional annotation following the initial sequencing of several reference genomes (The FAANG Consortium, 2015).[52] At that PAG conference, the Animal Biotechnology Working Group of the EU-US Biotechnology Research Task Force convened an "AgENCODE" workshop.[53] As the name suggests, the aim was to emulate ENCODE, and to that end, several speakers from that project contributed to the session and to subsequent workshops and conferences held by what became the Functional Annotation of Animal Genomes (FAANG) Consortium (Tuggle et al., 2016).

[51] https://www.isag.us/2012/docs/ISAG_2012_Abstracts.pdf (last accessed 19th December 2022).

[52] The reference genome completion dates were 2004 for the red jungle fowl (*Gallus gallus*), the wild progenitor of the domesticated chicken; 2009 for cattle (*Bos taurus*); 2011 for pig (*S. scrofa*); and 2014 for sheep (*Ovis aries*). Reference assemblies of other species (such as goat and salmon) were not deemed to be of sufficient quality or were not completed at this point.

[53] This working group was initially established in 2008. It aimed "to encourage an integrated program of US-EC collaboration combining research, training and dissemination activities" concerning animal genomics, animal health, and bioinformatics, with the added purpose of fostering "[i]nteractions among the agricultural science, life science and medical science communities" to enable the elucidation of phenotypes from genotypes, a key theme that we return to in this discussion. https://web.archive.org/web/20170918061400/ https://ec.europa.eu/research/biotechnology/eu-us-task-force/pdf/20th-meeting/ working_group_on_animal_biotechnology_en.pdf (last accessed 19th December 2022). The overall task force was founded in 1990 by the EC and the White House Office of Science and Technology: http://archive.euussciencetechnology.eu/bilat-usa/news/id/231 (last accessed 19th December 2022).

Presenting the outcomes of the AgENCODE workshop in a PAG conference session the following day were key figures from the genome mapping and sequencing of chicken, cattle and pig from the previous two decades: Gary Rohrer of USDA MARC, Alan Archibald of the Roslin Institute, Christine Elsik of the University of Missouri, Elisabetta Giuffra of the Animal Genetics and Integrative Biology unit (Génétique Animale et Biologie Intégrative, GABI) at the Jouy-en-Josas station of INRA and Martien Groenen of Wageningen University.[54] Reflecting the practices and careers of many livestock geneticists, these researchers worked on the genomes of multiple species. All this demonstrates the agriculturally-inclined origins of FAANG, which has shaped the aims and outputs of the project ever since. Although other potential applications such as the use of animals as biomedical models and understanding domestication and evolution have also been cited as motivations, these have not formed a substantial part of the published output or attention of the consortium.[55]

The aim of the FAANG Consortium (and its constituent steering committee and working groups) has been to "produce comprehensive maps of functional elements in the genomes of domesticated animal species based on common standardized protocols and procedures" (The FAANG Consortium, 2015, pp. 2-3). The Consortium (Table 7.3 lists pig genomicists who were founding members) narrowed its focus to the animals for which there were reference assemblies most amenable to functional annotation (chicken, cattle, pig and sheep), identified a small set of core assays and defined experimental protocols based on the experiences of ENCODE, established a Data Collection Centre based at the EBI to aid and validate submissions to the data portal hosted on the FAANG website, and defined a core set of tissues to be used. The collection and sharing of a limited set of tissues derived from populations of low genetic diversity was intended to aid the replicability and comparability of the data produced using them across the community and to ensure that associations between functional genomic annotations and quantitative phenotypic data could be made even in the early stages of the project (The FAANG Consortium, 2015).

---

[54] https://pag.confex.com/pag/xxii/webprogram/Paper9366.html (last accessed 19th December 2022).

[55] E.g., there are some mentions in The FAANG Consortium (2015), Tuggle et al. (2016), and Clark et al. (2020) of these applications, but they are largely incidental to the dominant agricultural focus.

Table 7.3 Pig genomicists who were initial members of the FAANG Consortium, identified through authorship of 'The FAANG Consortium' article (2015). This article indicated that "[a]ll authors are signatories of the FAANG Consortium" (p. 5). The Consortium included 30 other members who primarily worked on other livestock species such as cattle and chicken

Name	Institution
Leif Andersson	Department of Medical Biochemistry and Microbiology, Uppsala University & Department of Animal Breeding and Genetics, Swedish University of Agricultural Sciences
Alan L. Archibald	Roslin Institute, Royal (Dick) School of Veterinary Studies, University of Edinburgh
Elisabetta Giuffra	Animal Genetics and Integrative Biology unit (Génétique Animale et Biologie Intégrative, GABI), INRA Jouy-en-Josas
Martien A. Groenen	Animal Breeding and Genomics Centre, Wageningen University
Heebal Kim	Department of Agricultural Biotechnology, Seoul National University
Joan K. Lunney	Animal Parasitic Diseases Laboratory, USDA Agricultural Research Service, Beltsville, Maryland
Graham S. Plastow	Livestock Gentec Centre, Department of Agricultural, Food and Nutritional Science, University of Alberta
James M. Reecy	Department of Animal Science, Iowa State University
Gary A. Rohrer	USDA Meat Animal Research Center
Christopher K. Tuggle	Department of Animal Science, Iowa State University
Shuhong Zhao	Key Laboratory of Agricultural Animal Genetics, Breeding and Reproduction of Ministry of Education of China, Huazhong Agricultural University

Source: Elaborated by James Lowe

A key feature of FAANG has been its focus on defining and decomposing the phenotype, or the phenome. Phenome is a term that denotes the phenotypic equivalent of the genome, with phenomics constituting concerted phenotyping on the model of the major genomic sequencing projects. In the farm animal world, researchers have access to extensive gross phenotypic data (such as on coat colour, slaughter weight, number of eggs laid per day) on animals with well-defined pedigrees. This is due to the role that measuring phenotypes has played in the breeding industry, with which researchers have enjoyed close ties since at least the 1960s. The means by which to measure, analyse and interpret phenotypic data are long-established and have continually evolved as animal geneticists adopted more molecular approaches in the 1980s and then pursued genome mapping, sequencing and analysis from the 1990s onwards. Both molecular and genomic approaches have intersected with quantitative genetics research and methods.

The extent of this focus on phenotypic data eclipses the other two species we have examined throughout this book. Yeast biologists have paid close attention to the phenotypes of their organism, but these are phenotypes of far less complexity than those of farm animals. Concerning the human, concerted efforts to characterise large groups of humans in phenotypic terms, for example in the history of physical anthropology (Müller-Wille & Rheinberger, 2012, pp. 106-107) or more recent initiatives such as the UK Biobank project (Bycroft et al., 2018), constitute exceptions to the general trend in which phenotypic data collection—and the development of infrastructures and practices to enable this—has been far less extensive than for at least some farm animal species. One cannot control the breeding or environmental conditions experienced by humans or track multiple phenotypic measurements in such a continuous and intrusive way as can be done for an experimental herd or flock (or for plants; see: Müller-Wille, 2018).

One of the key aims of FAANG has been to decompose the gross phenotypes they and breeders had previously been working with into more proximate molecular phenotypes (biomarkers), and then to causally link variation in these proximate molecular phenotypes to variation in gross phenotypes. Alongside other intended outputs of the FAANG collaboration, the identification of molecular phenotypes and associated specific genomic variants has been intended to better model the relationship between genotype and phenotype, to advance their agenda of improving genomic prediction from a known genotype to an expected phenotype. This emphasis on genotype–phenotype relationships and being able to more accurately predict the phenotype from a given genotype is not unique to pig or wider farm animal genomics, but it does attain a distinct salience and inflection in this area.

Within five years of FAANG swinging into action, the participants were looking beyond the initial in-depth studies of a limited range of tissues with low genetic diversity. While this had helped the Consortium to identify and map functional elements and regions, it became clear that data derived from a more genetically-diverse range of animals, and more tissues, would be necessary to further analyse the relationship between genomic variation and phenotypic variation. Genes specific to particular populations could be identified through this, and then visualised in pangenome graphs depicting variation aligned to the reference sequence. This, in turn, could aid the identification of candidate variants to be implemented in programmes of genome editing of livestock species, and in the tracing of genetic diversity in and across populations to inform conservation efforts. Beyond individual species, the functional genomic and phenotypic data that FAANG compiled enabled them to identify evolutionary conservation across species. On this foundation, they could develop comparative analyses and approaches to inform cross-species inferences as to the functional genomic basis of phenotypic traits (Clark et al., 2020).

This transition from a narrow focus to a broader outlook was eased by the design of FAANG and the long-standing entanglement of systematic and functional modes of research in pig genomics. Among pig genomicists, that entwinement had fostered both versatility and an acute appreciation of the wide array of possibilities and potential applications presented by the rich and connected data generated by the FAANG Consortium.

Beyond FAANG, there have been two other ways in which functional and systematic modes have been entangled in pig post-reference genomics. A 2013 paper reporting studies of the genetic diversity of rare breed Chato Murciano pigs kept on eight farms in Spain instantiates one of these. This research used an inspection of the extent of variation that existed in these pigs to assess their (functional) viability in the light of inbreeding and crossbreeding (Herrero-Medrano et al., 2013). The second way is the kind of cycle (as identified by Ed Louis for yeast) between further functional annotation of the genome and an appreciation of conservation and syntenic breakpoints: either across pig breeds, between related species or drawing on a multi-species comparative approach to enrich knowledge of genome evolution more broadly (e.g. Anthon et al., 2014). This, again, often depended on the construction of new sequences based on older ones, in order to establish new connections between genomes, to identify relationships, changes over evolutionary time and examples of different forms of variation. As Martien Groenen of Wageningen University observed in a review of pig genome research in the systematic mode, however, though advances in this direction were enabled by the existence of an annotated reference genome, they were also inhibited by its limitations (Groenen, 2016). A new reference sequence and improved annotation using it and through FAANG has, therefore, proved a considerable boon to both systematic and functional studies.

We close this discussion of the relationship between functional and systematic modes of post-reference genome research concerning the pig by

Fig. 7.2 The second-generation SNP chip for pigs: PorcineSNP60v2 BeadChip. Photograph courtesy of Illumina, Inc.

exploring a tool that represents a powerful platform to enable both: the Illumina PorcineSNP60 SNP chip or microarray (see Fig. 7.2).

A SNP chip is a tool that enables the detection of the presence or absence of a particular set of DNA polymorphisms in a sample. In constructing them, DNA—of complementary sequence to the polymorphisms to be detected—is attached to the surface of the chip. The samples to be assayed are then labelled, typically with a fluorescent dye, and added to the chip. Any sequences complementary to the probes should attach to the chip's surface and, when stimulated, produce a detectable signal which is recorded and can then be processed to give the results of the assay. There are numerous technical details and options that go into the construction and use of a particular chip. We focus here on the choice of the DNA to be attached to the chip surface: the probes that are used to detect particular genomic variation at the single-nucleotide allele level.

When it became possible to do so, the value of identifying and generating data on SNPs was quickly recognised by the community of pig genomicists. They had long valued the creation and mapping of genetic markers of various kinds (including those with no putative functional or mechanistic role), for the identification and mapping of QTL. SNPs are polymorphic—albeit less so than microsatellites—and abundant across the genome, including in regions poorly-represented by markers such as microsatellites. They therefore represented an opportunity to identify markers at a higher resolution and more broadly across the genome.

This is particularly significant given the translational domain most members of the pig genome community were working towards: animal breeding. While there had been efforts to identify particular genes and variants thereof from the 1980s, in many cases actual functional genes were not necessarily needed for the purposes of breeding. In the 1990s, for instance, an approach called 'Marker-Assisted Selection' (MAS) was developed that only required that a genetic marker be identified, provided that it was closely associated with a gene of interest that a breeder may want to select for or against (e.g. Rothschild & Plastow, 2002). While identifying a gene would be imperative for transgenic improvement of livestock, or for medical genetics research, it is not for animal breeding. Because the aim is to improve a *population* in measurable ways, finding and using markers that are *good-enough* indicators is a viable strategy. If it is mistaken in individual cases, this is not a problem, as they can simply be removed from the breeding pool. By the turn of the millennium, quantitative geneticists were proposing new ways to develop MAS. One of these was 'genomic selection', in which many more markers would need to be genotyped across the genome to ensure that at least some of them were closely linked to any (probably unknown) loci with an actual causative effect on the eventual phenotype (Haley & Visscher, 1998; Meuwissen et al., 2001). This, therefore, created the demand for SNPs to be generated and incorporated into a chip to enable the genotyping of multiple sets of them (Lowe & Bruce, 2019).

Alongside this, industry was pursuing SNPs with the view to identifying candidate genes. Sygen (as PIC had been renamed) secured EC funds for PORKSNP, a project running from 2002 to 2006 to identify SNPs in genes expressed in pig muscle and then run association studies to search for loci involved in meal quality traits. Sygen provided the samples for subcontracted biotechnology companies to sequence.[56] Monsanto, who had entered the pig breeding market having bought into DeKalb in 1996 (completing the purchase in 1998), were also deeply interested in SNPs for performing genome-wide association studies. In November 2001, Monsanto's Swine Genomics Technical Lead John Byatt spoke with Jane Peterson from the NHGRI's Extramural Program (Chap. 3) about potential support for a pig genome sequencing project. In Peterson's notes on the event, she observed that "Really what they need are SNPs—denser needed".[57] However, as pig genome sequencing did not proceed at the NHGRI, Monsanto looked elsewhere: to the IHGSC's competitor, Celera. In addition to its primary biomedical focus, Celera had acquired an agriculturally-oriented biotechnology company from its parent company Perkin-Elmer, in what was effectively an internal transfer. The head of this company, Celera AgGen, was Stephen Bates. Bates persuaded Craig Venter to shotgun sequence pigs, cattle and chickens and create livestock databases using the data so generated. In February 2002, this unit was sold to MetaMorphix Inc., a biotechnology company founded in 1994 by researcher Se-Jin Lee of the Johns Hopkins University School of Medicine, who was the discoverer of the protein myostatin. As part of the deal, MetaMorphix licenced Celera's databases for pigs, cattle and chickens. In June 2004, they licenced what they called 'GENIUS—Whole Genome System™' for pigs to Monsanto for one million dollars and a share of royalties in the new breeding lines (and their hybrid offspring) developed by Monsanto using their data, which encompassed approximately 600,000 mapped SNPs and related intellectual property.[58] Despite the apparent fruitfulness of this association, MetaMorphix filed for bankruptcy in 2010, and Monsanto abandoned the pig breeding sector in 2007, selling Monsanto Choice Genetics to Newsham Genetics.[59]

---

[56] https://cordis.europa.eu/project/id/HPMI-CT-2002-00205 (last accessed 19th December 2022).

[57] Notes from Jane Peterson's meeting with John Byatt, 20th November 2001. NIHGR archives, Box031-014, obtained 7th December 2016.

[58] https://www.sec.gov/Archives/edgar/data/1289370/000093041306007147/c44432_10sb12g-a.htm (last accessed 19th December 2022).

[59] https://www.thepigsite.com/news/2007/09/newsham-genetics-acquires-monsanto-choice-1 (last accessed 19th December 2022); https://www.sec.gov/Archives/edgar/data/1289370/000093041311002078/c64859_ex99-1.htm (last accessed 19th December 2022).

Meanwhile, the pig genome community was also pursuing SNPs and the creation of a SNP chip. In addition to their potential utility in animal breeding, the geneticists believed that the generation of SNPs would enable the exploitation of mouse and human data for homing in on candidate genes, as well as aiding the refinement of genetic linkage maps (Rohrer et al., 2002; Schook et al., 2005). Creating the basis for the production of SNPs was to be an outcome of the project to sequence the reference genome. Martien Groenen obtained funding to perform next-generation sequencing on additional pigs to identify SNPs, brought in other members of a consortium—which became the International Porcine SNP Chip Consortium—to pursue this, and led the analysis group to identify putative SNPs (Archibald et al., 2010).

Alongside this, a commercial partner was needed to produce and distribute the chip. The consortium held what Alan Archibald has described as a "beauty contest" at the PAG conference in January 2008, between genomic services and tool manufacturers Illumina and their main competitor, specialist microarray producer Affymetrix. Both had previously produced chips for cattle, and the judges were swayed by Illumina's articulations of the lessons learned from it.[60] Illumina's cattle chip was produced at the behest of the USDA in 2007, with its 54,001 SNPs used in genomic evaluations of American dairy cattle. This was quickly deployed in genomic selection, a process that has produced considerable results on a short timescale and demonstrated the value of the approach (Wiggans et al., 2017). In addition to Illumina's lessons, a group at the USDA facility in Beltsville (Maryland) offered advice based on their own involvement in creating and using the cattle chip, with Curt Van Tassell in particular contributing valuable insights.

Martien Groenen had been involved in the development of a 20K chip (containing 20,000 SNPs) for the chicken, in collaboration with the breeding industry for that species.[61] It therefore made sense for him to play a leading role in the effort to create a pig SNP chip. For this, he leveraged existing relationships, such as with the Dutch pig breeding company Topigs, which provided genotype and sequencing data derived from their

[60] Interview with Alan Archibald, conducted by James Lowe, Roslin Institute, November 2016.

[61] Interview with Martien Groenen, conducted by James Lowe over Skype, September 2017.

breeding lines.[62] As with other pig genomics projects, each participant brought their own funding to enable them to make their contributions, which included the provision of samples, the sequencing and identification of SNPs, conducting the selection and validation of SNPs, bioinformatics work and networking with other organisations (such as the EBI) to assist in developing and publishing the data produced through the project.[63]

The commercial exigencies of the pig chip structured its contents. So too did the interests of the members of the pig genome community and the kinds of pigs—and therefore DNA samples and SNPs—that were available (see Table 7.4 for members of the 'International Porcine SNP Chip Consortium'). Marylinn Munson from Illumina participated in the weekly working group meetings of the Consortium conducted over Skype, which made the crucial decisions shaping the chip, for instance, how many SNPs were included, with roughly 60,000 chosen. Options of up to a million SNPs were floated, but this was deemed to be excessive when the trade-off between the number of SNPs and the cost of the chip was considered. For the chips designed to genotype humans, which needed to be able to identify rare alleles (possibly involved in rare diseases) and to sample a variety of different populations, a chip with as many SNPs as technically feasible was required. For the pig, however, to ensure the competitive pricing and commercial viability of the chip, advance orders of $5 million would have to be obtained. Breeders therefore had to be interested in the chip, and this meant including alleles of at least 5% prevalence that were present in a range of breeds that mainly reflected commercial populations used by the major breeders. Where possible, SNPs known to be of relevance to livestock traits were included. Proprietary SNPs were excluded.[64] The team narrowed down the approximately half-a-million SNPs to the selection of tens of thousands to be included on the chip.[65] The DNA samples used on the chip were obtained from the Duroc, Piétrain, Landrace

[62] Interview with Barbara Harlizius, conducted by Ann Bruce and James Lowe over Skype, December 2018; personal communication from Barbara Harlizius to James Lowe, January 2022.

[63] "Pig SNP Working Group" folder, Lawrence Schook's personal papers, obtained 6th April 2018.

[64] Interview with Martien Groenen, conducted by James Lowe over Skype, September 2017; interview with Lawrence Schook conducted by James Lowe over Skype, August 2017.

[65] "Pig SNP Working Group" folder, Lawrence Schook's personal papers, obtained 6th April 2018.

Table 7.4 List of members of the International Porcine SNP Chip Consortium and their institutional affiliations, from "Pig SNP Working Group" folder, Lawrence Schook's personal papers, obtained 6th April 2018. Note the continuity of personnel and institutions from prior mapping and sequencing projects (Chap. 5, Table 5.2). This illustrates the stability of actors in the pig genomics community and their involvement in the creation of multiple successive genomic resources, as well as the primarily agricultural orientation of the participants

Members of the International Porcine SNP Chip Consortium

Name	*Institution*
Alan Archibald	Roslin Institute
Jon Beever	University of Illinois
Mario Caccamo	European Bioinformatics Institute
Richard Clark	Sanger Institute
Richard Crooijmans	Wageningen University
Martien Groenen	Wageningen University
Lakshmi Matukumalli	USDA Beltsville (Bovine Functional Genomics Laboratory)
Denis Milan	INRA Castanet-Tolosan
Dan Nonneman	USDA Meat Animal Research Center
Gary Rohrer	USDA Meat Animal Research Center
Max Rothschild	Iowa State University
Robert Schnabel	University of Missouri
Lawrence Schook	University of Illinois
Tim Smith	USDA Meat Animal Research Center
Jerry Taylor	University of Missouri
Curt Van Tassell	USDA Beltsville (Bovine Functional Genomics Laboratory)
Ralph Wiedmann	USDA Meat Animal Research Center

Source: Elaborated by James Lowe

and Large White commercial breeds from Europe and North America and wild boar from Japan and Europe.[66]

SNPs were identified through a series of procedures, some of which used the latest versions of the reference assembly. The SNPs that passed validation were then put through a selection process which included assessment across a variety of parameters. The resulting PorcineSNP60 Genotyping BeadChip was released by the end of 2008 (Ramos et al., 2009). The advent of SNP chips made genomic selection in pigs feasible,

---

[66] There was a wider sampling including other domesticated breeds, including Asian ones, and related species to the pig as well, with the data from this sequencing and SNP discovery published in the publicly-available dbSNP database.

and it was adopted in the pig breeding industry as it had been in cattle (Knol et al., 2016; Samorè & Fontanesi, 2016).[67] A second version of the Illumina chip has since been developed, as well as other chips created with different selections of SNPs (Samorè & Fontanesi, 2016).

In addition to the direct use in genomic selection, the chip has also been extensively used in systematic studies, for instance concerning the diversity and patterns of domestication and geographic distributions of pigs. As with yeast, such research can reveal differences between populations and signatures of selection that enable candidate genes to be identified for further functional exploration (e.g. Diao et al., 2019; Yang et al., 2017).

A plethora of more direct functional analyses have been enabled by the chip, aiding researchers in finding and investigating genetic loci related to livestock production and welfare traits, for example through association studies (e.g. Maroilley et al., 2017). It has also helped researchers developing pigs as animal models of particular diseases (e.g. for muscular dystrophy: Selsby et al., 2015).[68] And finally, SNP chips can be used to produce and/or validate new reference resources, for instance in constructing a new high-density genetic linkage map (Tortereau et al., 2012) or assessing the completeness of the new reference sequence (Warr et al., 2020).

SNP chips, much like reference genomes and other reference resources, constitute platform tools that can be deployed for a variety of purposes. They enable new characterisations of variation and the creation of fresh resources based on them. In this, the variation imprinted in it, conditions its affordances as a platform tool. And in the case of the pig, the heavy involvement of the pig genomics community in the generation and selection of the SNPs to be included, and the commercial demands driving this process, affects what the SNP chip can do, and what new resources it can help seed. For example, the lack of representation of samples of DNA from African breeds and populations of pigs in the Illumina 60K chip makes it of limited usefulness for breeding applications in that continent. As a result, there has been a call for the creation of more Africa-specific

---

[67] Additional source: interview with Michael Goddard, conducted by James Lowe and Ann Bruce in Edinburgh, October 2018.

[68] For a list of all papers that have cited Ramos et al. (2009) that describes the creation and validation of the first-generation 60K Illumina SNP chip for pigs, see: https://pubmed.ncbi.nlm.nih.gov/?size=200&linkname=pubmed_pubmed_citedin&from_uid=19654876 (last accessed 19th December 2022).

livestock SNP chips, as well as breed or region-specific reference genomes (Ibeagha-Awemu et al., 2019).[69]

The development of genomic resources and the exploitation of them are therefore strongly conditioned by the historical paths taken. In the case of pig genomics, we have observed a close integration of functional and systematic modes of research from pre-reference genomics onwards, continuing even during the narrower and more concentrated endeavour to sequence the reference genome. The heavy involvement of the community of pig genomicists in the creation of genomic resources from the early-1990s onwards has enabled them to facilitate versatility in the wide use and applications of these resources once the pig reference genome was released. As we have seen though, this does not mean that the data and materials they have helped to generate lend themselves to an unlimited array of uses. It does mean, however, that they have a keen awareness of what these resources represent, how they can be built on and what they can be used for. The pig community has also benefited greatly from knowledge concerning the genomes and genomic research of other species. They have identified practices in human and cattle genomics, for example, and adapted them to their own ends and ways of working. They have also developed a comparative framework for making use of genomic data and other resources on mammals such as humans. As we have seen, the development of pig post-reference genomics differs considerably from that of human and yeast. We close the chapter by assessing the consequences of this, introducing the concept of webs of reference to help us to further characterise post-reference genomics and compare the historical trajectories of genomics across different species.

7.5   Seeding Webs of Reference

This chapter, together with elements of preceding ones, challenges existing views of postgenomics. By looking beyond human genomics and especially beyond the determination of the human reference sequence, we

---

[69] As well as the representation of particular alleles, this is also because of the differential genetic structure of livestock populations, as a result of different breeding and herd/flock management practices. There have been initiatives to sequence particular breeds and populations that were not included in the reference genome, combined with new methods of incorporating and displaying variation in reference assemblies (e.g., for one involving two African cattle breeds, see Talenti et al., 2022). However, many breeds—and species—of social and economic importance in the Global South remain uncharacterised (Ebenezer et al., 2022).

have shown that an emphasis on variation, multi-dimensionality and the contextualisation of sequence (and mapping) data has pre-existed reference genomics, and can be part of reference genomics itself, rather than simply succeeding and complementing reference genome sequences once they are produced.

Across the three species we have examined, the relationships between pre-reference genome research, reference genomics and post-reference genomics are affected by the differential involvement of particular communities in these efforts. In yeast and pig, there is a high-level of continuity across these phases, with the respective communities involved in constitutive aspects of the process of reference genome sequencing, and in enriching and improving the products. They have done this through engagement with large-scale sequencing centres (e.g. the Sanger Institute) and other centralised actors (e.g. MIPS), though in different ways. For example, the relationship of the pig community to the Sanger Institute was more like Mark Johnston's relationship to the Genome Sequencing Center at Washington University than it was equivalent to the role of the Sanger Institute as a contributor to the YGSP.

The yeast and pig communities also differed in their overall goals, the nature of their target organisms and the variation exhibited by these organisms. The yeast community were self-consciously curating a model organism with a panoply of linked datasets and experimental resources, with an eye towards comprehensiveness, permanence and accumulation. They worked with a highly-constructed laboratory strain of *S. cerevisiae* specifically designed to minimise variation within and between colonies. The pig community, on the other hand, often worked with a mixture of primarily commercial breeds of pig, reflecting the mainly agricultural aims of their research but also the ready availability of these creatures. But they also used wild boar, as well as crosses between breeds presumed to be genetically distinct due to their geographical distance. They created genetic markers, maps, mapping tools, QTL detection methods, families and pedigrees of pigs, reference assemblies, annotations of these, as well as masses of SNPs and the chips to genotype selections of them. They worked in a satisficing mode, with researchers, groups and institutions contributing to consortia and collaborations with their own pots of money from various funding sources, building on and using existing sets of resources they had produced for a prior purpose. In both species, we see a convergence between functional and systematic modes of practising genomics, involving considerable overlaps between actors pursuing both modes. Both communities realised that an investigation of diversity could aid functional analyses either directly through the identification and analysis of key physiological and genetic differences, or more indirectly by using the insights gained from systematic analysis to improve the functional annotation and characterisation of reference genomes and other reference resources associated with the species.

In human genomics, there has been more than one community at play. There is the IHGSC community, that through the mid-to-late 1990s and into the 2000s became increasingly narrow and concentrated. They emphasised the technical refinement of sequencing in large-scale centres and the development, advancement and integration of informatics pipelines. Then there has been the medical genetics community, focused on variation between individuals (and across populations more broadly) and in the sequences of particular genes. This latter community, as we have seen, became increasingly divorced from the IHGSC effort. Instead, they established connections with Celera and their activities, for instance through the annotation jamboree, the sequencing and analysis of chromosome 7 (Chap. 6), and in further developing the HGMD. This interaction constitutes a rapprochement between the medical genetics community and an institution that specialised in the sequence determination and informatics aspects of genomics to an exquisite degree, mediated by its own commercial strategies and responses to the actions of the IHGSC. A newer rapprochement between medical genetics and the mode of genomics characterised by centralised infrastructures and data repositories has been through ClinGen and ClinVar. These constitute an attempt to compile and interpret more richly-contextualised data on genetic variants of potential clinical import, and in so doing incorporate medical genetics practices and practitioners more into the centralised NCBI framework.

The community dynamics we have identified, in tandem with the way that pre-reference genomics and the creation of a reference genome proceeded, have affected how post-reference genome functional and systematic research related to each other. Throughout our examination of functional and systematic research, we have found that separately assessing the limitations of individual reference resources or tools fails to capture the inter-relations between them. Inter-relatedness has been a feature across the history of genomics, however, as existing resources are used for the construction of new ones, often through the deployment of comparative practices. Additionally, reference resources can relate to each other contemporaneously, through overlapping repertoires and data infrastructures, and by the ways in which one resource can inform the interpretation or validation of another.

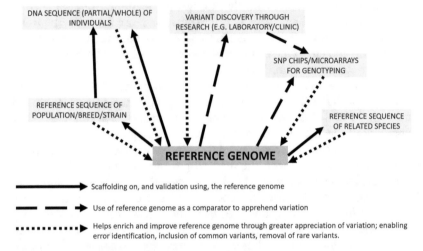

Fig. 7.3 A simplified depiction of a *web of reference* in which types of resources in the web are represented, rather than individual instantiations (there may be many different resources for each type). The development of webs of reference enables the exploration and characterisation of the extent, frequency, range and combinations of different types of genomic variation across a representational domain, such as a species. In so doing, the reference genome and other resources are further developed. The development of individual webs depends on the different historical trajectories leading to, and arising from, the creation of a reference genome. Elaborated by both authors

Through interpreting the products of genomic research as part of *webs of reference* that exhibit a range of connections (Fig. 7.3), we can better assess the infrastructural roles and consequences of reference resources. In the three species, post-reference genome work involved the creation of reference resources that identified and characterised more genomic variation. The reference resources refer to the reference genome, are explicitly intended to connect different manifestations of variation, and contain a surplus of possibilities for the further identification and characterisation of genomic variation and the translation of such data into a multitude of different working worlds.

Based on our examination of the different confluences of systematic and functional research, we can observe that post-reference genomics does not merely consist of increasing dimensionality: the recording and linking of additional genomic variation and other forms of biological variation in data infrastructures. It also involves the generation of these dimensions and the establishment of relations between them, in different concrete ways. Additional dimensions close to the level of the DNA sequence such as RNA sequences and protein sequences do not just exist in nature to be the next logical source of data to link to the reference sequence after its production. These forms of data are produced and catalogued for particular purposes and from particular sources: recall, in Chap. 6, the use of cDNA from the cloned offspring of TJ Tabasco in pig genome annotation. Other forms of data may derive from different origins, and be chosen for their practical utility rather than their representativeness of the species or particular biological processes. Here, we might consider the narrow range of genetically homogeneous tissue samples and assays used in the initial phases of FAANG. Furthermore, as FAANG shows, additional dimensions of data being arrayed on top of reference sequences may not only represent distinct kinds of macromolecules, but phenotypes as well.

Systematic studies entail and power comparative genomic approaches that generate dense sets of data and knowledge concerning the relationships between the genomes of different strains, populations or species. This helps researchers to characterise the extent and nature of genomic variation across populations, species and sets of related species. The extent of the potential variation (including different types of genomic and other biological variation) that can be apprehended and compared is limitless. Therefore, a selection of what is actually identified and represented from that limitless array of the potentially comparable is made either *a priori* or during the process of analysis. What dimensionality is added to the web of reference depends on the history and interests of the community producing a resource and how this community relates to the processes involved in producing and improving the reference genome. In other words, we cannot characterise this expansion of dimensionality as being a mere consequence of a simple transition from genomics to postgenomics (or even to post-reference genomics): there are different temporalities and models across (and within) yeast, human and pig genomics.

Across both functional and systematic studies separately, and even more acutely in their intersection, the variation that is measured, analysed and integrated into data infrastructures constitutes only some of the potential range that could be pursued and exploited. The dimensions that are explored, even if they are apparently of the same kind, may be directed towards distinct goals, use different materials and be related to other dimensions differently. We refer to this as a *variational surplus*, in analogy to the surplus possibilities open to researchers working on particular experimental systems, as characterised by Hans-Jörg Rheinberger (1997, p. 161). So, does all this just result in a blooming, buzzing confusion of different approaches to variation among distinct projects and communities? The construction of infrastructures to establish links and relationships between different forms of data and material objects, and efforts towards integration (e.g. Leonelli, 2013), suggests not.[70]

The history of post-reference genomics, elements of which we have examined in this chapter, suggests that there has been a shift in the kind of research on and using genomes. In the next chapter, we explore this in terms of "epistemic iteration", a term coined by philosopher Hasok Chang (2004). For now, we note that in the absence of direct access to the 'truth', the improvement of standards such as reference genomes is evaluated using epistemic virtues, values and goals as guides. This occurs through the correction and enrichment of these resources and builds on and supersedes prior standards. The past serves as a constraint or a condition but is not wholly determinative of the future course of the standard. Reference genomes and other reference resources can be seen as products of their history: the choices made by particular communities amongst those available to them, including objects, methods, and modes of validation and enrichment. These activities use and devise standards such as designated reference genomes and up-to-date maps. Each standard undergoes its own process of improvement, in which new versions succeed old ones. Linkages are made between different kinds of standards or reference resources, and such linkages are used in the construction and evaluation of one resource in terms of another. What makes the shift to post-reference genomics

---

[70] The notion of connection or linkage between resources that we use in this chapter, to describe the way that reference resources are related to each other in a web of reference, is more generic than the concept of *data linkage*. Data linkage entails the implementation of specific methods and infrastructures to allow data from different sources to be brought together on a common platform (e.g., Tempini, 2020). The kind of interoperability and data mobility that data linkage in this sense enables may play a role in establishing and exploiting connections between reference resources, such as the alignment of new sequence data to an existing reference genome, or being able to move from a representation of one kind of map to another in a browser (as discussed by de Chadarevian, 2004). However, connections need not require this kind of data linkage. For example, maps and reference genomes can be used interactively as visual sources by researchers who use established inferences in the production and evaluation of new reference resources (Chaps. 4, 5 and 6; Lowe et al., 2022).

significant depends on two related phenomena. One is the increase in the number of linkages that contributes to the improvement of individual standards/resources and their use in the improvement of other standards/resources. The other is the amplifying and ramifying effect of such improvements at the more global level of webs of reference.

Before we discuss this shift further, however, we should acknowledge that for the purposes of organising the narrative and our analysis, we have assessed the production and nature of reference genomes, their annotation, and post-reference genomics in separate chapters. This should not be taken to imply that these are discrete aspects of genomics or that they occur in a regular and linear sequence. Rather, as we have attempted to demonstrate throughout, the boundaries between any one particular set of practices that depend on the outcomes of another set are rarely sharply drawn. Conceptually later processes such as annotation may inform revisions of assemblies or even details of the sequence of reference genomes, for example, and the distinctions between structural and functional annotation, and manual and automated means of conducting it, are rarely clear-cut.

With that in mind, we consider how the aims and shape of genomic research changed following the release of reference sequences. These reference genomes were not themselves static, but were continually modified and improved according to widely held epistemic criteria. These improvement efforts were often informed by the results of post-reference genomic projects that themselves relied on and used an existing reference sequence.

Alongside the enrichment of the reference genome, a panoply of reference resources have been created for distinct populations and individuals, and the means to make comparisons within and between species has been further developed. These have fed functional analysis, but have also enabled the increasing exploration and mapping of the terrain of variation within species and the establishment of connections between different species. While this has led to concerns about the extent to which the reference sequence *represents* the increasingly mapped terrain, the new locales established throughout this land were still seeded from the reference genome, and related to it. The terrain is not three-dimensional like a geographical landscape, but more like a hyperdimensional state space. In this way, webs of reference have been constructed, exploring the variational space for a given type (the species, a sub-species, or a higher-level grouping or taxon) as new reference standards are created to capture specified types or sub-types. These webs of reference, in which each node is related to others, have developed iteratively and recursively. The more linked data there is concerning the variational space of the type, the more that further exploration can be conceived, and existing reference resources improved using the new linked data. This is where the development of population-specific resources, and ways of representing genomic variation such as pangenome graphs, have taken post-reference genomics: seeding the web.

The reference genome is useful to the extent that it is a viable origin of radiation that enables functional and systematic lines of investigation to bloom and produce linkages between different kinds of data and material. Genomics involves the creation of standards that improve over time relative to the epistemic aims of their creation and use, becoming more stable over time, though never achieving completion due to shifts in epistemic goals and the non-existence of even a theoretical absolute standard. But this is just a part of the picture, particularly for post-reference genomics, in which developments include the progressive exploration of the indefinitely-dimensional variation space for particular species (or other types) and the establishment of connections between these concerning different species (or between other types). The more the space is explored, the more connections can be made and the basis for further exploration—extensively across the space and intensively in particular regions of it—is created (Fig. 7.3).

The way this process unfolds, and the webs of reference that are constructed through it, is unlikely to be generic. The greater degrees of freedom offered compared with reference genomics indicates that the involvement of particular communities in the generation of genomic resources will be at least as salient to how these webs develop as they were to how reference genomes were produced. However, the historicity and contingency underlying these webs of reference should not distract from the potentially new emergent dynamics generated through them. The existence of a web of reference at a certain level of development lowers the threshold for adding—and connecting—new reference resources. New groups and communities can draw upon and link to existing resources to generate their own, and therefore to contribute towards and help shape the web. The wider context of reference resources should therefore be considered as a factor in enabling fresh participation and the connection of genomic data and resources to more specific research goals, in addition to the more widespread and distributed capacity to conduct sequencing that has emerged in the last 20 years.

REFERENCES

Agar, J. (2020). What is science for? The Lighthill report on artificial intelligence reinterpreted. *The British Journal for the History of Science, 53*(3), 289–310.

Ankeny, R. A., & Leonelli, S. (2011). What's so special about model organisms? *Studies in History and Philosophy of Science Part A, 42*(2), 313–323.

Ankeny, R. A., & Leonelli, S. (2015). Valuing data in postgenomic biology. In Richardson and Stevens (Ed.), *Postgenomics: Perspectives on biology after the genome* (pp. 126–149). Duke University Press.

Anthon, C., Tafer, H., Havgaard, J. H., Thomsen, B., Hedegaard, J., Seemann, S. E., et al. (2014). Structured RNAs and synteny regions in the pig genome. *BMC Genomics, 15*, 459.

Archibald, A. L., Bolund, L., Churcher, C., Fredholm, M., Groenen, M. A., Harlizius, B., et al. (2010). Pig genome sequence—Analysis and publication strategy. *BMC Genomics, 11*, 1.

Bassett Jr, D. E., Basrai, M. A., Connelly, C., Hyland, K. M., Kitagawa, K., Mayer, M. L., et al. (1996). Exploiting the complete yeast genome sequence. *Current Opinion in Genetics & Development, 6*(6), 763–766.

Bolker, J. (2012). There's more to life than rats and flies. *Nature, 491*, 31–33.

Bolotin-Fukuhara, M., Casaregola, S., & Aigle, M. (2005). Genome evolution: Lessons from Genolevures. In P. Sunnerhagen & J. Piškur (Eds.), *Topics in current genetics, Vol. 15: Comparative genomics* (pp. 165–196). Springer-Verlag.

Bork, P., Dandekar, T., Diaz-Lazcoz, Y., Eisenhaber, F., Huynen, M., & Yuan, Y. (1998). Predicting function: From genes to genomes and back. *Journal of Molecular Biology, 283*(4), 707–725.

Bycroft, C., Freeman, C., Petkova, D., Band, G., Elliott, L. T., Sharp, K., et al. (2018). The UK Biobank resource with deep phenotyping and genomic data. *Nature, 562*, 203–209.

Cambrosio, A., Campbell, J., Vignola-Gagné, E., Keating, P., Jordan, B. R., & Bourret, P. (2020). 'Overcoming the Bottleneck': Knowledge architectures for genomic data interpretation in oncology. In S. Leonelli & N. Tempini (Eds.), *Data Journeys in the Sciences* (pp. 305–327). Springer Nature.

Chang, H. (2004). *Inventing temperature: Measurement and scientific progress.* Oxford University Press.

Church, D. M., Schneider, V. A., Graves, T., Auger, K., Cunningham, F., Bouk, N., et al. (2011). Modernizing reference genome assemblies. *PLoS Biology, 9*(7), e1001091.

Clark, E. L., Archibald, A. L., Daetwyler, H. D., Groenen, M. A. M., Harrison, P. W., Houston, R. D., et al. (2020). From FAANG to fork: Application of highly annotated genomes to improve farmed animal production. *Genome Biology, 21*(1), 285.

Cliften, P. F., Hillier, L. W., Fulton, L., Graves, T., Miner, T., Gish, W. R., et al. (2001). Surveying *Saccharomyces* genomes to identify functional elements by comparative DNA sequence analysis. *Genome Research, 11*, 1175–1186.

Cliften, P., Sudarsanam, P., Desikan, A., Fulton, L., Fulton, B., Majors, J., et al. (2003). Finding functional features in *Saccharomyces* genomes by phylogenetic footprinting. *Science, 301*, 71–76.

de Chadarevian, S. (2004). Mapping the worm's genome. Tools, networks, patronage. In J.-P. Gaudillière & H.-J. Rheinberger (Eds.), *From molecular genetics to genomics: The mapping cultures of twentieth-century genetics* (pp. 95–110). Routledge.

Deplazes-Zemp, A. (2018). 'Genetic resources', an analysis of a multifaceted concept. *Biological Conservation, 222*, 86–94.

Diao, S., Huang, S., Xu, Z., Ye, S., Yuan, X., Chen, Z., et al. (2019). Genetic diversity of indigenous pigs from South China Area revealed by SNP array. *Animals, 9*, 361.

Dujon, B. (1998). European Functional Analysis Network (EUROFAN) and the functional analysis of the *Saccharomyces cerevisiae* genome. *Electrophoresis, 19*, 617–624.

Dunne, M. P., & Kelly, S. (2017). OrthoFiller: Utilising data from multiple species to improve the completeness of genome annotations. *BMC Genomics, 18*, 390.

Dwight, S. S., Balakrishnan, R., Christie, K. R., Costanzo, M. C., Dolinski, K., Engel, S. R., et al. (2004). *Saccharomyces* genome database: Underlying principles and organisation. *Briefings in Bioinformatics, 5*(1), 9–22.

Ebenezer, T. E., Muigai, A. W. T., Nouala, S., Badaoui, B., Blaxter, M., Buddie, A. G., et al. (2022). Africa: Sequence 100,000 species to safeguard biodiversity. *Nature, 603*, 388–392.

Engel, S. R., Dietrich, F. S., Fisk, D. G., Binkley, G., Balakrishnan, R., Costanzo, M. C., et al. (2014). The reference genome sequence of *Saccharomyces cerevisiae*: Then and now. *G3: Genes|Genomes|Genetics, 4*(3), 389–398.

Ewing, B., & Green, P. (1998). Base-calling of automated sequencer traces using Phred. II. Error probabilities. *Genome Research, 8*(3), 186–194.

Ewing, B., Hillier, L., Wendl, M. C., & Green, P. (1998). Base-calling of automated sequencer traces using Phred. I. Accuracy assessment. *Genome Research, 8*(3), 175–185.

Feldmann, H. (2000). Editorial: Génolevures—A novel approach to evolutionary genomics. *FEBS Letters, 487*, 1–2.

Felsenfeld, A., Peterson, J., Schloss, J., & Guyer, M. (1999). Assessing the quality of the DNA sequence from the Human Genome Project. *Genome Research, 9*, 1–4.

Fisk, D.G., Ball, C.A., Dolinski, K., Engel, S.R., Hong, E.L., Issel-Tarver, L., et al. (2006). *Saccharomyces cerevisiae* S288C genome annotation: A working hypothesis. *Yeast, 23*(12), 857–865.

Food and Agriculture Organization of the United Nations. (1999). *The global strategy for the management of farm animal genetic resources: Executive brief.* FAO.

García-Sancho, M., Leng, R., Viry, G., Wong, M., Vermeulen, N., & Lowe, J. W. E. (2022). The Human Genome Project as a singular episode in the history of genomics. *Historical Studies in the Natural Sciences, 52*(3), 320–360.

García-Sancho, M., Lowe, J. W. E., Viry, G., Leng, R., Wong, M., & Vermeulen, N. (2022). Yeast sequencing: 'Network' genomics and institutional bridges. *Historical Studies in the Natural Sciences, 52*(3), 361–400.

García-Sancho, M., & Lowe, J. W. E. (Eds.). (2022). The sequences and the sequencers: A new approach to investigating the emergence of yeast, human, and pig genomics. Special issue of *Historical Studies in the Natural Sciences, 52*(3).

Giaever, G., Chu, A. M., Ni, L., Connelly, C., Riles, L., Véronneau, S., et al. (2002). Functional profiling of the *Saccharomyces cerevisiae* genome. *Nature, 418*, 387–391.

Giaever, G., & Nislow, C. (2014). The yeast deletion collection: A decade of functional genomics. *Genetics, 197*(2), 451–465.

Goffeau, A. (2000). Four years of post-genomic life with 6000 yeast genes. *FEBS Letters, 480*, 37–41.

Goffeau, A., Aert, R., Agostini-Carbone, M., Ahmed, A., Aigle, M., Alberghina, L., et al. (1997). The yeast genome directory. *Nature, 387*(6632).

Grivell, L. A., & Planta, R. J. (1990). Yeast: The model 'eurokaryote'? *Trends in Biotechnology, 8*, 241–243.

Groenen, M. A. M. (2016). A decade of pig genome sequencing: A window on pig domestication and evolution. *Genetics Selection Evolution, 48*, 23.

Güldener, U., Münsterkötter, M., Kastenmüller, G., Strack, N., van Helden, J., Lemer, C., et al. (2005). CYGD: The Comprehensive Yeast Genome Database. *Nucleic Acids Research, 33*, D364–D368.

Guttinger, S. (2019). Beyond the genome: The transformative power of functional genomics. *Genomics in Context*, edited by James Lowe, published 2nd August 2019. Retrieved December 19, 2022, from https://genomicsincontext. wordpress.com/beyond-the-genome-the-transformative-power-of-functional-genomics/

Guttinger, S., & Dupré, J. (2016). The ENCODE project and the ENCODE controversy. In Zalta, E. N. (Ed.), *The Stanford Encyclopedia of Philosophy* (Winter 2016 Edition). Retrieved December 19, 2022, from https://plato. stanford.edu/entries/genomics/encode-project.html

Haley, C., & Visscher, P. M. (1998). Strategies to utilize marker-Quantitative Trait Loci Associations. *Journal of Dairy Science, 81*(2), 85–97.

Harrow, J., Frankish, A., Gonzalez, J. M., Tapanari, E., Diekhans, M., Kokocinski, F., et al. (2012). GENCODE: The reference human genome annotation for The ENCODE project. *Genome Research, 22*(9), 1760–1774.

Herrero-Medrano, J. M., Megens, H. J., Crooijmans, R. P., Abellaneda, J. M., & Ramis, G. (2013). Farm-by-farm analysis of microsatellite, mtDNA and SNP genotype data reveals inbreeding and crossbreeding as threats to the survival of a native Spanish pig breed. *Animal Genetics, 44*(3), 259–266.

Hilgartner, S. (2017). *Reordering life: Knowledge and control in the genomics revolution*. The MIT Press.

Hill, W. G. (1999). Advances in quantitative genetics theory. In J. C. M. Dekkers, S. J. Lamont, & M. F. Rothschild (Eds.), *From Jay Lush to genomics: Visions for animal breeding and genetics* (pp. 35–46). Iowa State University.

Hollingsworth, P. M., Li, D.-Z., Van der Bank, M., & Twyford, A. D. (2016). Telling plant species apart with DNA: From barcodes to genomes. *Proceedings of the Royal Society of London B, 371*, 20150338.

Ibeagha-Awemu, E. M., Peters, S. O., Bemji, M. N., Adeleke, M. A., & Do, D. N. (2019). Leveraging available resources and stakeholder involvement for improved productivity of African livestock in the era of genomic breeding. *Frontiers in Genetics, 10*, 357.

International Human Genome Sequencing Consortium. (2004). Finishing the euchromatic sequence of the human genome. *Nature, 431*, 931–945.

Kellis, M., Patterson, N., Endrizzi, M., Birren, B., & Lander, E. S. (2003). Sequencing and comparison of yeast species to identify genes and regulatory elements. *Nature, 423*(6937), 241–254.

Kellis, M., Wold, B., Snyder, M. P., Bernstein, B. E., Kundaje, A., Marinov, G. K., et al. (2014). Defining functional DNA elements in the human genome. *Proceedings of the National Academy of Sciences of the United States of America, 111*(17), 6131–6138.

Khamsi, R. (2022). The quest for an all-inclusive human genome. *Nature, 603*, 378–381.

Knol, E. F., Nielsen, B., & Knap, P. W. (2016). Genomic selection in commercial pig breeding. *Animal Frontiers, 6*(1), 15–22.

Kokocinski, F., Harrow, J., & Hubbard, T. (2010). AnnoTrack–A tracking system for genome annotation. *BMC Genomics, 11*, 538.

Landrum, M. J., & Kattman, B. L. (2018). ClinVar at five years: Delivering on the promise. *Human Mutation, 39*, 1623–1630.

Leonelli, S. (2013). Integrating data to acquire new knowledge: Three modes of integration in plant science. *Studies in History and Philosophy of Biological and Biomedical Sciences, 44*(4), 503–514.

Leonelli, S., & Ankeny, R. A. (2013). What makes a model organism? *Endeavour, 37*(4), 209–212.

Louis, E. (2011). *Saccharomyces cerevisiae*: Gene annotation and genome variability, state of the art through comparative genomics. In J. I. Castrillo & S. G. Oliver (Eds.), *Yeast systems biology, methods in molecular biology* (Vol. 759, pp. 31–40). Springer Science+Business Media.

Lowe, J. W. E. (2021). Adjusting to precarity: How and why the Roslin Institute forged a leading role for itself in international networks of pig genomics research. *The British Journal for the History of Science., 54*(4), 507–530.

Lowe, J. W. E., & Bruce, A. (2019). Genetics without genes? The centrality of genetic markers in livestock genetics and genomics. *History and Philosophy of the Life Sciences, 41*, 50.

Lowe, J. W. E., Leng, R., Viry, G., Wong, M., Vermeulen, N., & García-Sancho, M. (2022). The bricolage of pig genomics. *Historical Studies in the Natural Sciences, 52*(3), 401–442.

Mackenzie, A. (2015). Machine learning and genomic dimensionality: From features to landscapes. In Richardson and Stevens (Ed.), *Postgenomics: Perspectives on biology after the genome* (pp. 73–102). Duke University Press.

Maretty, L., Jensen, J. M., Petersen, B., Sibbesen, J. A., Liu, S., Villesen, P., et al. (2017). Sequencing and *de novo* assembly of 150 genomes from Denmark as a population reference. *Nature, 548*(7665), 87–91.

Maroilley, T., Lemonnier, G., Lecardonnel, J., Esquerré, D., Ramayo-Caldas, Y., Mercat, M. J., et al. (2017). Deciphering the genetic regulation of peripheral blood transcriptome in pigs through expression Genome-Wide Association Study and allele-specific expression analysis. *BMC Genomics, 18*, 967.

M'Charek, A. (2005). *The Human Genome Diversity Project: An ethnography of scientific practice*. Cambridge University Press.

Megens, H.-J., Crooijmans, R. P. M. A., San Cristobal, M., Hui, X., Li, N., & Groenen, M. A. M. (2008). Biodiversity of pig breeds from China and Europe estimated from pooled DNA samples: Differences in microsatellite variation between two areas of domestication. *Genetics Selection Evolution, 40*(1), 103–128.

Meuwissen, T. H. E., Hayes, B. J., & Goddard, M. E. (2001). Prediction of total genetic value using genome-wide dense marker maps. *Genetics, 157*, 1819–1829.

Müller-Wille, S. (2018). Making and unmaking populations. *Historical Studies in the Natural Sciences, 48*(5), 604–615.

Müller-Wille, S., & Rheinberger, H.-J. (2012). *A cultural history of heredity*. The University of Chicago Press.

Myelnikov, D. (2017). Cuts and the cutting edge: British science funding and the making of animal biotechnology in 1980s Edinburgh. *The British Journal for the History of Science, 50*(4), 701–728.

Nurk, S., Koren, S., Rhie, A., Rautiainen, M., Bzikadze, A. V., & Mikheenko, A. (2022). The complete sequence of a human genome. *Science, 376*, 44–53.

Oliver, S. (1996). A network approach to the systematic analysis of yeast gene function. *Trends in Genetics, 12*(7), 241–242.

Oliver, S. G. (1997). Yeast as a navigational aid in genome analysis. *Microbiology, 143*, 1483–1487.

Ollivier, L. (2009). European pig genetic diversity: A minireview. *Animal, 3*(7), 915–924.

Parolini, G. (2018). *Building human and industrial capacity in European biotechnology: The Yeast Genome Sequencing Project (1989–1996)*. Technische Universität Berlin. Retrieved December 19, 2022, from https://depositonce.tu-berlin.de/bitstream/11303/7470/4/parolini_guiditta.pdf

Proux-Wéra, E., Armisén, D., Byrne, K. P., & Wolfe, K. H. (2012). A pipeline for automated annotation of yeast genome sequences by a conserved-synteny approach. *BMC Bioinformatics, 13*, 237.

Rajagopalan, R. M., & Fujimura, J. H. (2018). Variations on a Chip: Technologies of difference in human genetics research. *Journal of the History of Biology, 51*, 841–873.

Ramos, A. M., Crooijmans, R. P. M. A., Affara, N. A., Amaral, A. J., Archibald, A. L., Beever, J. E., et al. (2009). Design of a high density SNP genotyping assay in the pig using SNPs identified and characterized by next generation sequencing technology. *PLoS ONE, 4*(8), e6524.

Reardon, J. (2004). *Race to the finish: Identity and governance in an age of genomics*. Princeton University Press.

Rehm, H. L., Berg, J. S., & Plon, S. E. (2018). ClinGen and ClinVar—Enabling genomics in precision medicine. *Human Mutation, 39*, 1473–1475.

Rheinberger, H.-J. (1997). *Toward a history of epistemic things: Synthesizing proteins in the test tube*. Stanford University Press.

Rheinberger, H.-J., & Müller-Wille, S. (Bostanci, A., Trans.) (2017). *The gene: From genetics to postgenomics*. The University of Chicago Press.

Richardson, S. S., & Stevens, H. (Eds.). (2015). *Postgenomics: Perspectives on Biology after the genome*. Duke University Press.

Richterich, P. (1998). Estimation of errors in "Raw" DNA sequences: A validation study. *Genome Research, 8*(3), 251–259.

Roberts, I. N., & Oliver, S. G. (2011). The yin and yang of yeast: Biodiversity research and systems biology as complementary forces driving innovation in biotechnology. *Biotechnology Letters, 33*, 477–487.

Rohrer, G., Beever, J. E., Rothschild, M. F., Schook, L., Gibbs, R., & Weinstock, G. (2002). *Porcine sequencing white paper: Porcine Genomic Sequencing Initiative*. Retrieved December 19, 2022, from https://www.animalgenome. org/pig/community/WhitePaper/2002.html

Rothschild, M. F., & Plastow, G. S. (2002). Development of a genetic marker for litter size in the pig: A case study. In M. F. Rothschild & S. Newman (Eds.), *Intellectual property rights in animal breeding and genetics* (pp. 179–196). CABI Publishing.

Samorè, A. B., & Fontanesi, L. (2016). Genomic selection in pigs: State of the art and perspectives. *Italian Journal of Animal Science, 15*(2), 211–232.

SanCristobal, M., Chevalet, C., Haley, C. S., Joosten, R., Rattink, A. P., Harlizius, B., et al. (2006). Genetic diversity within and between European pig breeds using microsatellite markers. *Animal Genetics, 37*, 189–198.

Scannell, D. R., Zill, O. A., Rokas, A., Payen, C., Dunham, M. J., Eisen, M. B., et al. (2011). The awesome power of yeast evolutionary genetics: New genome sequences and strain resources for the *Saccharomyces sensu stricto* genus. *G3: Genes|Genomes|Genetics, 1*, 11–25.

Scherer, S. W., Cheung, J., MacDonald, J. R., Osborne, L. R., Nakabayashi, K., Herbrick, J. A., et al. (2003). Human chromosome 7: DNA sequence and biology. *Science, 300*(5620), 767–772.

Schook, L. B., Beever, J. E., Rogers, J., Humphray, S., Archibald, A., Chardon, P., et al. (2005). Swine Genome Sequencing Consortium (SGSC): A strategic roadmap for sequencing the pig genome. *Comparative Functional Genomics, 6*, 251–255.

Selsby, J. T., Ross, J. W., Nonneman, D., & Hollinger, K. (2015). Porcine models of muscular dystrophy. *ILAR Journal, 56*(1), 116–126.

Souciet, J.-L., for the Génolevures Consortium (GDR CNRS 2354) (2011). Ten years of the Génolevures Consortium: A brief history (Les dix ans du consortium Génolevures: un bref historique). *Comptes Rendus Biologies, 334*, 580-584.

Stenson, P. D., Mort, M., Ball, E. V., Chapman, M., Evans, K., Azevedo, L., et al. (2020). The Human Gene Mutation Database (HGMD®): Optimizing its use in a clinical diagnostic or research setting. *Human Genetics, 139*, 1197–1207.

Stevens, H. (2013). *Life out of sequence: A data-driven history of bioinformatics.* The University of Chicago Press.

Stevens, H. (2015). Networks: Representations and tools in postgenomics. In S. S. Richardson & H. Stevens (Eds.), *Postgenomics: Perspectives on biology after the genome* (pp. 103–125). Duke University Press.

Stevens, H., & Richardson, S. S. (2015). Beyond the genome. In S. S. Richardson & H. Stevens (Eds.), *Postgenomics: Perspectives on biology after the genome* (pp. 1–8). Duke University Press.

Szymanski, E., Vermeulen, N., & Wong, M. (2019). Yeast: One cell, one reference sequence, many genomes? *New Genetics and Society, 38*(4), 430–450.

Talenti, A., Powell, J., Hemmink, J. D., Cook, E. A. J., Wragg, D., Jayaraman, S., et al. (2022). A cattle graph genome incorporating global breed diversity. *Nature Communications, 13*, 910.

Tempini, N. (2020). The reuse of digital computer data: Transformation, recombination and generation of data mixes in big data science. In S. Leonelli & N. Tempini (Eds.), *Data journeys in the sciences* (pp. 239–263). Springer Open. Retrieved December 19, 2022, from https://link.springer.com/book/10.1007/978-3-030-37177-7

The 1000 Genomes Project Consortium. (2010). A map of human genome variation from population-scale sequencing. *Nature, 467*, 1061–1073.

The FAANG Consortium, Andersson, L., Archibald, A. L., Bottema, C. D., Brauning, R., Burgess, S. C., et al. (2015). Coordinated international action to accelerate genome-to-phenome with FAANG, the Functional Annotation of Animal Genomes project. *Genome Biology, 16*, 57.

The Génolevures Consortium. (2009). Comparative genomics of protoploid *Saccharomycetaceae. Genome Research, 19*, 1696–1709.

The International HapMap Consortium. (2003). The International HapMap Project. *Nature, 426*, 789–796.

Thieffry, D., & Sarkar, S. (1999). Postgenomics? A conference at the Max Planck Institute for the History of Science in Berlin. *Bioscience, 49*(3), 223–227.

Tortereau, F., Servin, B., Frantz, L., Megens, H.-J., Milan, D., Rohrer, G., et al. (2012). A high density recombination map of the pig reveals a correlation between sex-specific recombination and GC content. *BMC Genomics, 13*, 586.

Tuggle, C. K., Giuffra, E., White, S. N., Clarke, L., Zhou, H., Ross, P. J., et al. (2016). GO-FAANG meeting: A Gathering On Functional Annotation of Animal Genomes. *Animal Genetics, 47*, 528–533.

Wach, A., Brachat, A., Pöhlmann, R., & Philippsen, P. (1994). New heterologous modules for classical or PCR-based gene disruptions in *Saccharomyces cerevisiae. Yeast, 10*(13), 1793–1808.

Warr, A., Affara, N., Aken, B., Beiki, H., Bickhart, D. M., Billis, K., et al. (2020). An improved pig reference genome sequence to enable pig genetics and genomics research. *GigaScience, 9*(6), giaa051.

Wiggans, G. R., Cole, J. B., Hubbard, S. M., & Sonstegard, T. S. (2017). Genomic selection in dairy cattle: The USDA experience. *Annual Review of Animal Biosciences, 5*, 309–327.

Winzeler, E. A., Shoemaker, D. D., Astromoff, A., Liang, H., Anderson, K., Andre, B., et al. (1999). Functional characterization of the *S. cerevisiae* genome by gene deletion and parallel analysis. *Science, 285*, 901–906.

Yang, B., Cui, L., Perez-Enciso, M., Traspov, A., Crooijmans, R. P. M. A., Zinovieva, N., et al. (2017). Genome-wide SNP data unveils the globalization of domesticated pigs. *Genetics Selection Evolution, 49*, 71.

CHAPTER 8

Conclusion

The heavy scholarly and media focus on the determination of the human reference sequence, popularly portrayed as the sequencing of 'the human genome', has had the effect of limiting the public perception of what constitutes genomics and its history. This perception concerns the characteristic practices, products and organisational configurations of genomics, and also locates genomics in a distinct era that closed with the 'completion' of the human reference genome in 2003. Genomics has become synonymous with a narrow set of practices and events associated with the creation of a reference sequence, and chiefly that of one species, the human. The industrial forms of production that this reference sequence required, as well as the possibilities it opened in biomedical research, has established rigid boundaries between what is conceived as pre-genomics, genomics and post-genomics. This periodisation foregrounds discontinuities and complicates any possible connection to be made between pre-genomic history and post-genomics. Throughout the preceding seven chapters, we have challenged this limited canonical view and argued that beyond *Homo sapiens* and the practice of large-scale sequencing, the historical vistas of genomics expand considerably.

In this final chapter, we reflect on the broader historiographical implications of our challenge. One of the consequences of the canonical view of genomics and its narrow historical lens has been that academic and policy appraisals of the nature and role of reference genomes treat them as

© The Author(s) 2023   327
M. García-Sancho, J. Lowe, *A History of Genomics across Species,
Communities and Projects*, Medicine and Biomedical Sciences in
Modern History, https://doi.org/10.1007/978-3-031-06130-1_8

isolated objects preserved in aspic. Instead, we have uncovered the dynamics of genomics as well as the changes arising from—and happening to—reference genomes over the course of: their establishment as scientific objects; the efforts to produce them; activities to refine, improve and enrich them; and their connection to related resources built using them. Our presentation of different communities of genomicists with distinct, historically-specific mechanisms of inclusion and exclusion and the active role that these communities played in configuring the affordances and ontological status of each reference genome, has been crucial for our dynamic approach to genomics and its history.

The reflections in this final chapter help us to elaborate on the main conceptual payoffs of our analysis, namely the portrayal of the reference genome as a dynamic and generative entity that shapes our understanding of the past, present and future prospects of genomics research. We outline this using our key distinction between post-genomics and post-reference genomics and assess what this differentiation can offer us analytically in considering the question of research translation. Finally, we close with a discussion concerning how our multi-species approach and emphasis on communities of genomicists as historical actors affects the historiography of genomics. With this, we attempt to marry specificity with being able to make more global claims about genomics and its history.

8.1   The Never-Ending Frontier: Querying the Limits of Genomics and Characterising Progress Within It

An ideal of completeness and comprehensiveness has guided some of the leading promoters of genomics. What constitutes completeness and comprehensiveness has been, though, a continually receding horizon. Understanding the nature of the reference genome and the multidimensionality of webs of reference following the production of a reference sequence, as detailed in Chap. 7, helps us to appreciate why. Even if it is possible to determine end-to-end sequences without gaps—as it is now for humans (Nurk et al., 2022)—there can be no absolute and final way of apprehending and characterising the variation in and of a particular type, such as a species. The goals shift as the available data and knowledge grows and new research aims are developed. A surplus of potential representations and instantiations of variation becomes available to genomicists and other researchers: the *variational surplus*. Variation is the measured or measurable differences of particular parameters within a defined type of object or process. In genomics, variants are detected by comparison of novel data with reference sequences or other standard resources. This variational surplus provides a plethora of potential routes through—and maps of—the variation, by which researchers can pursue their aims. There can therefore never be convergence on a final standard or ultimate set of linked standards.

One of the most compelling ways of interpreting the shifting frontier of what constitutes completeness or comprehensiveness is to consider that genomics manifests a particular open-ended version of what Hasok Chang has articulated as "epistemic iteration". This concept, and its particular features, was drawn by Chang from his studies of the establishment of standards for the measurement of temperature and development of thermometers across eighteenth and nineteenth-century physics. Chang defines epistemic iteration as "a process in which successive stages of knowledge, each building on the preceding ones, are created in order to enhance the achievement of certain epistemic goals" (Chang, 2004, p. 45). He emphasises that his characterisation of the key features of epistemic iteration, while abstract, cannot necessarily be conceived as general or universal, even across the physical sciences, let alone the biological. We do not intend to risk plunging into the deep waters—around which Chang has posted warning signs—by merely transposing or applying epistemic iteration as developed in the context of thermometry, to the establishment and development of reference resources in genomics. Instead, by assessing the historical development of genomics, we adapt this conceptual framework. In so doing, we intend to shed light on genomics and also examine how to extend epistemic iteration to domains that appear quite different to the precision measurement of physical parameters.

There are a number of features of epistemic iteration that Chang identifies. A "correct answer" may not be knowable. Different stages need not feature the same knowledge production processes, nor be reducible to prior stages (Chang, 2004, pp. 45–46). What guides the process of iteration, according to Chang, is an "imperative of progress" judged against certain epistemic virtues and values (Chang, 2004, p. 44). Furthermore, although there is evident conservatism based on a "principle of respect" for prior standards (Chang, 2004, p. 43), this manifests in a "pluralistic traditionalism" in which "each line of inquiry needs to take place within a tradition, but the researcher is ultimately not confined to the choice of one tradition, and each tradition can give rise to many competing lines of development" (Chang, 2004, p. 232).

In his discussion of thermometry, Chang details debates on the establishment of fixed points around which to base the temperature scale, the choice of substance (e.g. mercury or air) to incorporate into thermometers, the establishment of a theory-based absolute temperature and attempts to operationalise this by connecting it to concrete measurement methods. He demonstrates that some form of grounding on assumptions or imperfect empirical observations is necessary. Crucially, the improvement of standards—as evaluated against epistemic virtues, values and goals—often occurs through self-correction and enrichment, by building on and superseding prior standards.

There are some basic analogies between Chang's discussion of the development of thermometry and the history of genomics and reference genomes. The reference genome is indeed, at any one time, a fixed point, a contingent result of consensus. However, over time it changes; no reference genome, at least yet, has attained the near-permanency of the Celsius and Fahrenheit scales. The choice of thermometry substance is analogous to the selection of the source material to be sequenced in a project to determine a reference genome. In genomics, though, rather than measurements and arguments being conducted by a community around a material, the material itself is a community product: a result of opportunity and availability (pig genomics), the prior history of the genomicists involved in the sequencing effort (the use of the S288C strain of yeast) or an attempt to represent, quasi-metaphysically, the species in question (human genomics).

What, though, restricts the stipulative freedom when producing and presenting a reference genome? What is there to stop it from being arbitrary? In line with the conservatism of the processes of producing temperature standards, reference genomes must be consistent with previously-established antecedents and exhibit improvement according to metrics of validation and evaluation that allow comparison of quality. Robust processes for ironing out sources of error (e.g. through deep sequence coverage, as well as statistical and computational means) are especially important when post-hoc detection of 'errors' may not be possible, and where the status of something as an error may itself be questioned. Epistemic goals are crucial in shaping this iterative process; indeed, we can identify what a particular community is seeking through the metrics it uses to validate new versions of reference resources. For instance, an abstract idea of completeness and universality underlay the production of the human genome, whereas more specific agricultural and immunogenetic motives were behind the determination of the pig genome. Here, it may be observed that for partial or whole-genome sequence assemblies, using the quality of the assembly as some kind of context-free criterion, without reference to specific applications, may inhibit the use of it for other, translational purposes, and therefore complicate the development and usability of reference resources across communities.

It is important to note that epistemic goals in genomics are not merely subordinated to widely-held standards of quality or completeness. Throughout the history of genomics, different epistemic goals have motivated genomic research and data generation beyond just the creation of gold-standard reference genomes. And even for the creation of reference genomes, we have shown that maximising their completeness and quality according to certain metrics has not always been the sole or overriding concern of those promoting and conducting genomic projects. We have observed something distinctive about post-reference genomics, though, in that epistemic goals tend to shift towards the development and exploitation of reference resources built on and linked to the available reference genomes. These post-reference genomic resources characterise different forms of variation within the overall potential array of variation that can be apprehended and captured for a given species or across different species. Such aims to capture variation in this way existed before the advent of reference genomes, but once reference genomes are created, they present possibilities and opportunities to do this kind of work, ones that may not have been practical or conceivable before.

In the open-ended epistemic iteration characterising genomics, we therefore see the exploration of a particular *variational space*, by way of the creation of new genomic resources (data, materials, tools and infrastructures) that are based, in some respect, on the reference genome. This epistemic iteration constitutes a radiation from the fixed point of the reference genome, rather than a convergence to a fixed point as in Chang's thermometry. This explorative radiation is often conducted by a wider array of actors than were involved in the creation of reference genomes. It is shaped, though, by the initial conditions that are set by the processes by which the reference genome is produced, and is subsequently developed. In other words, the room for manoeuvre in post-reference genomics is shaped by the historicity of genomics: by the affordances and representativeness that different communities of genomicists envisioned and enacted.[1] This is why the inclusion or exclusion of particular communities in the production of a reference genome is so important.

8.2   A DYNAMIC VIEW OF REFERENCE GENOMES AND THEIR ROLE

Throughout the book, we have shown how the processes and differential involvement of particular communities in the generation of reference genomes affect their nature and exploitation. Typically, criticisms of reference genomes within the life sciences and philosophy of science focus on matters related to the extent to which they represent or stand-in for their target species in meaningful ways (e.g., Ballouz et al., 2019; Barnes & Dupré, 2008; Rosenfeld et al., 2012; Tauber & Sarkar, 1992). We have argued, however, that the question of what the reference genome represents, and the identification of alleged deficiencies in the processes of abstraction, misleads us by directing attention only to the reference genome as an object or end in itself. When its role as an active foundation for the seeding of webs of reference is considered, the ways in which reference genomes are produced becomes pertinent to appreciating their infrastructural role, and not merely their representative one. These ways, we have shown, include the thicker array of practices and configurations involved in the production of a reference genome, and not just the determination of a string of nucleotides and the absences and presences in these.

The webs of linked reference resources built on and around a reference genome, in turn, feed into the ongoing development and context of use of the reference genome to further seed explorations of variational space. The reference genome is therefore a dynamic entity, shaped and reformed by the very processes of production that generated it, and by the webs of linked resources it has helped to create. Later in this section, we reflect on the ontological implications of these dynamics for reference genomes.

---

[1] This is congruent with Soraya de Chadarevian's observation of the dynamic relationship between the setting of standards and exploration of human variation in the histories of cytogenetics and genomics. She has noted that "[t]he search for variation, then, seems to be built deeply into the study of heredity. Yet how variation is interpreted—as variation on a theme or deviation from a norm, in a hierarchical or inclusive manner—and how it is acted upon, is a matter of interpretation and historical contingency" (de Chadarevian, 2020a; see also 2020b).

First, we pursue some suggestions about the type of object that reference genomes constitute, or have been thought to constitute.

Leading figures in human genomics were adamant both before and after the publication of the human reference sequence that the resulting object would not constitute a "normal genome" in any respect. In 1989, for example, Victor McKusick, the co-founder of the journal *Genomics* (Chap. 3), emphasised that it was "well recognized by geneticists, that there is no single normal, ideal, or perfect genome". Interestingly, this was stated in justification of the idea that "the DNA can come from different persons chosen for study of particular parts of the genome. Such an approach is consistent with that of most biologic research, which depends on a few, and even on single individuals, to represent the whole". After all, if the reference genome was not presumed to be normative in some way or another, then why should it matter what it represented? McKusick did not, however, suggest a completely arbitrary basis for the reference genome. Writing more than a decade before its accomplishment, he argued that the DNA would need to come from actual human beings, and its assembly would be guided by prior standards such as maps, with the reference sequence constituting "the ultimate map", and validated according to other procedures to assess its quality and coverage (McKusick, 1989, p. 913).

Lisa Gannett (2003, pp. 179 and 182) identifies a range of positions on the idea of a "normal genome", from David Hull and Elliott Sober's "outright rejections of the notion of a normal genome and any treatment of genetic variation as deviation" to "the idea of a single genetic norm for the species from which all variation is deviation". Advocates of the latter position appeal to evolution or adaptation to the environment as the basis for such a norm. Within this variety of views, McKusick's contribution intended to present the reference genome as a kind of standard that abstracted from the genomic variation of the species but was not supposed to represent either the most common or the 'best' genome. To the extent that it is accepted as 'normal' by a community of practitioners—from genomicists involved in its production to other life scientists—it is a stipulated standard.

This conception of the human reference genome was to change. Writing a brief reflection on the tenth anniversary of the February 2001 draft sequence publication, with the benefit of the resulting knowledge gained about the human genome, Maynard Olson offered the view that "[a] model for human genetic individuality is emerging in which there actually is a 'wild-type' human genome—one in which most genes exist in an evolutionarily optimized form". He argued against this normative view on the grounds that "[t]here just are no 'wild-type' humans: we each fall short of this Platonic ideal in our own distinctive ways" (Olson, 2011, p. 872).

In his interpretation of the human reference genome, Olson—who played a crucial role in the mapping of the yeast *Saccharomyces cerevisiae* and devised tools to map larger genomes (Chap. 2)—referred to a particular concept, the wild type. This concept has been and is still used in medical genetics concerning a gene or functionally-relevant sequence that is not associated with a manifestation of disease or disorder. It therefore presumes that there are functional and non-pathogenic forms of genes. The wild type here is defined negatively, as not possessing certain forms of variation that would render the sequence non-functional or pathogenic. Since this is a function-first definition, what constitutes a wild type is not evident from the sequence itself: whatever deviates from the functional criteria used to assess the presence or absence of the wild type form of a gene or sequence is deemed to be a "mutation".

There are other meanings of wild type that have been used in the life sciences, dating back to the early nineteenth century. At the outset of the twentieth century, a variety of interpretations of wild type flowered. It became applied by William Bateson, for instance, to organisms that exhibited a "normal body" as a result of experiencing "normal development", as judged against the evolutionary history of the species. The wild type was therefore healthy and well-functioning, and a baseline against which variants could be assessed as beneficial or harmful. This was very much in line with the normative medical genetic version of it (Holmes, 2017; on normal development, see Lowe, 2016).

With the advent of what became known as 'classical genetics' in the laboratory of Thomas Hunt Morgan from the second decade of the twentieth century onwards, the wild type came to designate not just strains, individuals and genomes that represented the 'normal' as seen in nature, but also particular genes without evident mutant characters. So, for *Drosophila*, the wild type could refer to organisms with two symmetrical wings, the standard red eye colour, or other characteristics.

In this approach—that became prevalent in genetic experimentation—an organism or strain may be deemed to be a wild type, provided the characteristics pertinent to what is being investigated were themselves wild type. In this way, these characteristics serve as a baseline against which deviations from the wild type—variation—can be apprehended and then interpreted. This shift enabled the articulation of genes as difference makers whose effect could be discerned not by the presence or absence of particular characters or traits but through comparing observable variation to a standard. By this point, wild types could not be considered to be wild, though they were supposed to stand-in for nature in the laboratory, and thus function as a correlate within the laboratory of the nature outside.

This assumption that laboratory wild type strains were supposed to constitute a particular reflection of standard traits and provide a means to apprehend and measure variation outside the laboratory came under devastating attack by neo-Darwinian 'Modern Synthesis' theorists in the mid-twentieth century. This critique highlighted the limitations of some of the programmes of research conducted using wild types, and undermined their conceptual basis. The wild type endured in the life sciences, however, as embodied in "standard lab strains of experimental organisms […] [that] operate as controls to measure variation in model organism systems" (Holmes, 2017, p. 15). Indeed, the criticisms of the use of laboratory wild type strains also echo many of those levelled at the use of a small number of highly-standardised model organisms across biological research (Table 8.1).

Neither wild types nor model organisms account for the extent of natural variation. The very qualities that make a model organism useful for laboratory-based research also make them quite unlike even their wild cousins of the same species. Furthermore, the extent to which they possess the representational scope to capture biological processes and phenomena that occur in different species has been questioned (e.g. Bolker, 2017).

Table 8.1 The main model organisms used in biological research. Table elaborated by James Lowe

Escherichia coli (Bacteria)
Saccharomyces cerevisiae (Brewer's and baker's yeast)
Arabidopsis thaliana (Thale cress)
Caenorhabditis elegans (Nematode worm)
Drosophila melanogaster (Fruit fly)
Danio rerio (Zebrafish)
Mus musculus (Mouse)
Gallus gallus (Chicken)
Xenopus laevis (Frog)

Much of the recent concern over the translational gap between laboratory research and the clinic—e.g., relating to the development of new pharmaceutical products—has focused on the panoply of differences between laboratory workhorses such as the mouse *Mus musculus* and the humans who are supposed to benefit from such research (e.g. Garner, 2014).

Philosophical responses to such criticisms of the nature and use of model organisms have focused on their role as intensive hubs of resources concerning all aspects of the biology of the model organism species, which therefore function as a well-characterised basis for the generation of comparisons and the apprehension of variation across species (e.g. Ankeny & Leonelli, 2011; Leonelli & Ankeny, 2013; Ankeny, 2007, pp. 49–51; Leonelli, 2016, pp. 18–24, 145–148). Drawing on Rachel Ankeny's analysis of work on *Caenorhabditis elegans*, Lisa Gannett has observed that, like model organisms, reference genomes constitute a kind of descriptive model, in that they instantiate an abstraction that is used as a foundation for explanatory questions (Ankeny, 2000; Gannett, 2019). In this sense, they should be assessed in terms of how they ground further research—as infrastructures—rather than on the extent to which they alone sufficiently represent the genomic variation of a species or sub-species.

One criticism within genomics itself concerning the utility and representativeness of reference genomes is that they act as type specimens: reference samples that taxonomists use "to define the general class by example, often for a species". Reference genomes and type specimens share an "idiosyncratic" nature, in the sense that "[t]he data and assembly that made up the reference sequence reflect a highly specific process operating on highly specific samples". This means that, even if a reference genome is a useful and "good" type specimen of its target species—which some critics admit for the human reference sequence—it cannot adequately reflect the variational landscape of that species in nature (Ballouz et al., 2019, quotes from pp. 1–3).

How apt a designation is this for reference genomes, and what would interpreting them as type specimens mean for understanding the nature and function of reference sequences and other genomic reference resources?

Type specimens are defined and used in the fields of taxonomy and systematics as standards around which practices of classification, and apprehension and cataloguing of variation, can operate. In taxonomy and systematics, type specimens are material instantiations of an organism, on which the classification and name of a given type—such as a species—is anchored. This is vital for the enterprise of cataloguing and identification, and detailed specifications of different versions of type specimens have been developed by different communities. These kinds of designations, as well as the practices and rules governing them, have changed over time and also vary according to the kind of organism concerned, for example between animals and plants.

The use of type specimens in taxonomy has not been uncontroversial. It is intriguing how the questioning of their role has reflected some of the criticisms of reference genomes. For example, George Gaylord Simpson's critique of type specimens in the 1930s echoed the concern with how well they captured relevant variation to represent the type (Witteveen, 2018). Type specimens and reference genomes are indeed comparable, as both are fixed points of reference, at any one point in time. The representativeness of them in terms of biological variation is circumscribed, but they both enable variation to be apprehended, articulated, measured and recorded.

However, we emphasise that they are fixed points of reference *only at one point in time*. As the philosopher of biology Joeri Witteveen has noted (Witteveen, 2016), type specimens are not absolutely fixed as primary referents to particular species. They are, though, far less changeable than reference genomes have proven to be. We may speculate why this is the case. Possible reasons include the fact that reference genomes rely on already-designated species, and that they have a wider range and ever-changing set of epistemic goals that motivate continual iteration towards them. Furthermore, they have always been in digital form, allowing different versions to be designated and referred to far more easily. Reference genomes may offer a fixed point of reference, but serially rather than perpetually. By engaging with their historicity and the motivations of the communities of genomicists that created them—as we have done throughout the book—we can capture changes in their nature as references and as standards.

To introduce our assessment of that, we return to the determination of the human reference genome by the public and charitably-funded International Human Genome Sequencing Consortium (IHGSC) and the production of a whole-genome human sequence by the company Celera Genomics. They had different ways of generating their genomic data. Crucially, they also had different aims for the eventual product, which conditioned the strategies they pursued, but also their conception of the objects they were creating. The IHGSC aimed to release, into the commons, a record of the 'Book of Life', *the* genetic code of the human species. This universalist view of the human genome was buttressed with data that indicated that DNA sequence similarity between humans was 99.9% and therefore far closer than in other species. Therefore, it did not seem to matter that the selection of donors was largely arbitrary, conducted through a newspaper advertisement (Chap. 4). IHGSC members argued that it was unnecessary and meaningless to use DNA from people of different ethnicities and sexes, as the differences in the DNA of humans across the globe were minimal.

Celera's business model, on the other hand, was based on the identification and analysis of sequence variation. They wanted to sell that data to companies who would find it useful, for example in the development of diagnostic tools or therapeutic drugs. Later, they would try to exploit that data themselves for these purposes (García-Sancho, Leng, et al., 2022). Their emphasis was therefore on difference, rather than commonality or universality. Both efforts produced a comprehensive representation of the human genome, albeit one was a publicly released 'official' reference sequence, and the other was only available in full behind a paywall. Historians have already observed that these can indeed be regarded as two separate objects, because of the differential processes and configurations that went into producing them: Celera's whole-genome shotgun approach and the IHGSC's choice to construct physical maps and use these to help put the sequence together (Chap. 4; Bostanci, 2006).

Beyond that, we note that they constituted different forms of representation. For the IHGSC, their reference genome was representative of the species in the sense of faithfully depicting the genomes of humans across the world, except for a few minor and insignificant differences. For Celera, their genome was able to *stand-in* for the human species without substantially representing or reflecting its totality or diversity. At an event in August 2001, Gene Myers, a leading bioinformatician who worked at Celera from 1998 to 2002, pointed out that while there could be "no one single human genome", his company had indeed "determined a single reference sequence"—albeit an unofficial one (quotes in Bostanci, 2006).

Many of the criticisms of reference genomes we have observed involve some conflation of the ways in which a reference genome can represent or 'stand-in' for a species. The idea that the reference genome must be *representative* of the species rather than merely being *a representation* writes cheques that reference genomes often cannot cash. This problem arose when the basis for the IHGSC conception became untenable, as the extent of functionally-significant genetic variation across humans became apparent. This variation became possible to apprehend and record because of the advent of the reference genome, but undermined the idea that it represented the human species in a universal or metaphysical way. It did not undermine the conception proposed by Myers, in which the reference genome was something more like a type specimen. The appreciation of the extent of genomic variation—and the dissatisfaction with the reference genome occasioned by this growing knowledge and the increasing mismatch between this and the IHGSC's view—has helped effect a change in the nature of the human reference genome.

As a result, the ontological status of the human reference genome and those for other species such as *S. cerevisiae* has evolved. When the newer reference genome of *S. cerevisiae* was announced in 2014 (Chap. 7), subsequent revisions were supposed to incorporate more variation and better represent the species. In the case of the newer pig reference genome released in 2017 and published in 2020, the authors placed great emphasis on the benefits of the new assembly for finding and exploiting different forms of genomic variation. Developing the reference genome to incorporate more variation was less important to them, though, than it was to human and yeast genomicists. While there has been a general change in the ontological status and modality of reference genomes, with more focus on variability, these may not always be as fully realised for some species or carry the same weight relative to other avenues by which post-reference genomic resources can be developed.

In the case of the human reference genome, having originally been something more like a type specimen (an arbitrary extraction from the diversity of variation found in nature), it has been shifting to become something more like an idealised normal genome, reflecting common non-pathological variants found across populations. This transition constitutes one from the reference genome being an abstraction to becoming more of an idealisation.[2] What does this mean? As an abstraction, it has been based on the omission of genomic variation through a selective process that depended on multiple choices made throughout all of the stages resulting in the production of a reference sequence. This selectivity has

---

[2] There is a rich philosophical literature on abstraction and idealisation. Here, we have deployed some of the senses captured in Cartwright (1989), Godfrey-Smith (2009), Jones (2005), Levy (2018) and Love (2010).

not, though, necessarily been to create a product that is representative or normal in the sense of being only comprised of the most common or non-pathological variants. At the stage of the inception of reference genomes, these were not known, and therefore this was not possible to do. Only with the subsequent apprehension of variation and its functional significance can reference genomes be shaped to take account of—or even incorporate—the common and non-pathological.

Arising from this appreciation of genomic variation, in conjunction with existing ambitions to represent humankind, revisions of the reference genome increasingly tend towards idealisation. It now becomes possible to state that a reference genome is, to some degrees and in some respects, a misrepresentation, as there are now concrete epistemic goals directed towards a specific representational target. This signifies a shift from the dominant epistemic goals of the abstraction phase, which emphasised the contiguity, coverage and quality of assembly—and level of annotation—of the reference genomes. In the idealisation phase, due to the added normative dimensions and the new role that the reference genome is being asked to fulfil, a gap begins to be perceived between the genome itself and the representativeness that it is supposed to embody.

The implication of our transformed picture of the nature and role of the reference genome is not that equality and social justice concerns about the representational scope of genomic resources are invalid. Instead, we would direct such critiques from the reference genome towards the wider webs of reference and observe that such concerns become more salient as one enters deeper into the idealisation phase. Considering the reference genome as a dynamic object that is created and transformed through recursive and iterative processes involving—and sometimes excluding—particular communities of practitioners, is crucial if we are to avoid conflating different ways in which a reference genome can 'stand-in' as a representation of a species. As we observe in the next section, the alignments between the aims of genomic research and the concrete processes of idealisation are crucial to effecting translation.

8.3   Genomics and Translation

The advent of a reference genome is a significant event for any community concerned with the genetics of a particular species. In providing researchers with a comprehensive consensus sequence of the target species, the reference genome constitutes a resource to which existing and newly-determined genomic data can be related and aligned. It informs the assembly and annotation of new sequences—such as of specific pig breeds or human populations, or different microorganismal strains—and also provides a basis for intra and inter-species comparison.

Particular configurations of pre-reference genomics, and the decisions made in them and in the determination of the reference sequence, affect how readily certain forms of variation can be explored in post-reference genomics. In yeast, there was a pragmatic decision to focus on one particular strain, and this shaped the trajectory of research after the release of the reference genome: participants in the EUROFAN project to functionally annotate the reference sequence were largely drawn from the prior Yeast Genome Sequencing Project. For the human, the gap between the producers of the reference sequence and the medical genetics community led to problems in squaring variation—at least the variation on which medical geneticists had worked before and during the production of the reference genome—with the reference sequence. In pig, although the 'thin' compilation of the reference sequence was delegated to the Sanger Institute, significant community continuity happened at the level of the thick sequencing: mapping, assembly and annotation practices. This allowed pig genomicists to appreciate what variation was incorporated in the reference genome, what was missing and what further work needed to be done to characterise different kinds of variation across the species.

The kind of epistemic iteration concerning the development of new genomic resources implies that the characterisation of variation beyond the reference sequence is the central epistemic task of post-reference genomics. Post-reference genomics research involves, in one way or another, the identification, cataloguing, control and use of variation. What variation is being compared, over what time-frame, how it is to be measured, and for what purpose, is up to the researchers involved, who work within various material, theoretical and technical constraints. There are, conceivably, unlimited ways in which comparisons between two (or more) parts, individuals or groups can reveal variation. The particular means by which variation is generated, apprehended, identified, measured, recorded and integrated with other types of variation, conditions (but does not fully determine) the further use that may be made of it.

Sufficiently rich webs of well-connected resources represent different kinds of variation. Key here is the creation of the resources, the processes by which they instantiate particular kinds and ranges of variation, and the data and material linkages and connections established between them. Capturing extra dimensionalities of data—for instance, through annotation, cataloguing of sequence variants and generating non-genomic biological data— ensures this. So too does apprehending diversity by using the reference sequence, in whole or in part, to characterise specific breeds, strains, populations or even individuals. These practices seed comparisons that enable further functional analysis and the apprehension and detection of variation.

Following on from the points made in Chap. 7 about the development and intersection between the functional and systematic aspects of post-reference genomics, we suggest that translation involves the establishment of means to integrate, link and compare data of these different kinds: those that are associated with phenotypic effects as well as those that pertain to intra- or inter-specific patterns in the sequences. This need not involve a collapsing of distinctions between these modes, but require the alignment and commensuration between resources representing—and derived from—different sources and kinds of data.

In foregrounding alignment in this way, we therefore present a concept of translation that echoes previous social scientific scholarship (Lowe et al., 2020; Sunder Rajan & Leonelli, 2013). This is an interpretation that has more in common with Michel Callon's sociology of translation (Callon, 1986) than with the common use of translation as a policy category concerning the strategy and governance of scientific research. For Callon, achieving translation involves the shaping of a network of actors in a way that structures relations and actions around particular problems and solutions that are posed by one (or more) of the actors. In this way, translation involves "creating convergences and homologies by relating things that were previously different" (Callon, 1980, p. 211; as cited in Wæraas & Nielsen, 2016), be they biological objects themselves or the scientific groups and communities oriented towards them, and their ongoing practices and organisations. These convergences and homologies of biological entities and communities require alignments and commensurations of norms, organisational models and genomic resources.

To adapt Callon's analytical framework into the domain of genomics, we can say that the process of translation consists in defining the epistemic goals (the problems) and determining the means by which these epistemic goals are worked towards (the solutions). In genomics, these processes operate at multiple levels and present different casts and configurations of actors, although there are undoubtedly multiple different overlaps and relations between them.

A main level of operation is the creation of a reference genome, both as a generic object and in specific instances. Generically, the process involves the creation of the category of reference genome (with different designated levels of quality and completeness) by large-scale data infrastructures, such as the RefSeq database of the US National Center for Biotechnology Information, and control over revisions to reference genomes by bodies such as the Genome Reference Consortium. This work creates objects that are commensurate with other forms of genomic (and other omic) data and with other reference sequences (Chap. 1). Yet, before their entrance and commensuration within these centralised infrastructures, genomic data has been produced by different processes involving distinct modes of interaction between the target species and specific communities of genomicists.

Here, we can make sense of some of the different trajectories we have observed for the three species we examined. For the human genome, Callon-style translation was achieved by a small group of actors, primarily at the US Department of Energy, the National Institutes of Health, the Wellcome Trust and some large-scale sequencing centres. They successfully designated the quick generation of a common, accessible reference sequence as the main problem—the epistemic goal of whole-genome sequencing—and so sidelined medical geneticists. An alternative attempt at translation around the sequence produced by Celera provided a more amenable alignment with the interests, practices and norms of the medical geneticists. However, while this enabled some medical geneticists to advance their research, and to produce some genomic resources of use to the wider community (García-Sancho, Leng, et al., 2022), the way that Celera's data was released—in terms of both access and format—restricted the availability and linkage opportunities around their sequence. Only recently, through initiatives such as ClinGen/ClinVar and the 100,000 Genomes Project, has there been a concerted effort to align large-scale genomics data infrastructures with the interests of, and data produced by, medical geneticists.[3]

The situation, as we have seen, was quite different in the cases of pig and yeast. There, existing communities working on those organisms achieved translation mostly on their own terms, and this has enabled them

---

[3] On the 100,000 Genomes Project, see Jarmo de Vries's blogpost concerning his ongoing PhD research at: https://genomicsincontext.wordpress.com/2021/06/11/the-100000-genomes-project-shaping-genomic-medicine-in-the-nhs/ (last accessed 20th December 2022).

to pursue post-reference genome research and alignment with their working world concerns—domains of application such as medicine, agriculture and biotechnology (Agar, 2020)—more-or-less seamlessly, and successfully. For them, the defined epistemic goals shared some commonalities with human genomics, but presented distinct problems that required different solutions to be provided by the reference genome and subsequent resources built using it and relating to it.

In the case of yeast, there needed to be an immediate connection to the experimental practices and aims of the researchers involved, and this meant generating a sufficiently well-annotated reference sequence to enable further exploration through extensive deletions—knockout experiments—and laboratory assays to functionally analyse the genome and its products. This went well beyond the functional annotation that the genome centres initially pursued on the human reference sequence.

For the pig community, the genome simply had to be good enough to enable the selective annotation and further biological explorations of certain regions known to be associated with traits of interest for breeding, developing the pig as an animal model, and furthering the utility of the pig in transplantation biology and xenotransplantation. Additionally, it had to provide the basis for the identification of multitudes of genetic markers such as Single Nucleotide Polymorphisms (SNPs) that would constitute the foundation for new methods of breeding based on the use of these masses of markers. These variants and tools, such as the SNP chips, also furthered the characterisation of the genetic diversity and patterns of distribution of pigs, contributing towards the synergistic relationship between functional and systematic modes that had been a part of pig genomics since the mid-1990s (Chap. 7).

This brings us to another level at which the processes of genomic research operates: concerning the relation of a reference genome to wider webs of reference, and these to forms of biological variation that are pertinent to various working worlds. Alignments and commensurations of the reference genome to various forms of variation enable data, information and interpretations of all kinds to travel through networks of inference and meaning. Working with and beyond reference genomes engenders a greater appreciation of the extent and biological significance of different forms of genomic variation. It also leads to the collection and analysis of data concerning other forms of variation: transcriptomic, metabolomic, all the way to phenotypic, population and community-level. Data concerning these kinds of variation can be linked and related to each other, and to the web of genomic data. We have seen that this post-reference genomics has informed revisions to reference sequences, and even a shift in the nature of the object of the reference genome. In turn, however, the content of the reference genome conditions what and how new forms of variation can be apprehended and made sense of.

The processes of abstraction of variation involved in the creation of the reference genome, therefore, shape the subsequent idealisation of it and its connections to other reference resources and biological data and materials. The interests of the genomicists that were involved in the production of reference genomes affect their capacity for seeding and influencing the development of subsequent webs of reference. It is within the affordances of the data and materials that result from the historical development of reference genomics that new interconnected nodes can be placed in the abstract variational space that the web of reference 'explores'. This placement and evolving topology of the web depends on which forms of variation (which new abstractions or idealisations) the communities involved want to generate, to aid the purposes of their research goals and tackling of working world problems. It is easy to see, based on this, that continuity between those actors that successfully seeded and shaped the early development of the web and those actors connected to working world concerns (e.g. in agriculture, biotechnology or medicine) increases the chances of effecting agricultural, biotechnological or medical translation. In other words, Callon-style translation in the production of a reference genome is an important factor in easing or hindering the translation of genomic data and other resources towards addressing practical research problems.

We cannot, of course, consider such webs only in isolation: though they may be furnished with rich internal connections, they undoubtedly also connect to other webs. These connections may be between webs pertaining to different species, but not necessarily, as there may be distinct webs closely associated with particular working worlds. Here, some of the key alignments consist in forming comparative relationships and interoperability between the resources in each of the webs. Again, this is historically conditioned. The extensive development of a comparative inferential architecture between pig and human genomics (Lowe, 2022) aids alignments between webs representing those species. The types and densities of connections and the topology of the ecology of webs depend on socio-historical factors and the nature of the organisms being worked with. On the basis of pan-species projects such as Génolevures, for example, we may expect stronger connections and perhaps fuzzier distinctions between the webs of reference of different species of yeast.

Finally, there is the more general level of the overall infrastructure and norms of genomics. At this level, the actors who successfully achieved and built on their Callon-style translations at other levels may have more subordinate roles or at least less dominant ones. This level, consisting of the data infrastructures and their associated rules and norms, institutions, funding and publication policies, and even a certain vision of what genomics is and should be, has been strongly bent in the direction of the problems and solutions presented by those core actors that directed the production of the human reference sequence. Because of this, some species such as yeast with the resources and disposition of a model organism community, as well as a history of genomics that precedes the completion of the human reference genome, may exhibit more independence than pig genomics, which conducted its sequencing afterwards and always had strong connections with the mainstream of human genomics. The pig genomics community, though, has been able to shape genomics in a way congenial to the aims and interests of the genomicists comprising it. Their existing working world ties to the breeding industry have provided a means for the recapitulation of the pre-genomic norms of animal genetics into farm animal genomics.

Sociological translations are involved at all these levels, each of which have been shaped by the ways that different communities of genomicists have been formed and their attempts—and differential success—at effecting the translation of their interests. All these levels and factors have conditioned the development of reference genomes and subsequent webs of reference. These genomes and webs of reference, in turn, affect how the tools for the further characterisation of data concerning variation can align with working world problems, be it medical genetics, livestock breeding or the investigation of a model organism. As well as furnishing the socio-historical conditions affecting the chances of successful scientific translation, these processes also shape what medical and agricultural applications are considered doable or desirable.

Further characterising webs of reference and the nature of post-reference genomics is a vital task. It will require working across methods and disciplines, combining more conventional historical and philosophical inquiry alongside qualitative and quantitative methods in the social sciences. It also will require an engagement with, and sensitivity to, the concrete paths of research developed across different domains of species and working world orientations. We close this concluding chapter with reflections on some methodological aspects that future research should take account of, concerning the periodisation and demarcation of genomic research.

8.4   Periodisation, Multispecies Approaches and Communities as Historical Actors of Genomics

One of the main arguments of the book has been to distinguish between a historical periodisation that strictly identifies an age of genomics (roughly 1990 to 2003, with post-genomics succeeding it) and our narrative in which genomics is an ongoing enterprise, albeit featuring distinctive shifts in the organisation and nature of the endeavour following the production of a reference genome. Our interpretation takes fuller account of the differential historical trajectories of genomics concerning different species and the communities that worked on them. It also stresses the fact that the practices and outputs of genomics continue into the so-called 'post-genomic' era. Furthermore, in certain communities and genomic enterprises concerning particular species, constitutive features that scholars have attributed to post-genomics (e.g., in Richardson & Stevens, 2015) were also present in genomic research.

Genomics is not a discrete—nor complete—phase of scientific endeavour. It is continually transformed and enters into new combinations and relations with other data being generated and handled in particular ways. Our notion of post-reference genomics captures this, but also encapsulates the situatedness and historicity of particular strands of post-reference genomics that deal with specific objects, such as species or groups of related species. While post-reference genomics represents a category, it can be manifested in distinct ways by different communities, in different time periods and with differing consequences. We can therefore observe diverse historical trajectories, other than those of the canonical periodisation of genomics centred on the 'completion' of the human reference genome and the alleged start of a new post-genomic era (Fig. 8.1).

As we have shown in Chaps. 3 and 4, even for *H. sapiens* there were a plethora of initiatives that, while directed to the human genome as an object, did not pursue the production of a full reference sequence. Because of this, these initiatives did not adopt the genome centre model or industrial forms of organisation aimed at the rapid production of a whole-genome sequence. Rather than deploying large-scale approaches, these initiatives sought to map and sequence targeted areas of the genome. The distance between the communities of medical and human geneticists that undertook these initiatives and the producers of the reference genome created a perceived 'translational gap' around the exploitation of the clinical and scientific potential of the full sequence. In other words, the distinct

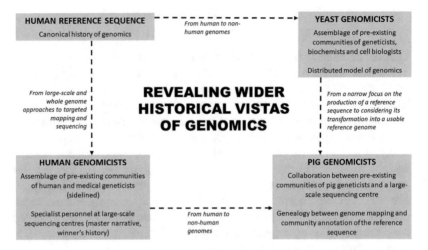

Fig. 8.1   A diagrammatic representation of how an emphasis on the interactions between different communities and their target genomes expands the historical vistas of genomics. Dotted lines represent our historiographical de-centring from the production of the human reference sequence. Below each community of genomicists, we outline how their trajectory diverges from the canonical history of genomics. Elaborated by both authors. For a larger version of this figure that can be zoomed in and out, see https://www.pure.ed.ac.uk/ws/portalfiles/portal/290406893/Fig_8_1_increased_final.pdf

historicities and motivations of two different communities of genomicists—human and medical geneticists, on the one hand, and more specialised operatives at genome centres, on the other—created a disjunction between the reference sequence produced by just one of them, but that was intended for use by the other.

If we shift from the human to non-human species, we observe that while yeast and pig genomics sought the production of a full reference sequence, their historical trajectories differ from the canonical one. For yeast, a long-established, tight-knit community working on a specific strain of S. cerevisiae decisively contributed to the production of the reference sequence in what we called the distributed model of genomics, as opposed to the concentrated determination of the human reference genome at specialised sequencing centres (Chap. 2). Pig genomics squares with the canonical trajectory if we consider the 'thin' production of the reference sequence; after all, this endeavour was modelled on the plans and methods of the international consortium that produced the equivalent sequence for *H. sapiens*, and it was largely undertaken by the Sanger Institute. Yet if we consider the 'thicker' practices that were involved in making this sequence a robust reference resource, other gene-alogies become apparent and challenge the rigid periodisation of genom-ics and post-genomics. For instance, the agriculturally-inclined geneticists and immunogeneticists involved in the prior mapping of the pig genome were crucial in its community annotation, which required collaboration between the Sanger Institute and those long-established pig genomicists (Chaps. 5 and 6).

Although our perspective de-centres the human reference sequence as the paradigmatic—even definitional—instantiation of genomics, it does not necessarily remove it from an important role in the shaping of the his-tory of genomics more broadly. Instead, it calls attention to examining the concrete ways in which this reference sequence, and more specifically the idea of one Human Genome Project that produced it, generated a gravi-tational attraction around the version of genomics it embodied. As we have shown throughout the book, this centripetal force was associated with broader socio-political processes and, crucially, established retrospec-tively in the accounts of James Watson and other prominent participants. The master narrative of genomics, centred on the idea of a single and suc-cessful Human Genome Project, was—and is still—influential because of its alignment with other influential historical forces, not because it repre-sents an intrinsically superior or dominant way of conducting science.

There is a tension implicit in our de-centring and alternative periodisa-tion of genomics. Through identifying the advent a reference genome as an inflection point, rather than a transition to a wholly new post-genomic endeavour, we appear to suggest that the structure of the history of genomics differs according to the species. After all, while yeast entered our proposed post-reference genomic period in 1996, the human did not do so until 2003, and the pig until 2011.

This historiographical transition from a human-centred periodisation towards one based on species-specific designations of pre-reference, refer-ence and post-reference genomic phases constitutes an advance in appre-ciating the heterogeneities and continuities we have observed in this book. It, however, still constitutes an incomplete and patchy picture. This is because, in spite of the distinct periodisations for each species, the overall development of genomics—its infrastructures, norms, data, materials, methods and techniques—possesses its own rhythm and historicity. These may have developed out of one or a few distinct initiatives—such as

Ensembl and the Human and Vertebrate Analysis and Annotation (HAVANA) group being born out of the human reference genome sequencing programme—but once created have had a life, development and impact beyond them. It matters for understanding some of the differences between the histories of pig and yeast genomics that an existing sequencing, assembly and annotation infrastructure was in place at the Sanger Institute for pig genomics but not for yeast, for example. And in turn, it is consequential that the particular way in which pig genomics developed affected the way that HAVANA, in particular, changed in the post-human reference genome era.

Our approach to the history of genomics has enabled us to identify this relationship between more global and local repertoires, processes and configurations. As well as de-centring from the illusion that one model—the Human Genome Project—is generalisable, it has helped us to unpick the commensuration work of administrative agencies and large-scale infrastructures, such as the RefSeq database. It has also enabled us to reveal the historical trajectories that give the products of genomics research different affordances and limitations. The usefulness of our approach is not restricted to being merely comparative; it also enables connections to be identified. It remains an open question how best to harmonise—or, at least, operationalise—the always conflicting tension between histories that are strongly species-specific and those that concern the more general development of genomics as an infrastructural and data-centred endeavour. We have, we hope, now opened up the space for such questions to be asked and explored.

References

Agar, J. (2020). What is science for? The Lighthill report on artificial intelligence reinterpreted. *The British Journal for the History of Science, 53*(3), 289–310.

Ankeny, R. A. (2000). Fashioning descriptive models in biology: Of worms and wiring diagrams. *Philosophy of Science, 67*(Proceedings), S260–S272.

Ankeny, R. A. (2007). Wormy logic: Model organisms as case-based reasoning. In A. N. H. Creager, E. Lunbeck, & M. N. Wise (Eds.), *Science without laws: Model systems, cases, exemplary narratives* (pp. 46–58). Duke University Press.

Ankeny, R. A., & Leonelli, S. (2011). What's so special about model organisms? *Studies in History and Philosophy of Science Part A, 42*(2), 313–323.

Ballouz, S., Dobin, A., & Gillis, J. A. (2019). Is it time to change the reference genome? *Genome Biology, 20,* 159.

Barnes, B., & Dupré, J. (2008). *Genomes and what to make of them*. The University of Chicago Press.

Bolker, J. A. (2017). Animal models in translational research: Rosetta Stone or stumbling block? *BioEssays, 39*(12), 1700089.

Bostanci, A. (2006). Two drafts, one genome? Human diversity and human genome research. *Science as Culture, 15*(3), 183–198.

Callon, M. (1980). Struggles and negotiations to define what is problematic and what is not; the socio-logic of translation. In K. D. Knorr, R. Krohn, & R. Whitley (Eds.), *The social process of scientific investigation* (pp. 197–219). D. Reidel.

Callon, M. (1986). Some elements of a sociology of translation: Domestication of the scallops and the fishermen of St Brieuc Bay. In J. Law (Ed.), *Power, action and belief: New sociology of knowledge?* (pp. 196–229). Routledge & Kegan Paul.

Cartwright, N. (1989). *Nature's capacities and their measurement*. Clarendon Press.

Chang, H. (2004). *Inventing temperature: Measurement and scientific progress*. Oxford University Press.

de Chadarevian, S. (2020a). Normalization and the search for variation in the human genome. *Historical Studies in the Natural Sciences, 50*(5), 578–595.

de Chadarevian, S. (2020b). *Heredity under the microscope: Chromosomes and the study of the human genome*. The University of Chicago Press.

Gannett, L. (2003). The normal genome in twentieth-century evolutionary thought. *Studies in History and Philosophy of Biological and Biomedical Sciences, 34*, 143–185.

Gannett, L. (2019). The Human Genome Project. In E. N. Zalta (Ed.), *The Stanford Encyclopedia of Philosophy* (Winter 2019 Edition). Retrieved December 20, 2022, from https://plato.stanford.edu/archives/win2019/entries/human-genome

García-Sancho, M., Leng, R., Viry, G., Wong, M., Vermeulen, N., & Lowe, J. W. E. (2022). The Human Genome Project as a singular episode in the history of genomics. *Historical Studies in the Natural Sciences, 52*(3), 320–360.

Garner, J. P. (2014). The significance of meaning: Why do over 90% of behavioral neuroscience results fail to translate to humans, and what can we do to fix it? *ILAR Journal, 55*(3), 438–456.

Godfrey-Smith, P. (2009). Abstractions, idealizations, and evolutionary biology. In A. Barberousse, M. Morange, & T. Pradeu (Eds.), *Mapping the future of biology: Evolving concepts and theories* (pp. 47–56). Springer.

Holmes, T. (2017). The wild type as concept and in experimental practice: A history of its role in classical genetics and evolutionary theory. *Studies in History and Philosophy of Biological and Biomedical Sciences, 63*, 15–27.

Jones, M. R. (2005). Idealization and abstraction: A framework. In M. R. Jones, N. Cartwright, & N. (Eds.), *Idealization XII: Correcting the model; idealisation and abstraction in the sciences* (pp. 173–218). Rodopi.

Leonelli, S. (2016). *Data-centric biology: A philosophical study*. The University of Chicago Press.

Leonelli, S., & Ankeny, R. A. (2013). What makes a model organism? *Endeavour, 37*(4), 209–212.

Levy, A. (2018). Idealization and abstraction: Refining the distinction. *Synthese, 198*, 5855–5872.

Love, A. C. (2010). Idealization in evolutionary developmental investigation: A tension between phenotypic plasticity and normal stages. *Philosophical Transactions of the Royal Society B, 365*, 679–690.

Lowe, J. W. E. (2016). Normal development and experimental embryology: Edmund Beecher Wilson and *Amphioxus*. *Studies in History and Philosophy of Biological and Biomedical Sciences, 57*, 44–59.

Lowe, J. W. E. (2022). Humanising and dehumanising pigs in genomic and transplantation research. *History and Philosophy of the Life Sciences, 44*, 66.

Lowe, J. W. E., Leonelli, S., & Davies, G. (2020). Training to translate: Understanding and informing translational animal research in pre-clinical pharmacology. *Tecnoscienza, 10*(2), 5–30.

McKusick, V. A. (1989). Mapping and sequencing the human genome. *The New England Journal of Medicine, 320*, 910–915.

Nurk, S., Koren, S., Rhie, A., Rautiainen, M., Bzikadze, A. V., & Mikheenko, A. (2022). The complete sequence of a human genome. *Science, 376*, 44–53.

Olson, M. V. (2011). What does a "normal" human genome look like? *Science, 331*(6019), 872.

Richardson, S. S., & Stevens, H. (Eds.) (2015). *Postgenomics: Perspectives on biology after the genome*. Duke University Press.

Rosenfeld, J. A., Mason, C. E., & Smith, T. M. (2012). Limitations of the human reference genome for personalized genomics. *PLoS ONE, 7*(7), e40294.

Sunder Rajan, K., & Leonelli, S. (2013). Introduction: Biomedical trans-actions, postgenomics, and knowledge/value. *Public Culture, 25*(3), 463–475.

Tauber, A. I., & Sarkar, S. (1992). The Human Genome Project: Has blind reductionism gone too far? *Perspectives in Biology and Medicine, 35*(2), 220–235.

Wæraas, A., & Nielsen, J. A. (2016). Translation theory 'translated': Three perspectives on translation in organizational research. *International Journal of Management Reviews, 18*(3), 236–270.

Witteveen, J. (2016). Suppressing synonymy with a homonym: The emergence of the nomenclatural type concept in nineteenth century natural history. *Journal of the History of Biology, 49*, 135–189.

Witteveen, J. (2018). Typological thinking: Then and now. *Journal of Experimental Zoology Part B: Molecular and Developmental Evolution, 330*, 123–131.

Appendix A: Oral Histories

Name	Role	Location/date
Mark Johnston	Yeast geneticist and member of the *S. cerevisiae* mapping team at Washington University	Skype, September 2020
Rolf Stucka	Molecular geneticist based at Ludwig-Maximilian University of Munich	Adolf Butenandt Institute, Munich, Germany, November 2019
Horst Domdey	Early-career and then senior researcher at Genzentrum in Munich	Genzentrum, Munich, Germany, November 2019
Brigitte Obermaier	Head of the genome analysis team at Genzentrum and of sequencing services at MediGene and MediGenomix in Munich	IZB Building, Munich, Germany, November 2019: Telephone, June 2021
Karl Kleine	Bioinformatician at Martinsried Institute for Protein Sequences	Telephone, October 2019
Thomas Pohl	Director of sequencing company GATC	Telephone, September 2019
Jane Peterson	Administrator in charge of distributing genome mapping grants during the early days of the US Human Genome Project	National Human Genome Research Institute, Bethesda, Maryland, USA, November 2018

(*continued*)

© The Author(s) 2023
M. García-Sancho, J. Lowe, *A History of Genomics across Species, Communities and Projects*, Medicine and Biomedical Sciences in Modern History, https://doi.org/10.1007/978-3-031-06130-1

(continued)

Name	Role	Location/date
Mark Guyer (extended personal communication)	Director of the extramural (grant-funding) programme of the US National Human Genome Research Institute	National Human Genome Research Institute, Bethesda, Maryland, USA, November 2018
Keith Peters	Physician researching and practicing immunology at London's Hammersmith Hospital and then Regius Professor of Physic at the University of Cambridge	London, UK, October 2013 Telephone, December 2013
Ross Sibson	Biological manager, UK Human Genome Mapping Project Resource Centre	Royal Liverpool University Hospital, Liverpool, UK, March 2014
Martin Bobrow	Medical geneticist and member of the Wellcome advisory group	Cambridge, UK, June 2015
Nick Hastie	Member of the Wellcome advisory group	Edinburgh, UK, July 2015
David Bentley	Researcher at Bobrow's group in the Division of Medical and Molecular Genetics of Guy's Hospital (London, UK) and key member of the human genome mapping and sequencing team at the Sanger Institute	Cold Spring Harbor Laboratory, New York, US, July 2015
Jane Rogers	Senior administrator and then Head of Sequencing at the Sanger Institute	Cold Spring Harbor Laboratory, New York, US, July 2015
Tony Vickers	Manager, UK Human Genome Mapping Project	Email, September and December 2013
Claire Rogel-Gaillard	Member of CEA-INRA team that developed genome libraries	Skype, May 2017
Patrick Chardon, Christine Renard and Marcel Vaiman	Members of CEA-INRA team that developed genome libraries	Paris, France, November 2017
Lawrence Schook	Co-director of the Swine Genome Sequencing Consortium	Skype, January and August 2017
Jennifer Harrow	Led HAVANA team at Sanger Institute	Cambridge, UK, October 2017
Peter Li	Head of Chromosome Team at Celera, involved in manual annotation of the human genome	Skype, September 2020
Kerstin Howe	Led team that analysed, validated and improved genome assemblies at Sanger Institute	Wellcome Trust Genome Campus, Hinxton, Cambridgeshire, UK, October 2017

(*continued*)

(continued)

Name	Role	Location/date
Craig Beattie	Quantitative geneticist who worked at the USDA Meat Animal Research Center	Skype, March 2017
James Reecy	Iowa State University bioinformatician working on livestock genetics	Zoom, May 2021
Jane Loveland	Member of HAVANA team at Sanger Institute	Wellcome Trust Genome Campus, Hinxton, Cambridgeshire, UK, October 2017
Christopher Tuggle	Livestock geneticist at Department of Animal Science, Iowa State University	Skype, March 2017
Alan Archibald	Principal Investigator, Roslin Institute	Roslin Institute, Edinburgh, UK, November 2016
Chris Haley	Roslin Institute quantitative geneticist involved in 1990s mapping projects	Institute of Genetics and Molecular Medicine, Edinburgh, UK, February and December 2017
Martien Groenen	FAANG member and livestock geneticist at Animal Breeding and Genomics Centre, Wageningen University, Wageningen, Netherlands	Skype, September 2017
Barbara Harlizius	Member of Groenen's Animal Breeding and Genomics Centre, Wageningen University, Wageningen, Netherlands	Skype, December 2018
Michael Goddard	Quantitative geneticist based at the University of Melbourne	Ashworth Laboratories, King's Buildings, Edinburgh, UK, October 2018
Sydney Brenner	*Molecular biology pioneer who proposed the creation of the UK Human Genome Mapping Project*	*Ely (Cambridgeshire), July 2013*
Michael Kemp	*Administrator at the Medical Research Council in charge of starting the Human Genome Mapping Project (mid-to-late 1980s)*	*Email, January 2014*
Werner Mewes	*Director, Martinsried Institute for Protein Sequences (MIPS)*	*Munich, Germany, November 2019*
Louis Ollivier	*Quantitative geneticist at INRA Jouy-en-Josas who led the PigBioDiv projects*	*Jouy-en-Josas, France, November 2017*

(*continued*)

(continued)

Name	Role	Location/date
Peter Philippsen	*Coordinator of chromosome XIV for the Yeast Genome Sequencing Project, based at Biozentrum, University of Basel, Switzerland*	*Telephone, September 2019*
Max Rothschild	*United States Department of Agriculture (USDA) extramural National Swine Genome Coordinator*	*Skype, January 2017*
David Weatherall	*Clinician and medical geneticist who established the Institute of Molecular Medicine at the John Radcliffe Hospital, Oxford*	*Institute of Molecular Medicine, John Radcliffe Hospital, April 2014*

Note: All reasonable efforts were made to obtain permission from the interviewees when their oral histories are directly quoted or referred to in this book. Oral histories are listed in the order in which they were initially cited in the book. The unquoted oral histories—italicized at the foot of the table above—were used as general background

Appendix B: Archival Sources

Name	Location	Date of initial access
Personal archive of Robert Waterston	University of Washington, Seattle	April 2021
Papers and Correspondence of Sir John Sulston	Wellcome Library, London, UK	June 2015
Hoechst Archives	Frankfurt, Germany	November 2019
Personal papers of Karl Kleine	Munich, Germany	November 2019
Medical Research Council Series	National Archives of the UK at Kew, London, UK	February 2013
Papers and Correspondence of Sir Walter Bodmer	Bodleian Library, Oxford, UK	March 2014
Wellcome Trust Corporate Archive (including uncatalogued files)	Wellcome Library, London, UK	July 2013
Papers and Correspondence of Michael Ashburner	Wellcome Library, London, UK	September 2020
Personal archive of Alan Archibald	Roslin Institute, University of Edinburgh, Edinburgh UK	March and May 2017
NHGRI History Archive	Bethesda, MD, USA	November 2016

(*continued*)

© The Author(s) 2023
M. García-Sancho, J. Lowe, *A History of Genomics across Species, Communities and Projects*, Medicine and Biomedical Sciences in Modern History, https://doi.org/10.1007/978-3-031-06130-1

(continued)

Name	Location	Date of initial access
Personal papers of Lawrence Schook	University of Illinois Urbana–Champaign, IL, USA	April 2018

Note: All reasonable efforts were made to obtain permission from the archivists or owners of the records when these are directly quoted, referred to or reproduced in this book. Archival sources are listed in the order in which they were initially cited in the book

INDEX[1]

---

[1] Note: Page numbers followed by 'n' refer to notes.

© The Author(s) 2023   361
M. García-Sancho, J. Lowe, *A History of Genomics across Species,*
Communities and Projects, Medicine and Biomedical Sciences in
Modern History, https://doi.org/10.1007/978-3-031-06130-1

Systematics (mode of genomics
research)
  biodiversity, 273–275, 278
  evolution/evolutionary biology,
    31, 242n41

T

Taxonomy and systematics, 336
The Institute for Genomic Research
  (TIGR), 126, 142, 144
TJ Tabasco, 187–189, 233, 315
Translation (of genomic data), 31,
  298, 336, 347
Transplantation/xenotransplantation,
  4, 160, 163, 168, 169, 240, 344
  Porcine Endogenous Retroviruses
    (PERVs), 168

U

UK Government, 42, 43, 47,
  123, 127n9
University of Illinois at Urbana-
  Champaign, 185–187
  The Keck Center for Comparative
    and Functional Genomics, 187
University of Toronto Hospital for
  Sick Children, 49, 108, 149–152
University of Washington in
  Seattle, 48, 266
  Department of Molecular
    Biotechnology, 48
US Department of Agriculture
  Agricultural Research Service Meat
    Animal Research Center
    (USDA MARC), 160, 161,
    166, 175, 178n18, 179, 183,
    186, 189, 270, 271, 300
  Beltsville Agricultural Research
    Center, 241, 307

Cooperative State Research,
  Education, and Extension
  Service (CSREES), 273
Pig Genome Coordination Program
  (PGCP), 160
US Department of Energy (DoE)
  Joint Genome Institute (and
    constituent National
    Laboratories), 144n34,
    145, 295
  Lawrence Berkeley National
    Laboratory, 86
  Lawrence Livermore National
    Laboratory, 86, 147
  Los Alamos National
    Laboratory, 86, 211
  Office of Biological and
    Environmental Research
    [formerly Office of Health and
    Environmental Research],
    87, 87n9
Users and user communities, 13, 23,
  27, 31, 59n9, 64, 101n31, 105,
  106, 141, 141n28, 142n30, 191,
  195, 206, 207, 213, 219–221,
  223, 227, 229, 236, 241, 247,
  262, 262n4, 283n32, 288, 295,
  296, 298
US Human Genome Project
  (US-HGP), 17n13, 81–85,
  87–90, 89n11, 93, 93n15, 94,
  97, 98, 100, 104, 105, 106n36,
  110, 112, 119, 121, 121n4, 123,
  124, 126, 128, 135, 136,
  139, 145–147

V

Vaiman, Marcel, 169, 178
Variation/variants, 329
  catalogues of variation, 105–112, 297

Printed in the United States
by Baker & Taylor Publisher Services